SOLUTIONS MANUAL
TO ACCOMPANY

ORGANIC
CHEMISTRY
A BRIEF COURSE

ROBERT C. ATKINS
Department of Chemistry
James Madison University

FRANCIS A. CAREY
Department of Chemistry
University of Virginia

McGRAW-HILL PUBLISHING COMPANY
New York   St. Louis   San Francisco   Auckland   Bogotá   Caracas
Hamburg   Lisbon   London   Madrid   Mexico   Milan
Montreal   New Delhi   Oklahoma City   Paris   San Juan   São Paulo
Singapore   Sydney   Tokyo   Toronto

SOLUTIONS MANUAL
TO ACCOMPANY
ORGANIC CHEMISTRY: A BRIEF COURSE

1 2 3 4 5 6 7 8 9 0   RMK   RMK   8 9 4 3 2 1 0 9

ISBN 0-07-009921-9

The editors were Randi B. Rossignol and Kirk Emry;
the production supervisor was Leroy A. Young.
Rand McNally and Company was printer and binder.

CONTENTS IN BRIEF

PREFACE

To the student...

You have already demonstrated your intent to master the material in your organic chemistry course by obtaining this *Solutions Manual*. It can be a very effective tool that will not only familiarize you with the kind of reasoning and analysis that one applies in attacking organic chemistry problems, but will also help you focus your attention on the most important topics of each chapter. Courses in organic chemistry build continuously from a relatively simple conceptual base and differ from courses in general chemistry where successive topics often bear little relationship to one another. No doubt your instructor has already told you the twin secrets of success in organic chemistry— *Do not fall behind! Do lots of problems!* Let us elaborate on these two admonitions.

(1) *Read the text before lecture.* Much of the terminology in organic chemistry will be new to you, so it is prudent to acquaint yourself with the language prior to hearing it for the first time. Just as important, is the knowledge of where the instructor is headed that day. You can be selective in your taking of notes if you have already examined the topics and recognize what material is included for orientation and motivation, and what is the core material.

(2) *Do the problems within the body of the chapter while learning the chapter.* None of the individual facts that one uses to solve introductory organic chemistry problems are very difficult by themselves, but a typical problem requires that you string several thoughts, ideas, and concepts together in a logical sequence. The problems within the chapter normally test the single concept discussed just prior to the presentation of the problem.

(3) *Do end-of-chapter problems as soon after completion of the chapter as possible.* These problems serve to integrate the most important aspects of the chapter. Too long a delay in attempting these problems will cause the principal points of the chapter to fade from your memory and require it to be relearned, rather than reinforced.

(4) *Make sure that when you "do problems" that you are, in fact, doing them and not just reading them.* It is all to easy to look at a problem and say "Oh, I can do that one," and never write anything down. **You must write out your answers**. The most common statement heard by instructors from students who are having difficulty is "I don't know why I don't do better on tests. I understand your lectures and I read the book, and it all seems very clear but I seem to make a lot of silly mistakes on the tests." We all make silly mistakes. But we can minimize them by practicing, and practicing in organic chemistry is nothing more than writing out the answers to problems.

(5) *Make sure you understand the correct answer to a problem.* Most organic chemistry problems require you to write a structural formula, or a chemical equation using structural formulas. Check your answer immediately against the answer given in this *Solutions Manual.* If they are different, evaluate yours critically to see if it is incorrect, and why it is incorrect. There are many ways to write chemical formulas and your answer may differ from the answer given here only in style. If it differs in substance, then you need to rethink the problem, reread the text, or seek help from a friend or from the instructor.

Each chapter in this *Solutions Manual* begins with a **Glossary of Terms** designed to place the most important terms of a particular chapter in context. The **Solutions to Text Problems** includes answers to all of the problems in *Organic Chemistry. A Brief Course* except for those for which answers were already provided in the text [part (a)of multipart in-chapter problems]. We have tried to present more than answers, by including a brief discussion of the reasoning behind the answers.

We thank...

We gratefully acknowledge the assistance of Jeff Buterbaugh, Charles Riddle, George Nowacek, and David Meyer, undergraduate organic chemistry students at the University of Virginia who helped produce the camera-ready manuscript. Structural formulas were drawn using the ChemIntosh desk accessory available from SoftShell International Ltd., 2004 North 12th Street, Grand Junction, CO 81501. We also thank Andrew Carey for checking many of the problems, and Randi Rossignol our editor at McGraw-Hill for her support.

Robert C. Atkins   Francis A. Carey
James Madison University   University of Virginia

CHAPTER 1

INTRODUCTION TO ORGANIC CHEMISTRY: CHEMICAL BONDING

GLOSSARY OF TERMS

Antibonding orbital An orbital in a molecule in which an electron is less stable than when localized on an isolated atom.

Atomic Number The number of protons in the nucleus of a particular atom. The symbol for atomic number is Z, and each element has a unique atomic number.

Bonding orbital An orbital in a molecule in which an electron is more stable than when localized on an isolated atom. All of the bonding orbitals are generally doubly occupied in stable neutral molecules.

Constitution The order of atomic connections that defines a molecule.

Constitutional isomers Isomers (see below) that differ in respect to the order in which the atoms are connected.

Covalent bond A chemical bond between two atoms that results from their sharing of two electrons.

Electronegativity A measure of the ability of an atom to attract the electrons in a covalent bond toward itself. Fluorine is the most electronegative element.

Empirical formula An expression of a chemical formula in which the experimentally determined elemental composition is presented as simplest whole number ratios. Thus, the empirical formula of glucose is determined to be CH_2O while its molecular formula (*see below*) is $C_6H_{12}O_6$.

Formal charge The charge, either positive or negative, on an atom calculated by subtracting from the number of valence electrons in the neutral atom a number equal to the sum of its unshared electrons plus one-half of the electrons in its covalent bonds.

Hund's rule When two orbitals are of equal energy they are populated by electrons so that each is half-filled before either one is doubly occupied.

Ionic bond A chemical bond between oppositely charged particles that results from the electrostatic attraction between them.

Isomer Different compounds that have same molecular formula are described as isomers of one another.

Lewis structures Chemical formulas in which electrons are represented by dots. Two dots (or a line) between two atoms represents a covalent bond in a Lewis structure. Unshared electrons are explicitly shown, and stable Lewis structures are those in which the octet rule (see below) is satisfied.

Molecular formula A chemical formula in which subscripts are used to indicate the number of atoms of each element present in one molecule. In organic compounds, carbon is cited first, hydrogen second, and the remaining elements in alphabetical order.

1

Molecular orbital theory A theory of chemical bonding in which electrons are assigned to orbitals in molecules. The molecular orbitals are described as combinations of the orbitals of all of the atoms that comprise the molecule.

Noble gases The elements in Group VIIIA of the periodic table (helium, neon, argon, krypton, xenon, radon). Also known as the "rare gases," they are, with few exceptions, chemically inert.

Octet rule When forming compounds, atoms gain, lose, or share electrons so that the number of their valence electrons is the same as that of the nearest noble gas. For the elements carbon, nitrogen, oxygen, and the halogens, this number is 8.

Orbital A region in space near an atom where there is a high probability (90-95%) of finding an electron.

Pauli exclusion principle No two electrons can have the same set of four quantum numbers. An equivalent expression is that only two electrons can occupy the same orbital, and then only when they have opposite spins.

Period The horizontal row of the periodic table in which a particular element appears.

Polar covalent bond A shared-electron pair bond in which the electrons are drawn more closely to one of the component atoms than the other.

Principal quantum number The quantum number (n) of an electron that describes its energy level. An electron with $n = 1$ must be an s electron, one with $n = 2$ has s and p states available.

Resonance A method by which electron delocalization may be shown using Lewis structures. The true electron distribution in a molecule is regarded as a hybrid of the various Lewis structures which can be written for a molecule.

σ bond A connection between two atoms in which the electron probability distribution has rotational symmetry along the internuclear axis. A cross section perpendicular to the internuclear axis is a circle.

Spin quantum number One of the four quantum numbers that describe an electron. An electron may have either of two different spin quantum numbers, +1/2 or -1/2.

Structural isomer Synonymous with *constitutional isomer*.

Valence electrons The outermost electrons of an atom. For second-row elements these are the $2s$ and $2p$ electrons.

Valence-shell electron pair repulsion model A method for predicting the shape of a molecule based on the notion that electron pairs surrounding a central atom will repel each other. Four electron pairs will arrange themselves in a tetrahedral geometry, three will assume a trigonal planar geometry, and two electron pairs will adopt a linear arrangement.

Vitalism A 19th century theory which divided compounds into two main classes, organic and inorganic, according to whether they originated in living (animal or vegetable) or nonliving (mineral) matter respectively. Vitalist doctrine held that the conversion of inorganic substances to organic ones could only be accomplished through the action of some "vital force."

SOLUTIONS TO TEXT PROBLEMS

1.1 The atomic number Z of carbon = 6, so its electron configuration is $1s^2, 2s^2, 2p^2$. The valence electrons are those in the $2s$ and $2p$ levels, so the number of valence electrons is **four**. Another way of reaching the same answer is to remember that main-group elements (those in the "A" groups) have the same number of valence electrons as their group number. Carbon is in group IVA, so has four valence electrons.

1.2 Electron configurations of elements are derived by applying the following principles:

- The number of electrons in a neutral atom is equal to its atomic number Z.

- The maximum number of electrons in any orbital is two.

- Electrons are added to orbitals in order of increasing energy, filling the $1s$ orbital before any electrons occupy the $2s$ level. The $2s$ orbital is filled before any of the $2p$ orbitals, and the $3s$ orbital is filled before any of the $3p$ orbitals.

- All of the $2p$ orbitals are of equal energy, and each is singly occupied before any is doubly occupied. The same holds for the $3p$ orbitals.

With this as background, the electron configurations of the third-row elements are derived as follows:

Na (Z=11) $1s^2\,2s^2\,2p^6\,3s^1$ · P (Z=15) $1s^2\,2s^2\,2p^6\,3s^2\,3p_x^1\,3p_y^1\,3p_z^1$

Mg (Z=12) $1s^2\,2s^2\,2p^6\,3s^2$ · S (Z=16) $1s^2\,2s^2\,2p^6\,3s^2\,3p_x^2\,3p_y^1\,3p_z^1$

Al (Z=13) $1s^2\,2s^2\,2p^6\,3s^2\,3p_x^1$ · Cl (Z=17) $1s^2\,2s^2\,2p^6\,3s^2\,3p_x^2\,3p_y^2\,3p_z^1$

Si (Z=14) $1s^2\,2s^2\,2p^6\,3s^2\,3p_x^1\,3p_y^1$ · Ar (Z=18) $1s^2\,2s^2\,2p^6\,3s^2\,3p_x^2\,3p_y^2\,3p_z^2$

1.3 Hydrogen has one valence electron and fluorine has seven. The covalent bond in hydrogen fluoride arises by sharing the single electron of hydrogen with the unpaired electron of fluorine.

Combine H · and · F̈ : to give the Lewis structure for hydrogen fluoride   H : F̈ :

1.4 (b) Nitrogen is a goup VA element and so contributes five valence electrons, while three hydrogen atoms contribute one each.

Combine :N · and three H ·   to write a   H
                                 Lewis structure   : N : H
                                 for ammonia   H

In the Lewis structure for ammonia, nitrogen has eight electrons in its valence shell; two are unshared and six are grouped in three shared pairs.

(c) There are 26 valence electrons to be accounted for in NF_3; each fluorine contributes seven and nitrogen contributes five.

Combine :N · and three · F̈ :   to write a   : F̈ :
                                   Lewis structure   : N : F :
                                   for NF_3 : F̈ :

Both nitrogen and fluorine are surrounded by eight electrons in NF_3.

(d) Phosphorus, like nitrogen is in group VA of the periodic table and so contributes five valence electrons. Each chlorine contributes seven.

Combine :P· and three :Cl·

to write a
Lewis structure
for PCl₃

: Cl :
: P : Cl :
: Cl :

The octet rule is satisfied for both phosphorus and chlorine.

(e) There is only one way to combine the atoms in CH₃Cl so as to produce a valid Lewis structure. Carbon can form four bonds while hydrogen and chlorine form one each in their stable compounds.

Combine ·C· with ·Cl: and three H·

to write the
Lewis structure
for CH₃Cl

H
H : C : Cl :
H

(f) In order to satisfy the Lewis rules, C₂H₆ must have a carbon-carbon bond.

Combine two ·C· and six H·

to write
Lewis structure
of ethane

H H
H : C : C : H
H H

There are a total of 14 valence electrons distributed as shown. Each carbon is surrounded by eight electrons.

1.5   (b) The electron count of carbon is five; there are two electrons in an unshared pair, and three electrons are counted as carbon's share of the three covalent bonds to hydrogen.

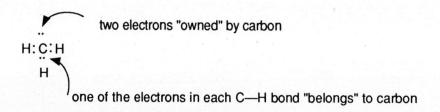

two electrons "owned" by carbon

one of the electrons in each C—H bond "belongs" to carbon

An electron count of five is one more than that for a neutral carbon atom. The formal charge on carbon is -1, as is the net charge on this species.

(c) This species has one less electron than that of the preceding problem. None of the atoms bears a formal charge. The species is neutral.

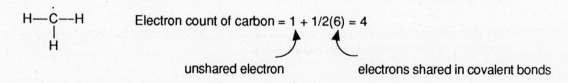

Electron count of carbon = 1 + 1/2(6) = 4

unshared electron   electrons shared in covalent bonds

4

(d) The formal charge of carbon in this species is +1. Its only electrons are those in its three covalent bonds to hydrogen, so its electron count is three. This corresponds to one less electron than in a neutral carbon atom, giving it a unit positive charge.

$$H—\overset{\displaystyle |}{\underset{\displaystyle |}{C}}—H$$
$$H$$

(e) In this species the electron count of carbon is four or, exactly as in (c), that of a neutral carbon atom. Its formal charge is zero and the species is neutral.

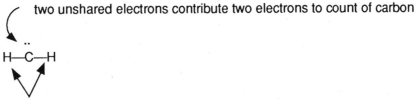

two unshared electrons contribute two electrons to count of carbon

half of the four electrons in these two covalent bonds contribute two electrons to count of carbon

1.6   First calculate the molecular weight corresponding to the empirical formula $C_4H_5N_2O$.

4C	=	4 x 12	= 48
5H	=	5 x 1	= 5
2N	=	2 x 14	= 28
1O	=	1 x 16	= 16
$C_4H_5N_2O$			= 97

The problem specifies that the molecular weight of caffeine is 194, or twice that of $C_4H_5N_2O$. Therefore, the molecular formula of caffeine must be $C_8H_{10}N_4O_2$.

1.7   While the statement of the problem does not explicitly require you to show all of the unshared electron pairs in the structures, they are included in the following solutions for completeness.

(b) Since carbon normally has four bonds, place two hydrogens on carbon and one on oxygen to give the constitution:

$$\begin{matrix} H & \\ \diagdown & \\ & C{=}N \\ \diagup & \diagdown \\ H & O{-}H \end{matrix}$$

The structure contains 6 bonds and these bonds account for 12 electrons. The neutral molecule CH3NO requires 18 valence electrons (4 from carbon, 1 each from 3 hydrogens, 5 from nitrogen, and 6 from oxygen). Therefore, 6 electrons (3 pairs) need to be added. Assign one pair (2 electrons) to nitrogen and two pairs (4 electrons) to oxygen so as to complete their octets.

$$\begin{matrix} H & \\ \diagdown & \ddot{} \\ & C{=}N \\ \diagup & \diagdown \\ H & :\ddot{O}{-}H \end{matrix}$$

Any other disposition of electron pairs would either violate the octet rule or would lead to two atoms bearing opposite formal charges and the structure would be less stable than the one shown.

(c)   Placing one hydrogen on oxygen, one on carbon, and one on nitrogen gives the structure:

$$\begin{array}{c} \text{H—O} \\ \diagdown \\ \text{C}=\text{N} \\ \diagup \quad \diagdown \\ \text{H} \qquad \text{H} \end{array}$$

Since the covalent bonds account for 12 electrons, 6 more (3 pairs) are needed. Place two electrons on nitrogen and four on oxygen.

$$\begin{array}{c} \text{H—}\ddot{\text{O}}: \\ \diagdown \\ \text{C}=\ddot{\text{N}} \\ \diagup \quad \diagdown \\ \text{H} \qquad \text{H} \end{array}$$

(d) Two of the hydrogens are on nitrogen, one is on carbon. There are two unshared electron pairs on oxygen and one pair on nitrogen.

$$\begin{array}{c} :\ddot{\text{O}} \qquad \text{H} \\ \| \qquad \diagup \\ \text{C—N}: \\ \diagup \quad \diagdown \\ \text{H} \qquad \text{H} \end{array}$$

1.8   (b) Move electrons from the negatively charged oxygen, as shown by the curved arrows.

The ion shown is bicarbonate (hydrogen carbonate) ion. The resonance interaction shown is more important than an alternative one involving delocalization of lone pair electrons in the OH group.

Not an equivalent structure: not as

stable because of charge separation

(c) All three oxygens are equivalent in carbonate ion. Either negatively charged oxygen can serve as the donor atom.

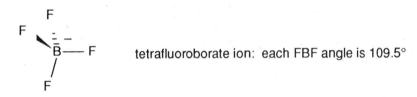

1.9 Boron is surrounded by four bonded pairs of electrons and has a tetrahedral arrangement of its four bonds.

tetrafluoroborate ion: each FBF angle is 109.5°

1.10 (b) Nitrogen in ammonium ion is surrounded by eight electrons in four covalent bonds. These four bonds are directed toward the corners of a tetrahedron.

each HNH angle is 109.5°

(c) Double bonds are treated as single bonds when deducing the shape of a molecule. Thus azide ion is linear.

the NNN angle is 180°

(d) Since the double bond in carbonate ion is treated as if it were a single bond, the three sets of electrons are arranged in a trigonal planar arrangement around carbon.

the OCO angle is 120°

1.11 (a) Li+ has one less electron than Li ($Z = 3$). Therefore the electron configuration of Li+ is $1s^2$.

(b) Mg+ has one less electron than Mg ($Z = 12$). Therefore the electron configuration of Mg+ is $1s^2\ 2s^2\ 2p_x^2\ 2p_y^2\ 2p_z^2\ 3s^1$.

(c) Mg^{2+} has two less electrons than Mg ($Z = 12$) and has the electron configuration of the noble gas neon $1s^2\ 2s^2\ 2p_x^2\ 2p_y^2\ 2p_z^2$.

(d) K^+ has one less electron than K ($Z = 19$) and has the electron configuration of the noble gas argon $1s^2\ 2s^2\ 2p^6\ 3s^2\ 3p_x^2\ 3p_y^2\ 3p_z^2$.

(e) H^- has one more electron than H ($Z = 1$). Its electron configuration is $1s^2$.

(f) O^- has one more electron than O ($Z = 8$). Its electron configuration is $1s^2\ 2s^2\ 2\ p_x^2\ 2p_y^2\ 2p_z^1$.

(g) O^{2-} has two more electrons than O ($Z = 8$). Its electron configuration is the same as the noble gas neon $1s^2\ 2s^2\ 2p_x^2\ 2p_y^2\ 2p_z^2$.

(h) S^{2-} has two more electrons than S ($Z = 16$). Its electron configuration is the same as the noble gas argon $1s^2\ 2s^2\ 2p^6\ 3s^2\ 3p_x^2\ 3p_y^2\ 3p_z^2$.

1.12   The atomic number Z identifies the element, and the difference between Z and the number of electrons gives its net charge.

(a)   Since $Z = 9$ the element is fluorine. The total number of electrons is 10, so the ion is F^-.

(b)   The atomic number $Z = 12$ tells us the element is magnesium. Since there are 11 electrons, the ion is Mg^+.

(c)   The atomic number is the same as in (b), so again the element is magnesium. With only 10 electrons, the ion is Mg^{2+}.

(d)   The atomic number $Z = 13$ reveals the element to be aluminum, and since there are 11 electrons the ion is Al^{2+}.

(e)   Again, as in (d), the element is aluminum. With 10 electrons the ion is Al^{3+}.

(f)   The atomic number $Z = 16$ corresponds to sulfur. Since there are 18 electrons the ion is S^{2-}.

1.13   The species most likely to have an ionic bond is the one in which the connected atoms differ in electronegativity by the greatest amount.

(a) All of the species contain an atom bonded to oxygen. Oxygen is an electronegative element, so the most ionic bond will be between oxygen and the most electropositive (least electronegative) element. Calcium oxide (CaO) is the compound most lilkely to have an ionic bond.

(b) Fluorine is the most electronegative element. Lithium is the most electropositive of the elements bonded to fluorine in this problem; LiF has the most ionic bond.

(c) Magnesium chloride has the most ionic bond of the group CCl_4, $MgCl_2$, Cl_2O, and Cl_2 because Mg is the only electropositive element in the group.

(d) Both aluminum and potassium are electropositive, but potassium is more electropositive than aluminum, so KBr is more likely to have an ionic bond than $AlBr_3$.

1.14   All these species are characterized by the formula $: X \equiv Y :$ and each atom has an electron count of five.

triple bond contributes half of its six electrons,
or three electrons each, to individual electron counts of X and Y

$$: X \equiv Y :$$

unshared electron pair
contributes two electrons
to electron count of X

unshared electron pair contributes two
electrons to electron count of Y

Electron count **X** = electron count **Y** = 2 + 3 = **5**

(a) $: N \equiv N :$ A neutral nitrogen atom has five valence electrons. Therefore, each atom is electrically neutral in molecular nitrogen.

(b) $: C \equiv N :$ Nitrogen, as before, is electrically neutral. A neutral carbon atom has four valence electrons so carbon in this species, with an electron count of five, has a unit negative charge. The species is cyanide anion; its net charge is -1.

(c) $: C \equiv C :$ There are two negatively charged carbon atoms in this species. It is a dianion; its net charge is -2.

(d) $: N \equiv O :$ Here again is a species with a neutral nitrogen atom. Oxygen, with an electron count of five, has one less electron in its valence shell than a neutral oxygen atom. Oxygen has a formal charge of +1; the net charge is +1.

(e) $: C \equiv O :$ Carbon has a formal charge of -1; oxygen has a formal charge of +1. Carbon monoxide is a neutral molecule.

1.15   All these species are of the type $: Y = X = Y :$ Atom **X** has an electron count of four, corresponding to half of the eight shared electrons in its four covalent bonds. Each atom **Y** has an electron count of six; comprising it are four unshared electrons plus half of the four electrons in the double bond of each **Y** to **X**.

(a) $: \ddot{O} = C = \ddot{O} :$ Oxygen with an electron count of six, and carbon, with an electron count of four, both correspond to the respective neutral atoms in the number of electrons they "own." Carbon dioxide is a neutral molecule, and neither carbon nor oxygen has a formal charge in this Lewis structure.

(b) $: \ddot{N} = N = \ddot{N} :$ The two terminal nitrogens each ahve an electron count (6) that is one more than a neutral atom and thus each has a formal charge of -1. The central N has an electron count (4) that is one less than a neutral nitrogen; it has a formal charge of +1. The net charge on the species is (-1+1-1), or -1.

(c) $: \ddot{O} = N = \ddot{O} :$ As in (b), the central nitrogen has a formal charge of +1. As in (a), each oxygen is electrically neutral. The net charge is +1.

9

1.16   Species **A**, **B**, and **C** have the same molecular formula, the same atomic positions, and the same number of electrons. They differ only in the arrangement of their electrons. They are resonance forms of the same compound. They do, however, differ in respect to their formal charge disposition.

$$H-\overset{-}{\underset{..}{C}}=\overset{+}{N}=\overset{..}{\underset{..}{O}}:\qquad\qquad H-C\equiv\overset{+}{N}-\overset{..}{\underset{..}{\underset{..}{O}}}:\qquad\qquad H-\overset{+}{C}=N-\overset{..}{\underset{..}{\underset{..}{O}}}:$$

A   **B**   **C**

(a)   Structure **C** has a formal charge of +1 on carbon.

(b)   Structures **A** and **B** both have a formal charge of +1 on nitrogen.

(c)   None of the structures has a formal charge of +1 on oxygen.

(d)   Structure **A** has a formal charge of -1 on carbon.

(e)   None of the structures has a formal charge of -1 on nitrogen.

(f)   Structures **B** and **C** both have a formal charge of -1 on oxygen.

(g)   All of the structures are electrically neutral.

1.17   Species **D**, **E**, and **F** are resonance forms of the same compound.

D   **E**   **F**

(a)   Structure **D** has a formal charge of +1 on carbon.

(b)   Structures **E** and **F** both have a formal charge of +1 on nitrogen.

(c)   None of the structures has a formal charge of +1 on oxygen.

(d)   Structure **F** has a formal charge of -1 on carbon.

(e)   None of the structures has a formal charge of -1 on nitrogen.

(f)   All of the structures have at least one oxygen with a formal charge of -1; structure **D** has two.

(g)   None of the structures are electrically neutral.

1.18   The direction of a bond dipole is governed by the electronegativity of the atoms it connects. In each of the parts to this problem, the more electronegative atom is partially negative and the less electronegative atom is partially positive. Electronegativites of the elements are given in Table 1.2 of the text.

(a) Chlorine is more electronegative than hydrogen.   (b) Chlorine is more electronegative than iodine.

(c) Iodine is more electronegative than hydrogen.   (d) Oxygen is more electronegative than hydrogen.

H—I

(e) Oxygen is more electronegative than either hydrogen or chlorine.

1.19 (a) Phosphorus is in group VA of the periodic table, so has five valence electrons. Each of the three hydrogens contributes one valence electron. The correct Lewis structure is:

$$\begin{array}{c} H \\ \cdot\cdot \\ H:P:H \\ \cdot\cdot \end{array}$$

(b) Aluminum contributes three valence electrons and each of the four hydrogens contributes one for a total of seven. Since the species has a net charge of -1, the total number of electrons must be eight. Each of the four hydrogens is covalently bonded to aluminum, and the octet rule is satisfied.

$$\begin{array}{c} H \qquad - \\ \cdot\cdot \\ H:Al:H \\ \cdot\cdot \\ H \end{array}$$

(c) The total number of valence electrons is 24; carbon contributes 4, oxygen 6, and each of the two chlorines contributes 7. The problem specifies that all the atoms are connected to carbon, so connect each to carbon by a two-electron covalent bond. These bonds account for six electrons, leaving 18 to be assigned. Place six electrons (three unshared pairs) on oxygen and each chlorine. The species so generated satisfies the octet rule for all atoms except carbon. It has a formal charge of +1 on carbon and -1 on oxygen. Reorganize electrons as shown to give the most stable Lewis structure. It is more stable because it has one more bond and no separation of opposite charges.

Use an electron pair of oxygen to form a double bond to carbon

(d) Even though this species is very much different from that of part (c), like the species in part (c) it too has 24 valence electrons. Three oxygens contribute 18 (6 each), carbon 4, and hydrogen 1, for a total of 23. Add one more electron because the species is negatively charged to give a grand total of 24. We can model this structure on the preceding one by starting with a carbon that has single bonds to two oxygens and a double bond to the third oxygen.

(e) Nitrogen contributes five valence electrons and each oxygen contributes six for a total of 17. But since the species given in the problem is positively charged, we need to substract one electron to give a final total of 16. The problem additionally states that the order of atoms is ONO. Therefore, connect

11

nitrogen to each oxygen by a two-electron covalent bond to give O : N : O. The remaining 12 electrons can be divided equally between the two oxygens giving each one three unshared pairs. The structural formula which results:

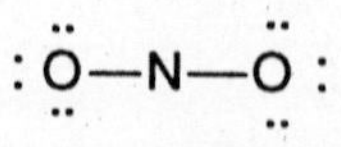

$$: \overset{\cdot\cdot}{\underset{\cdot\cdot}{O}} - N - \overset{\cdot\cdot}{\underset{\cdot\cdot}{O}} :$$

has only four electrons around the central nitrogen. The formal charge on each oxygen in this structure is -1, and the formal charge on nitrogen is +3. Use an electron pair on each oxygen to form a double bond to nitrogen. The resulting species satisfies the octet rule for all three atoms and has a formal charge of +1 on nitrogen.

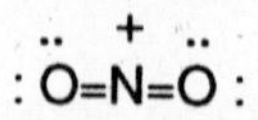

$$: \overset{\cdot\cdot}{O} = \overset{+}{N} = \overset{\cdot\cdot}{O} :$$

(f)   The species NO_2^- has two more electrons than $^+NO_2$ (preceding problem). Convert one of the N=O double bonds to a single bond by moving two of the electrons in the double bond to become an unshared pair on oxygen. Add the two additional electrons in NO_2^- to nitrogen. The structure shown satisfies the octet rule for all three atoms, and has a formal charge of -1 on one of the oxygens.

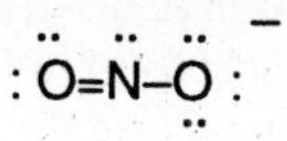

$$: \overset{\cdot\cdot}{O} = \overset{\cdot\cdot}{N} - \overset{\cdot\cdot}{\underset{\cdot\cdot}{O}} :^{-}$$

1.20   (a)   Phosphorous is surrounded by four electron pairs. Three are bonded pairs and one pair is unshared. The four pairs of electrons are directed toward the corners of a tetrahedron, making the overall geometry of a molecule **trigonal pyramidal**.

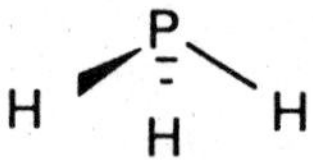

The unshared electron pair is not shown in the structural drawing because we describe molecular geometry according to the positions of the atoms.

(b)   Aluminum is surrounded by four bonded pairs of electrons and has a **tetrahedral** arrangement of those bonds.

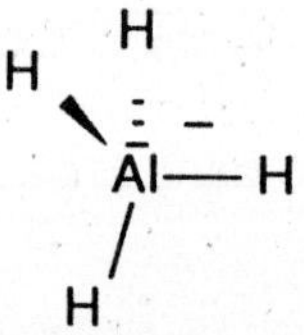

Each H-Al-H angle is exactly 109.5°

(c)   Carbon is bonded to three groups in $COCl_2$; it is connected to oxygen by a double bond. For the purposes of the VSEPR method, a double gond is considered to be like a single bond. The maximum separation of electron pairs is achieved in a **trigonal planar** geometry.

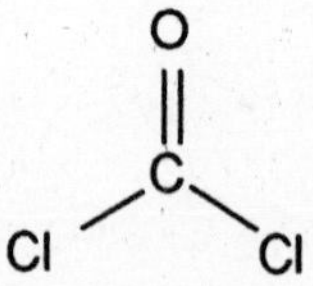

Bond angles approximately 120°

12

(d) As in (c), carbon is bonded to three groups in HCO_3^-. It, too, is a **trigonal planar** species.

O
‖
C
HO   O–

Bond angles approximately 120°

(e) The central nitrogen in $^+NO_2$ has its eight electrons divided equally in double bonds to two oxygens. Each double bond is treated as if it were a single bond. A central atom that bears two bonds and no unshared pairs has a **linear** arrangement of its attached groups.

+
:O=N=O:

ONO angle = 180°

(f) The central nitrogen in NO_2^- has an unshared electron pair to consider along with bonds to two oxygens. The electrons in the bonds to the two oxygens repel each other and repel the unshared electron pair on nitrogen to give a trigonal planar disposition of electron pairs. The geometry as defined by the atomic positions is **bent**.

N
O   O –

ONO angle is approximately 120°

1.21 (a) Each carbon has four valence electrons, each hydrogen one, and chlorine has seven. Hydrogen and chlorine each can only form one bond, so the only stable structure must have a carbon-carbon bond. Of the 20 valence electrons, 14 are present in the seven covalent bonds and 6 reside in the three unshared electron pairs of chlorine.

H H
H:C:C:Cl:
H H

or

H H
│ │
H—C—C—Cl
│ │
H H

(b) As in part (a) the single chlorine as well as all of the hydrogens must be connected to carbon. There are 18 valence electrons in C_2H_3Cl and the framework of five single bonds accounts for only 10 electrons. Six of the remaining 8 are used to complete the octet of chlorine as three unshared pairs and the last two are used to form a carbon-carbon double bond.

H H
H:C::C:Cl:

or

H   H
\   /
C=C
/   \
H   Cl

(c) All of the atoms except carbon (H, Br, Cl, and F) are monovalent. Therefore they can only be bonded to carbon. The problem states that all three fluorines are bonded to the same carbon so one of the carbons is present as a CF_3 group. The other carbon must be present as a CHBrCl group. Connect these groups together to give the structure of halothane.

(d) As in part (c) all of the atoms except carbon are monovalent. Since each carbon bears one chlorine, two $ClCF_2$ groups must be bonded together.

1.22 In each part of this problem first calculate the molecular weight associated with the given empirical formula, then determine how many of this molecular weight units are required to give the molecular weight of the specified compound.

(a) The empirical formula C_2H_4O corresponds to a molecular weight of 44. Since the molecular weight of dioxane is given as 88, its molecular formula must be two times C_2H_4O or $C_4H_8O_2$. The molecular formula of paraldehyde is 132, so its molecular formula must be three times C_2H_4O or $C_6H_{12}O_3$.

(b) The empirical formula is C_5H_6O and its molecular weight 82. The molecular weight of eugenol is given as 164, so its molecular formula must be twice C_5H_6O or $C_{10}H_{12}O_2$. Since the molecular formula of ambrosin is 246, or 3 x 82, its molecular formula must be $3(C_5H_6O)$ or $C_{15}H_{18}O_3$.

(c) Since the molecular weight given for thymol (150) corresponds to the molecular weight of the empirical formula $C_{10}H_{14}O$, then $C_{10}H_{14}O$ must also be the molecular formula of thymol. The molecular weight of retinoic acid is 300 or twice that of of $C_{10}H_{14}O$ so the molecular formula of retinoic acid must be $C_{20}H_{28}O_2$.

1.23 **(a)** Recognize that two types of carbon chains are possible. One isomer has an unbranched continuous chain of four carbons, the other isomer has a continuous chain of three carbons with a one-carbon branch off of the central carbon.

or $CH_3CH_2CH_2CH_3$

and

or CH_3CHCH_3 or $(CH_3)_3CH$
$\quad\quad\;\; CH_3$

14

(b) The three carbons can only form a continuous chain. The chlorine may be attached to a carbon at the end of the chain or at the carbon in the middle of the chain. When chlorine is at the end of the chain it does not matter whether it is at the left end or the right end; both positions are equivalent.

$$H-\underset{H}{\overset{H}{C}}-\underset{H}{\overset{H}{C}}-\underset{H}{\overset{H}{C}}-Cl \qquad \text{equivalent to} \qquad Cl-\underset{H}{\overset{H}{C}}-\underset{H}{\overset{H}{C}}-\underset{H}{\overset{H}{C}}-H \qquad \text{or} \qquad CH_3CH_2CH_2Cl$$

Similarly, it does not matter whether chlorine points up or down.

$$H-\underset{H}{\overset{H}{C}}-\underset{Cl}{\overset{H}{C}}-\underset{H}{\overset{H}{C}}-H \qquad \text{equivalent to} \qquad H-\underset{H}{\overset{H}{C}}-\underset{H}{\overset{Cl}{C}}-\underset{H}{\overset{H}{C}}-H \qquad \text{or} \qquad \underset{Cl}{CH_3CHCH_3} \qquad \text{or} \quad (CH_3)_2CHCl$$

(c) The chain of atoms may be in the order C-C-O or C-O-C and gives the two isomers shown. Remember that carbon has four bonds in its neutral compounds and oxygen has two.

$$H-\underset{H}{\overset{H}{C}}-\underset{H}{\overset{H}{C}}-O-H \qquad \text{or} \qquad CH_3CH_2OH$$

$$H-\underset{H}{\overset{H}{C}}-O-\underset{H}{\overset{H}{C}}-H \qquad \text{or} \qquad CH_3OCH_3$$

(d) Here the order may be C-C-N or C-N-C. An uncharged nitrogen normally has three bonds.

$$H-\underset{H}{\overset{H}{C}}-\underset{H}{\overset{H}{C}}-\overset{H}{N}-H \qquad \text{or} \qquad CH_3CH_2NH_2$$

$$H-\underset{H}{\overset{H}{C}}-\overset{H}{N}-\underset{H}{\overset{H}{C}}-H \qquad \text{or} \qquad CH_3NHCH_3 \qquad \text{or} \ (CH_3)_2NH$$

(e) The two chlorines may be on the same carbon, or on different carbons.

$$\begin{array}{c} \text{H} \quad \text{H} \\ | \quad\quad | \\ \text{H}-\text{C}-\text{C}-\text{Cl} \\ | \quad\quad | \\ \text{H} \quad \text{Cl} \end{array} \qquad \text{or} \qquad CH_3CHCl_2$$

$$\begin{array}{c} \text{H} \quad \text{H} \\ | \quad\quad | \\ \text{Cl}-\text{C}-\text{C}-\text{Cl} \\ | \quad\quad | \\ \text{H} \quad \text{H} \end{array} \qquad \text{or} \qquad ClCH_2CH_2Cl$$

(f) With four chlorines and two carbons we may have a compound in which each carbon bears two chlorines or an isomer in which three chlorines are on one carbon and one on the other.

$$\begin{array}{c} \text{H} \quad \text{Cl} \\ | \quad\quad | \\ \text{Cl}-\text{C}-\text{C}-\text{Cl} \\ | \quad\quad | \\ \text{H} \quad \text{Cl} \end{array} \qquad \text{or} \qquad ClCH_2CCl_3$$

$$\begin{array}{c} \text{Cl} \quad \text{Cl} \\ | \quad\quad | \\ \text{Cl}-\text{C}-\text{C}-\text{Cl} \\ | \quad\quad | \\ \text{H} \quad \text{H} \end{array} \qquad \text{or} \qquad Cl_2CHCHCl_2$$

1.24 (a) Approach the problem systematically by first writing the structural formula for the C_5H_{12} isomer with an unbranched carbon chain. Next shorten the chain by one carbon while adding a one-carbon branch. Finally, write a three-carbon chain with two one-carbon branches.

$$\text{C}-\text{C}-\text{C}-\text{C}-\text{C} \qquad \begin{array}{c} \text{C}-\text{C}-\text{C}-\text{C} \\ | \\ \text{C} \end{array} \qquad \begin{array}{c} \text{C} \\ | \\ \text{C}-\text{C}-\text{C} \\ | \\ \text{C} \end{array}$$

Fill in the number of hydrogens so that each carbon has four bonds and the molecular formula is C_5H_{12}.

$$\begin{array}{c} \text{H} \quad \text{H} \quad \text{H} \quad \text{H} \quad \text{H} \\ | \quad | \quad | \quad | \quad | \\ \text{H}-\text{C}-\text{C}-\text{C}-\text{C}-\text{C}-\text{H} \\ | \quad | \quad | \quad | \quad | \\ \text{H} \quad \text{H} \quad \text{H} \quad \text{H} \quad \text{H} \end{array}$$

These structural formulas are more conveniently presented in a condensed version (to be discussed in Section 2.3).

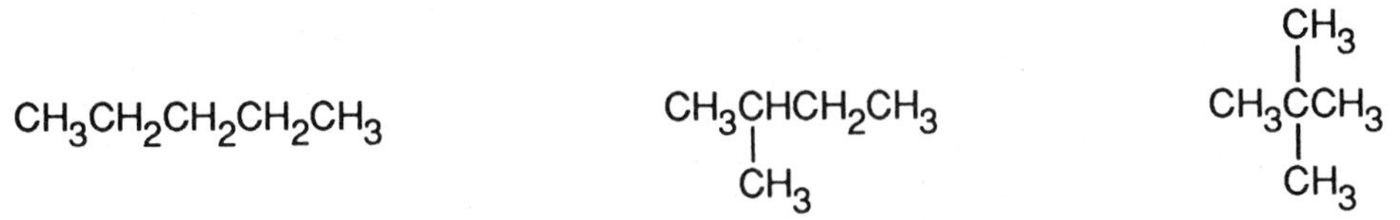

$$CH_3CH_2CH_2CH_2CH_3 \qquad \underset{\underset{\displaystyle CH_3}{|}}{CH_3CHCH_2CH_3} \qquad \underset{\underset{\displaystyle CH_3}{|}}{\overset{\overset{\displaystyle CH_3}{|}}{CH_3CCH_3}}$$

These are the only three isomers of C_5H_{12}. Other structural formulas are simply alternative representations of the compounds shown here.

(b) All of the atoms other than carbon in $C_2H_3Cl_2F$ are *monovalent*; they form only one bond in their stable compounds. Thus, all three isomers are characterized by a C-C structural unit.

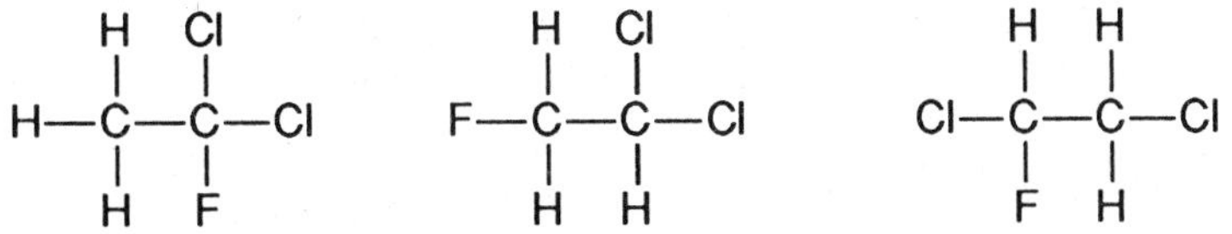

(c) The three isomers are as shown. In the first two, the main chain is C-C-C-O or C-O-C-C. In the third, the main chain is C-C-O and there is a one-carbon branch of this chain.

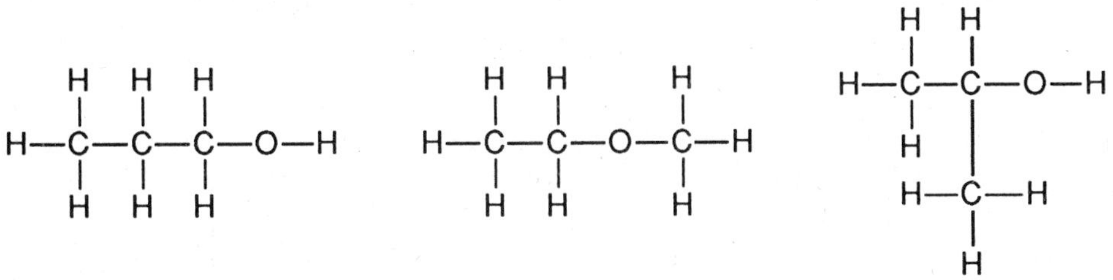

(d) On trying to write structural formulas for C_2H_4O, you will find that the only structure which includes only single bonds must contain a ring.

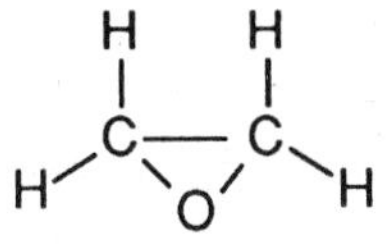

Rings can only be absent if double bonds are present. The double bond can be between two carbons or between one carbon and oxygen.

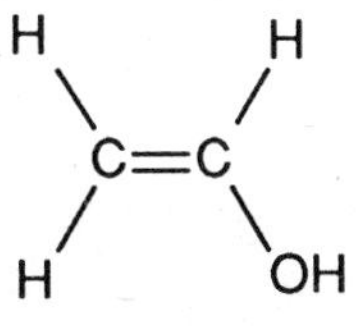

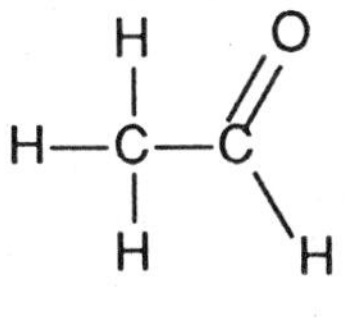

1.25 (a) These two structures are resonance forms of azide anion since they have the same atomic positions and the same number of electrons.

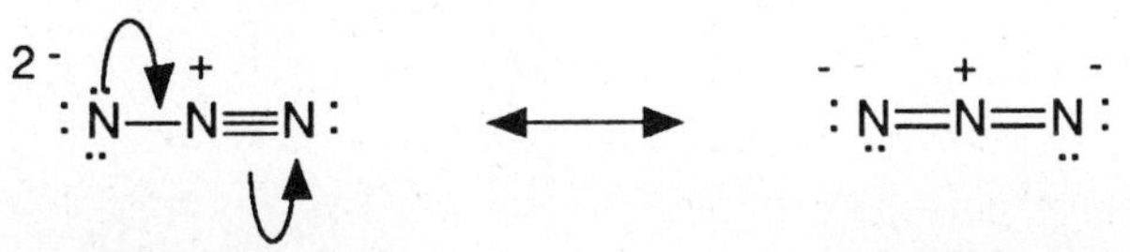

16 valence electrons   16 valence electrons
(net charge = -1)   (net charge = -1)

(b) The two structures have different numbers of electrons and therefore they are not resonance forms.

$$:\overset{2-}{\underset{..}{\overset{..}{N}}}\!-\!N\!\equiv\!\overset{+}{N}:$$

16 valence electrons
(net charge = -1)

$$:\overset{..}{N}\!-\!\overset{2+}{N}\!=\!\overset{..}{\underset{..}{N}}\overset{-}{}$$

14 valence electrons
(net charge = +1)

(c) These two structures have different numbers of electrons; they are not resonance forms.

$$:\overset{2-}{\underset{..}{\overset{..}{N}}}\!-\!N\!\equiv\!\overset{+}{N}:$$

16 valence electrons
(net charge = -1)

$$:\overset{2-}{\underset{..}{\overset{..}{N}}}\!-\!\overset{:}{\underset{..}{N}}\!-\!\overset{2-}{\underset{..}{\overset{..}{N}}}:$$

20 valence electrons
(net charge = -5)

1.26 (a) In this resonance form, the carbon atom of carbon dioxide has only six electrons surrounding it. Move electrons as indicated by the curved arrow so as to generate more stable Lewis structure in which all the atoms have complete octets.

$$:O\!=\!\overset{+}{C}\!-\!\overset{..}{\underset{..}{O}}:{}^{-} \longleftrightarrow :\overset{..}{O}\!=\!C\!=\!\overset{..}{O}:$$

(b) By moving electron pairs toward oxygen, the negative charge may be placed on a more electronegative atom. Oxygen is more electronegative than nitrogen.

$$:\overset{..}{\underset{..}{O}}\!=\!C\!=\!\overset{..}{N}:{}^{-} \longleftrightarrow {}^{-}:\overset{..}{\underset{..}{O}}\!-\!C\!\equiv\!N:$$

(c) One of the oxygen atoms in this structure of ozone has only six electrons surrounding it. A more stable Lewis structure, derived by moving electrons as shown, has complete octets at all three oxygens. Notice also that there is less charge separation in the more stable structure.

$$:\overset{+}{\underset{..}{O}}\!-\!\overset{..}{\underset{..}{O}}\!-\!\overset{..}{\underset{..}{O}}:{}^{-} \longleftrightarrow :\overset{..}{\underset{..}{O}}\!=\!\overset{+}{\underset{..}{O}}\!-\!\overset{..}{\underset{..}{O}}:{}^{-}$$

(d) By moving electrons toward positive charge and away from negative charge, a structure with complete octets around all second-row elements, as well as no charge separation, is generated.

$$\begin{array}{c} H \\ \diagdown \\ {}_{/}\overset{+}{C}\!-\!\overset{..}{C}\!-\!\overset{..}{\underset{..}{O}}:{}^{-} \\ H \end{array} \longleftrightarrow \begin{array}{c} H \\ \diagdown \\ {}_{/}C\!=\!C\!=\!\overset{..}{O}: \\ H \end{array}$$

18

(e) Negative charge can be placed on the most electronegative atom (oxygen) in this molecule by moving electrons as indicated.

(f) This exercise is similar to (e). Electron reorganization allows the negative charge to be placed on nitrogen.

(g) Octets of electrons are present around both carbon and oxygen if an oxygen unshared pair is moved toward the positively charged carbon to give an additional covalent bond.

(h) This exercise is similar to (g): move electrons from oxygen to carbon so as to produce an additional bond and satisfy the octet rule for each atom.

(i) By moving electrons from the site of negative charge toward the positive charge, a structure that has no charge separation is generated.

CHAPTER 2

ALKANES AND CYCLOALKANES

GLOSSARY OF TERMS

Alkane   A hydrocarbon in which all the bonds are single bonds.   Alkanes have the general formula C_nH_{2n+2}.

Angle strain   The strain a molecule possesses because its bond angles are distorted from their normal values.   This normal value is 109.5° for sp^3-hybridized carbon.

Axial bond   A bond to a carbon in the chair conformation of cyclohexane oriented as shown below:

Boat conformation   The unstable conformation of cyclohexane depicted below:

Carbon-skeleton formula   A method of representing structural formulas in which carbon atoms and hydrogen atoms attached to carbon are not

shown explicitly.   The pattern of carbon-carbon bonds is shown along with atoms other than carbon or hydrogen.   Also called "bond-line" or "line-segment" formulas.

Chair conformation   The most stable conformation of cyclohexane.

Combustion   The burning of a substance in the presence of oxygen.   All hydrocarbons yield carbon dioxide and water when they undergo combustion.

Common nomenclature   Names given to compounds on some basis other than a comprehensive, systematic set of rules.

Condensed structural formula   A standard way of representing structural formulas in which subscripts are used to indicate replicated atoms or groups.

Conformations   Non-identical representations of a molecule generated by rotation about single bonds.

Conformational analysis   The study of the conformations available to a molecule, their relative stability, and the role they play in defining the properties of the molecule.

Cycloalkane　An alkane in which a ring of carbon atoms is present.

Eclipsed bonds　The C—H bonds indicated in the structure shown are eclipsed.

Equatorial bond　A bond to a carbon in the chair conformation of cyclohexane oriented as shown below:

Ethyl group　The group CH_3CH_2—.

Exothermic reaction　A reaction that proceeds with the evolution of heat.

Hydrocarbon　A compound that contains only carbon and hydrogen.

Isobutyl group　The group $(CH_3)_2CHCH_2$—.

Isopropyl group　The group $(CH_3)_2CH$—.

IUPAC nomenclature　The most widely used method of naming organic compounds. It uses a set of rules proposed and periodically revised by the International Union of Pure and Applied Chemistry.

Methyl group　The group CH_3—.

n-Butyl group　The group $CH_3CH_2CH_2CH_2$—.

Newman projection　A method for depicting conformations in which one sights down a carbon-carbon bond and represents the front carbon by a point and the back carbon by a circle.

Paraffin hydrocarbons　An old name for alkanes and cycloalkanes.

Ring inversion　The process by which a chair conformation of cyclohexane is converted to a mirror-image chair. All of the equatorial substituents become axial, and *vice versa*.

***sec*-Butyl group**　The group $CH_3CH_2CHCH_3$ in which the point of attachment is the carbon of the CH group (indicated in bold face).

***sp*^{3} hybridization**　A model to describe the bonding of a carbon attached to four other atoms or groups. The carbon 2*s* orbital and the three 2*p* orbitals are combined to give a set of four equivalent orbitals having 25% *s* character and 75% *p* character. These orbitals are directed toward the corners of a tetrahedron.

Staggered conformation　A conformation of the type shown in which the bonds on adjacent carbons are as far away from one another as possible.

Stereoisomers　Isomers that have the same constitution but which differ in respect to the arrangement of their atoms in space.

Systematic nomenclature　Names for chemical compounds which are developed on the basis of a prescribed set of rules. Usually the IUPAC system is meant when the term "systematic nomenclature" is used.

***tert*-Butyl group**　The group $(CH_3)_3C$—.

Torsional strain　The decreased stability of a molecule that results from the eclipsing of bonds.

Trivial nomenclature　A term synonymous with *common nomenclature*.

van der Waals strain　The destabilization that results when two atoms or groups approach one another too closely. Also known as *van der Waals repulsion*.

SOLUTIONS TO TEXT PROBLEMS

2.1 Each of the carbons in propane is bonded to four other atoms. All the carbon atoms in propane are sp^3 hybridized.

$$H-\overset{\overset{\displaystyle H}{|}}{\underset{\underset{\displaystyle H}{|}}{C}}-\overset{\overset{\displaystyle H}{|}}{\underset{\underset{\displaystyle H}{|}}{C}}-\overset{\overset{\displaystyle H}{|}}{\underset{\underset{\displaystyle H}{|}}{C}}-H$$

All 10 of the bonds in propane are σ bonds. The eight C—H σ bonds arise from $C(sp^3)$—$H(1s)$ overlap. The two C—C bonds arise from $C(sp^3)$-$C(sp^3)$ overlap.

2.2 A Newman projection of isobutane is constructed by sighting along the C(1)—C(2) bond:

$$H-\overset{\overset{\displaystyle H}{|}}{\underset{\underset{\displaystyle H}{|}}{C}}-\overset{\overset{\displaystyle H}{|}}{\underset{\underset{\displaystyle CH_3}{|}}{C}}-CH_3$$

sight along this bond

The "front" carbon in the Newman projection bears three H's; the "back" carbon has one H and two CH_3 groups.

The remaining two isomers have two methyl branches on a four-carbon chain.

2.3 An unbranched alkane (*n*-alkane) of 28 carbons has 26 methylene (CH_2) groups flanked by a methyl (CH_3) group at each end. The condensed formula is $CH_3(CH_2)_{26}CH_3$.

2.4 The alkane represented by the carbon-skeleton formula has 11 carbons. The general formula for an alkane is C_nH_{2n+2}, and thus there are 24 hydrogens. The molecular formula is $C_{11}H_{24}$; the condensed structural formula is $CH_3(CH_2)_9CH_3$.

2.5 A second isomer exists having a five-carbon chain with a one-carbon (methyl) branch:

$$CH_3CH_2\overset{\overset{\displaystyle CH_3}{|}}{C}HCH_2CH_3 \qquad \text{or}$$

The remaining two isomers have two methyl branches on a four-carbon chain.

$$CH_3\overset{\overset{\displaystyle CH_3}{|}}{\underset{\underset{\displaystyle CH_3}{|}}{C}}HCHCH_3 \qquad \text{or} \qquad\qquad\qquad CH_3CH_2\overset{\overset{\displaystyle CH_3}{|}}{\underset{\underset{\displaystyle CH_3}{|}}{C}}CH_3 \qquad \text{or}$$

2.6 The alkane has an unbranched chain of 11 carbon atoms and is named **undecane**.

2.7 The ending "hexadecane" means that the longest continuous carbon chain of phytane has 16 carbon atoms. There are four methyl groups ("tetramethyl") and they are located at carbons 2, 6, 10, and 14.

2.8 (b) The structures and common names of the three C_5H_{12} isomers are given in the text. The systematic name of the unbranched isomer $CH_3CH_2CH_2CH_2CH_3$ is **pentane**.

The isomer $(CH_3)_2CHCH_2CH_3$ contains a methyl branch at C-2 of a four-carbon chain. Thus it is named as a derivative of butane. It is **2-methylbutane**.

The isomer $(CH_3)_4C$ has three carbons in its longest chain, and so is named as a derivative of propane. There are two methyl groups at C-2, so the isomer is named **2,2-dimethylpropane**.

(c) Begin by writing the structure in more detail (without parentheses) to reveal the longest continuous chain of carbon atoms.

write $(CH_3)_2CH_2CH(CH_3)_2$ as

The compound is named as a derivative of pentane; it is **2,2,4-trimethylpentane**.

(d) Writing the structural formula $(CH_3)_3CC(CH_3)_3$ in sufficient detail to identify the longest continuous chain reveals it to have four carbons; thus, the alkane is named as a derivative of butane.

The butane chain bears four methyl substituents; two at C-2, and two at C-3. The alkane is **2,2,3,3-tetramethylbutane**.

2.9   (b)   The longest continuous chain is 6 carbons so this compound is named as a derivative of hexane.

(CH$_3$CH$_2$)$_2$CHCH$_2$CH(CH$_3$)$_2$ is

$$\begin{array}{cccccc} 6 & 5 & 4 & 3 & 2 & 1 \end{array}$$
$$\text{CH}_3\text{CH}_2\text{CHCH}_2\text{CHCH}_3$$
$$\text{CH}_3\text{CH}_2 \quad \text{CH}_3$$

Substituents are listed in alphabetical order in the name; ethyl precedes methyl. The chain is numbered from the end which gives the lower number to the first appearing substituent. When the chain is numbered from the left, the first substituent (ethyl) is at C-3; when numbered from the right, the first substituent (methyl) is at C-2. Therefore, number the chain from right to left to give the name **4-ethyl-2-methylhexane**.

(c)   The longest continuous chain is 10 carbons so the alkane is named as a derivative of decane. There is an ethyl group (on C-8), an isopropyl group (C-4), and two methyl groups (C-2 and C-6). The name is **8-ethyl-4-isopropyl-2,6-dimethyldecane**.

2.10   (b)   There are 10 carbon atoms comprising the ring of this cycloalkane. The name of the unbranched alkane with 10 carbons is decane. Therefore this cycloalkane is **cyclodecane**.

(c)   There are *cyclopropyl* groups attached to C-1 and C-2 of a four-carbon chain. The compound is **1,2-dicyclopropylbutane**.

(d)   When two cycloalkyl groups are connected by a single bond, the compound is named as a cycloalkyl-substituted cycloalkane. This compound is **cyclohexylcyclohexane**.

2.11   (b)   The disposition of axial and equatorial bonds on a cyclohexane ring is shown below. The equatorial bond to the carbon which corresponds to C-3 in the problem is darkened. Notice that this bond is parallel to the ring bonds on carbons 2 and 4.

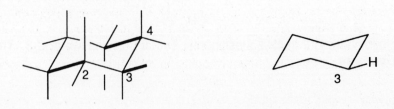

(c) and (d)   The disposition of axial and equatorial bonds on a cyclohexane ring oriented as given in the problem is as indicated. The equatorial bond to C-1 and the axial bond to C-5 are shown as darkened bonds.

2.12   A *tert*-butyl group is larger than a methyl group and has a greater preference for the equatorial position in the most stable conformation.

2.13   The two methyl groups are closer together in the cis isomer. The source of the van der Waals strain is repulsion between them.

cis-1,2-Dimethylcyclopropane

trans-1,2-Dimethylcyclopropane

2.14   The larger *tert*-butyl group has a preference for the equatorial position in the most stable conformation of each stereoisomer.

trans-1-*tert*-Butyl-2-methylcyclohexane

cis-1-*tert*-Butyl-2-methylcyclohexane

2.15   Each carbon of prismane is bonded to three other carbons, and thus bears one hydrogen. The molecular formula of prismane is C_6H_6. The carbons of cubane are bonded in a similar manner; the molecular formula is C_8H_8. Four of the carbon atoms of adamantane bear one hydrogen and six of the carbons have two hydrogens each. The molecular formula of adamantane is $C_{10}H_{16}$.

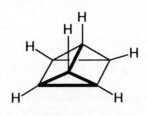

Prismane

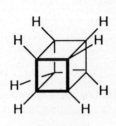

Cubane

Adamantane

2.16 The two unbranched-chain alkanes have the greatest surface area, and the highest boiling points. Nonane has the higher boiling point of the two (151° C), followed by octane (126° C). The alkane with the highest degree of branching has the lowest boiling point. Thus 2,2,3,3-tetramethylbutane boils at 106° C, and the remaining alkane, 2-methylheptane, boils at 116° C.

2.17 Alkane combustion is the reaction with oxygen to produce carbon dioxide and water. The balanced equation is:

$$C_5H_{12} \ + \ 8\,O_2 \longrightarrow 5\,CO_2 \ + \ 6\,H_2O$$

2.18 (a) There are two methylhexanes. They are 2-methylhexane and 3-methylhexane.

$$\begin{array}{c} CH_3 \\ | \\ CH_3CHCH_2CH_2CH_2CH_3 \end{array}$$

2-Methylhexane

$$\begin{array}{c} CH_3 \\ | \\ CH_3CH_2CHCH_2CH_2CH_3 \end{array}$$

3-Methylhexane

(b) Four isomers of heptane are named as dimethylpentanes. In two of the isomers the methyl substituents are on different carbon atoms.

$$\begin{array}{c} CH_3 \\ | \\ CH_3CHCHCH_2CH_3 \\ | \\ CH_3 \end{array}$$

2,3-Dimethylpentane

$$\begin{array}{c} CH_3 \\ | \\ CH_3CHCH_2CHCH_3 \\ | \\ CH_3 \end{array}$$

2,4-Dimethylpentane

The methyl substituents of the other two isomers are on the same carbon atom:

$$\begin{array}{c} CH_3 \\ | \\ CH_3CCH_2CH_2CH_3 \\ | \\ CH_3 \end{array}$$

2,2-Dimethylpentane

$$\begin{array}{c} CH_3 \\ | \\ CH_3CH_2CCH_2CH_3 \\ | \\ CH_3 \end{array}$$

3,3-Dimethylpentane

(c)   There is one isomer of heptane named as an ethyl-substituted pentane:

$$CH_2CH_3$$
$$|$$
$$CH_3CH_2CHCH_2CH_3$$

3-Ethylpentane

2.19 (a)   Each line of a skeleton formula represents a bond between two carbon atoms. Hydrogens are added so the number of bonds to each carbon atom totals four. The molecular formula can be seen by first writing the condensed structural formula.

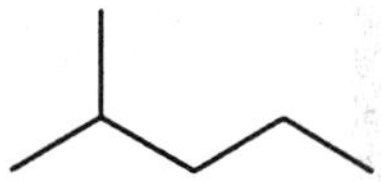

is the same as

$$CH_3$$
$$|$$
$$CH_3CHCH_2CH_2CH_3$$

which is C_6H_{14}

The IUPAC name is **2-methylpentane**.

(b)

is the same as

$$CH_3CH_2CHCH_2CHCH_2CH_3$$
$$|\qquad\quad|$$
$$CH_3\quad\ CH_3$$

which is C_9H_{20}

The IUPAC name is **3,5-dimethylheptane**.

(c)

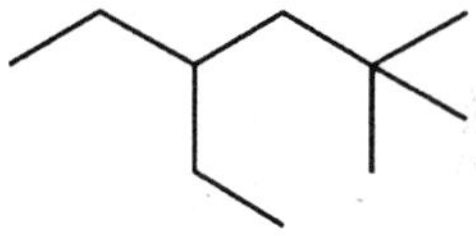

is the same as

$$CH_3CH_2CHCH_2C(CH_3)_3$$
$$|$$
$$CH_2CH_3$$

which is $C_{10}H_{22}$

The IUPAC name is **4-ethyl-2,2-dimethylhexane**.

(d)

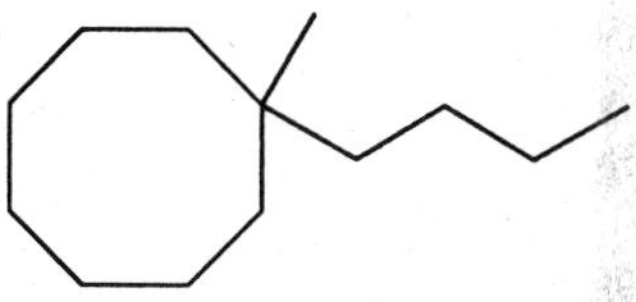

is the same as

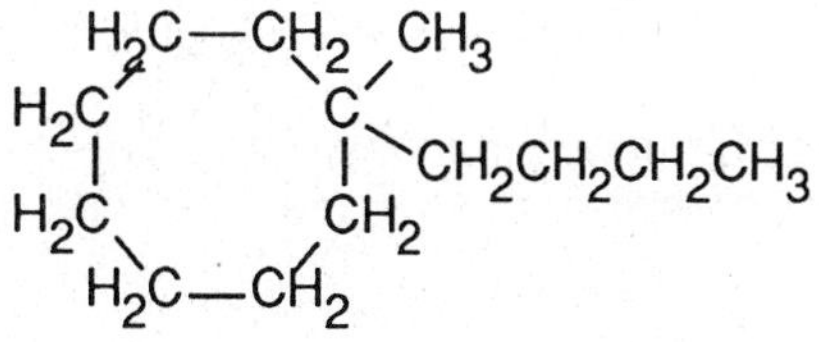

$$\begin{array}{c} H_2C{-}CH_2\ \ CH_3 \\ H_2C^{\nearrow}\qquad\quad{}^{\searrow}C^{\nearrow} \\ |\qquad\qquad\quad|^{\searrow}CH_2CH_2CH_2CH_3 \\ H_2C\qquad\quad\ \ CH_2 \\ {}^{\searrow}H_2C{-}CH_2^{\nearrow} \end{array}$$

which is $C_{13}H_{26}$

The IUPAC name is **1-butyl-1-methylcyclooctane**.

27

2.20 (a) Carbon-skeleton formulas are drawn by representing each carbon-carbon bond by a line. Thus $(CH_3)_2CCH_2CH_2CH_3$ becomes

To name the alkane, find the longest continuous chain of carbon atoms. Locate the substituents by numbering the chain in the direction that gives the substituent the lowest number at the first point of branching. The alkane has two methyl substituents on a five-carbon chain. It is named **2,2-dimethylpentane**.

(b) The alkane contains an ethyl branch on a hexane chain. It is named **3-ethylhexane**.

(c) The longest chain has 8 carbon atoms, thus the compound is named as an octane derivative. It is **5-ethyl-3-methyloctane**.

(d) The longest continuous chain has 7 carbon atoms, and bears an ethyl and three methyl substituents. The compound is named **4-ethyl-2,2,6-trimethylheptane**.

2.21 (a) The name 3-ethylhexane describes a compound whose longest chain is six carbons (hexane), having an ethyl substituent ($-CH_2CH_3$) on C-3.

(b) 6-Isopropyl-2,3-dimethylnonane has a nine-carbon chain (nonane) bearing an isopropyl group [$-CH(CH_3)_2$] on C-6 and two methyl groups on C-2 and C-3.

28

(c)   To the carbon skeleton of heptane (seven carbons) add a *tert*-butyl group to C-4 and a methyl group to C-3 to give 4-*tert*-butyl-3-methylheptane.

(d)   An isobutyl group is —$CH_2CH(CH_3)_2$. The structure of 4-isobutyl-1,1-dimethylcyclohexane is as shown.

(e)   A *sec*-butyl group is $CH_3CHCH_2CH_3$. *sec*-Butylcycloheptane has a *sec*-butyl group on a seven-membered ring.

(f)   Dicyclopropylmethane has two cyclopropyl groups as substituents on the same carbon.

(g)   A cyclobutyl group is a substituent on a five-membered ring in cyclobutylcyclopentane.

2.22   Isomers are different compounds that have the same molecular formula. In all parts of this problem the safest approach is to write a structural formula, then count the number of carbons and hydrogens.

(a)   Butane and isobutane have the formula C_4H_{10} and are isomers.

$$CH_3CH_2CH_2CH_3 \qquad\qquad (CH_3)_2CHCH_3$$

Butane   Isobutane

Cyclobutane (C_4H_8) and 2-methylbutane (C_5H_{12}) are not isomers.

29

(b)   Two compounds of this group are isomers, 2,2-dimethylpentane and 2,2,3-trimethylbutane. Both have the molecular formula (C_7H_{16}).

CH₃
|
CH₃CCH₂CH₂CH₃
|
CH₃

2,2-Dimethylpentane

CH₃
|
CH₃C———CHCH₃
| |
CH₃ CH₃

2,2,3-Trimethylbutane

(c)   Three compounds are isomers and have the formula C_6H_{12}.

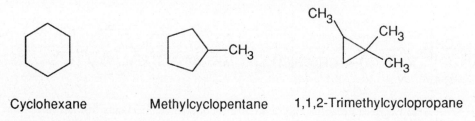

Cyclohexane   Methylcyclopentane   1,1,2-Trimethylcyclopropane

Hexane has the molecular formula C_6H_{14}; it is not an isomer of any of these three.

(d)   The three compounds that are isomers all have the molecular formula C_5H_{10}.

Ethylcyclopropane   1,1-Dimethylcyclopropane   Cyclopentane

1-Cyclopropylpropane is not an isomer of the others. Its molecular formula is C_6H_{12}.

(e)   Only 4-methyltetradecane and pentadecane are isomers. Both have the molecular formula $C_{15}H_{32}$.

CH₃(CH₂)₂CH(CH₂)₉CH₃
|
CH₃

CH₃(CH₂)₁₃CH₃

4-Methyltetradecane   Pentadecane

2,3,4,5-Tetramethyldecane has the molecular formula $C_{14}H_{30}$. The molecular formula of 4-cyclobutyldecane is $C_{14}H_{28}$.

2.23   Alkanes are characterized by the molecular formula C_nH_{2n+2}. The value of n can be calculated on the basis of the fact that the molecular weight is 240, and the atomic weights of carbon and hydrogen are 12 and 1, respectively.

$$12n + 1(2n + 2) = 240$$
$$14n = 238$$
$$n = 17$$

The molecular formula of the alkane is $C_{17}H_{36}$. Since the problem specifies that the carbon chain is unbranched, the hydrocarbon is heptadecane, $CH_3(CH_2)_{15}CH_3$.

2.24 Since it is an alkane, the sex attractant of the tiger moth has a molecular formula of C_nH_{2n+2}. The number of carbons and hydrogens may be calculated from its molecular weight.

$$12n + 1(2n + 2) = 254$$
$$14n = 252$$
$$n = 18$$

The molecular formula of the alkane is $C_{18}H_{38}$. In the problem it is stated that the sex attractant is a 2-methyl-branched alkane. It is therefore 2-methylheptadecane, $(CH_3)_2CHCH_2(CH_2)_{13}CH_3$.

2.25 (a) The IUPAC name for pristane reveals that the longest chain contains 15 carbon atoms (-pentadecane). The chain is substituted with four methyl groups at the positions indicated in the name.

Pristane (2,6,10,14-tetramethylpentadecane)

(b) The molecular formula of pristane is $C_{19}H_{40}$.

2.26 (a) The general formula for an alkane is C_nH_{2n+2}. Thus an alkane having 31 carbon atoms has 64 hydrogens, and the formula of hentriacontane is $C_{31}H_{64}$.

(b) Hentriacontane is the IUPAC name of a straight-chain alkane. The condensed structural formula is $CH_3(CH_2)_{29}CH_3$.

2.27 All single bonds are σ bonds. Draw a structural formula of each compound to reveal the number of bonds in the molecule.

Pentane; 16 σ bonds

Cyclopentane; 15 σ bonds

2.28 (a) and (b) An alkane having 100 carbon atoms has 2(100) + 2 = 202 hydrogens. The molecular formula of hectane is $C_{100}H_{202}$ and the condensed structural formula is $CH_3(CH_2)_{98}CH_3$.

(c) The 100 carbon atoms are connected by 99 σ bonds. The total number of σ bonds is 301 (99 C-C bonds + 202 C-H bonds).

(d) Unique compounds are formed by methyl substitution at carbons 2 through 50 on the 100-carbon chain (C-51 is identical to C-49, and so on). There are 49 methylhectanes.

(e) Compounds of the type 2,X-dimethylhectane can be formed by substitution at carbons 2 through 99. There are 98 of these compounds.

2.29 Torsional strain arises from the repulsion of parallel C—H bonds on adjacent carbon atoms. The torsional strain of cyclopentane is greater than that of cyclopropane; ten C—H bonds contribute to the torsional strain of the former and six to the latter. Cyclopropane has the greater angle strain due to the smaller internal angles of a triangle (60°) compared to a pentagon (108°).

2.30 Begin by drawing a structural formula for **2,2-dimethylpropane**.

A Newman projection of the most stable conformation is constructed by drawing the groups attached to the "front" and "back" carbons in a staggered arrangement.

2.31 Proceeding as in the previous problem, two different staggered conformations can be drawn for **2,3-dimethylbutane**.

2.32 The most stable conformations of alkanes are the staggered ones. Of the three staggered conformations of 2-methylbutane, two are equivalent. The third conformation has a gauche relationship between the methyl groups and is less stable than the others.

Equivalent; most stable conformations   Least stable staggered conformation

2.33 (a) By rewriting the structures in a form that shows the order of their atomic connections, it is apparent that the two structures are constitutional isomers.

is equivalent to $CH_3C(CH_3)_3$

is equivalent to $CH_3CH_2CH(CH_3)_2$

(b) The two compounds have the same constitution; both are $(CH_3)_2CHCH(CH_3)_2$. The Newman projections represent different staggered conformations of the same molecule: in one the hydrogen substituents are anti to each other, whereas in the other they are gauche.

and

Hydrogens at C-2
and C-3 are anti

Hydrogens at C-2
and C-3 are gauche

(c) The methyl and ethyl groups are cis in the first structure but trans in the second. The two compounds are stereoisomers; they have the same constitution but differ in the arrangement of their atoms in space.

cis-1-Ethyl-4-methylcyclohexane
(both alkyl groups are up)

trans-1-Ethyl-4-methylcyclohexane
(ethyl group is down; methyl group is up)

Do not be deceived because the six-membered rings look like ring-flipped forms. Remember, chair-chair interconversion converts all the equatorial bonds to axial and vice versa. Here the ethyl group is equatorial in both structures.

2.34 (a) Since the isomers must contain a four-membered ring, the ring must bear two methyl groups or an ethyl group. This part of the problem specifies that the compounds be *constitutional* isomers of *cis*-1,2-dimethylcyclobutane, so *trans*-1,2-dimethylcyclobutane is not included. The isomers are:

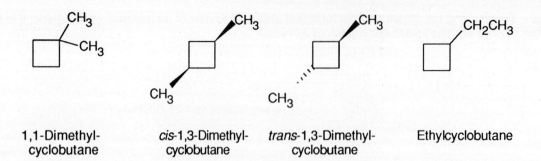

1,1-Dimethyl-
cyclobutane

cis-1,3-Dimethyl-
cyclobutane

trans-1,3-Dimethyl-
cyclobutane

Ethylcyclobutane

(b)   Each isomer must have the molecular formula C_6H_{12}. There are two compounds of this molecular formula which incorporate rings larger than four-membered, methylcyclopentane and cyclohexane.

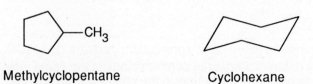

Methylcyclopentane   Cyclohexane

(c)   The stereoisomer of *cis*-1,2-dimethylcyclobutane is *trans* -1,2-dimethylcyclobutane.

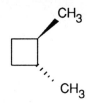

2.35   The most stable conformation of a cyclohexane derivative is the chair conformation in which the maximum number of substituent groups are in the equatorial position.

(a) 1,1,3-Trimethylcyclohexane   (b) 1,1,4-Trimethylcyclohexane

2.36   The less stable chair conformations of the molecules in the preceding problem have a methyl group in the axial position.

(a) 1,1,3-Trimethylcyclohexane   (b) 1,1,4-Trimethylcyclohexane

2.37 (a) The ethyl group is equatorial in the more stable chair conformation of ethylcyclohexane.

Axial ethyl group (less stable)   Equatorial ethyl group (more stable)

(b) Both methyl groups are equatorial in the more stable chair conformation of *trans*-1,4-dimethylcyclohexane.

Both methyl groups axial (less stable)   Both methyl groups equatorial (more stable)

(c) The two chair conformations of *cis*-1,4-dimethylcyclohexane are equivalent. In each, one methyl group is axial and one is equatorial.

(d) In order for the substituents to be cis to one another in a 1,4-disubstituted cyclohexane, one must be axial and the other equatorial. The larger substituent group has a greater preference for the equatorial position. Thus the more stable chair conformation of 1-*tert*-butyl-4-methylcyclohexane has an equatorial *tert*-butyl substituent and an axial methyl group.

tert-Butyl group is axial (less stable)   Methyl group is axial (more stable)

35

2.38 (a, b) For a stereoisomer of 1,3-dimethylcyclohexane to have two equivalent chair conformations, one methyl group must be equatorial while the other is axial. When the cyclohexane ring flips, the axial methyl becomes equatorial, and vice versa. This isomer is *trans*-1,3-dimethylcyclohexane.

(c) The other stereoisomer, *cis*-1,3-dimethylcyclohexane, has two nonequivalent chair conformations. In the less stable one, both methyl groups are axial. In the more stable conformation, the methyl groups are equatorial.

Both methyl groups axial (less stable)   Both methyl groups equatorial (more stable)

2.39 (a) An isopropyl group is larger than a methyl group and so will occupy an equatorial site in the most stable conformation of *cis*-1-isopropyl-3-methylcyclohexane. The methyl group at C-3 is also up in the cis stereoisomer, and is also equatorial.

(b) As in (a), the isopropyl group will occupy an equatorial position in *trans*-1-isopropyl-3-methylcyclohexane. In the trans stereoisomer, unlike the cis, the methyl group occupies an axial site.

(c) In order for two substituents to be cis and located on carbons 1 and 4 of a cyclohexane ring, one group must be axial and the other equatorial (see problem 2.37d). The larger *tert*-butyl group has a stronger preference for the equatorial site in *cis*-1-*tert*-butyl-4-ethylcyclohexane than the ethyl group.

36

2.40 Write each of the compounds in its most stable conformation. Compare them by examining their conformations for sources of strain, such as axial substituent groups.

(a) The cis stereoisomer of 1-isopropyl-2-methylcyclohexane has an axial methyl group and is less stable than the trans.

cis-1-Isopropyl-2-methylcyclohexane
(less stable; one alkyl group axial)

trans-1-Isopropyl-2-methylcyclohexane
(more stable; both alkyl groups equatorial)

(b) Both groups are equatorial in the cis stereoisomer of 1-isopropyl-3-methylcyclohexane; one group (the methyl) is axial in the trans stereoisomer. The cis is more stable.

cis-1-Isopropyl-3-methylcyclohexane
(more stable; both alkyl groups equatorial)

trans-1-Isopropyl-3-methylcyclohexane
(less stable; one alkyl group axial)

(c) The more stable stereoisomer of 1,4-disubstituted cyclohexanes is the trans; both alkyl groups are equatorial in trans-1-isopropyl-4-methylcyclohexane.

trans-1-Isopropyl-4-methylcyclohexane
(more stable; both alkyl groups equatorial)

cis-1-Isopropyl-4-methylcyclohexane
(less stable; one alkyl group axial)

(d)   The first stereoisomer of 1,2,4-trimethylcyclohexane is the more stable one. All its methyl groups are equatorial in its most stable conformation. The most stable conformation of the second stereoisomer has one axial and two equatorial methyl groups.

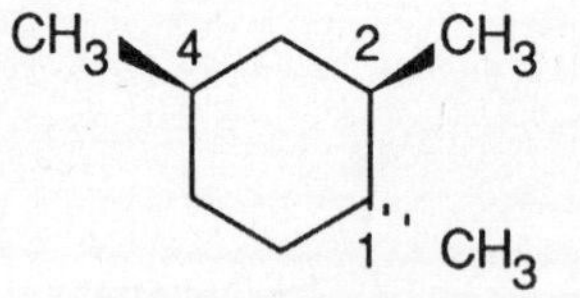

More stable stereoisomer

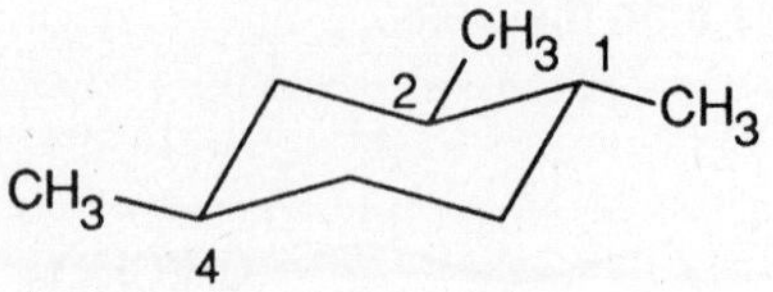

All methyl groups equatorial in most stable conformation

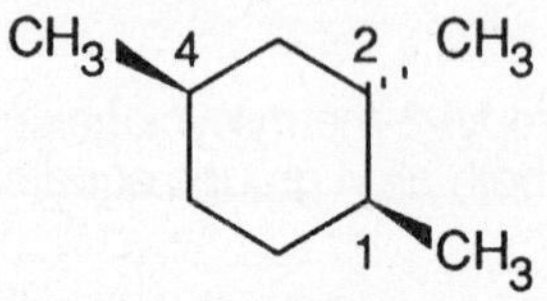

Less stable stereoisomer

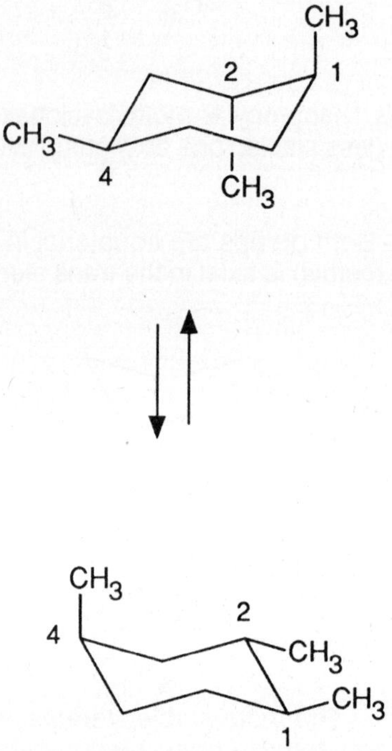

One axial methyl group in most stable conformation

CHAPTER 3

INTRODUCTION TO ORGANIC CHEMICAL REACTIONS

GLOSSARY OF TERMS

Acid According to the Arrhenius definition an acid is a substance which ionizes in water to produce protons. According to the Bronsted-Lowry definition an acid is a substance which donates a proton to some other substance. According to the Lewis definition an acid is an electron-pair acceptor.

Acid dissociation constant, K_a The equilibrium constant for dissociation of an acid defined as:

$$K_a = \frac{[H_3O^+][A^-]}{[HA]}$$

for the reaction:

$$HA + H_2O \rightleftharpoons H_3O^+ + A^-$$

Alcohol A compound of the type ROH.

Alkene A hydrocarbon that contains a carbon-carbon double bond (C=C).

Alkyl group A structural unit related to an alkane by replacing one of the hydrogens by a potential point of attachment to some other atom or group. The general symbol for an alkyl group is R-.

Alkyl halide A compound of the type RX in which X is a halogen substituent (F, Cl, Br, I).

Alkyloxonium ion A positive ion of the type ROH_2^+.

Alkyne A hydrocarbon that contains a carbon-carbon triple bond.

Amine A molecule in which a nitrogen-containing group of the type $-NH_2$, -NHR, or $-NR_2$ is attached to an alkyl or aryl group.

Ar- Symbol for an aryl group.

Arene A hydrocarbon that contains a benzene ring as a structural unit.

Base According to the Arrhenius definition a base is a substance which ionizes in water to produce hydroxide ions. According to the Bronsted-Lowry definition a base is a substance which accepts a proton from some suitable donor. According to the Lewis definition a base is an electron-pair donor.

Bronsted acid (See "Acid.")

Bronsted base (See "Base.")

Bronsted-Lowry definitions See "Acid" and "Base."

Carbocation A positive ion in which the charge resides on carbon. An example is *tert*-butyl cation $(CH_3)_3C^+$. Carbocations are unstable species which, though they cannot normally be isolated, are be-

39

lieved to be intermediates in certain types of organic reactions.

Chain reaction A reaction mechanism in which a sequence of individual steps repeat themselves many times, usually because a reactive intermediate consumed in one step is regenerated in a subsequent step. The halogenation of alkanes is a chain reaction proceeding *via* free radical intermediates.

Chlorofluorocarbon Low molecular weight compounds of the type $C_wH_xCl_yF_z$ used as refrigerants and aerosol propellants. Chlorofluorocarbons accumulate in the upper atmosphere and undergo photochemical reactions which degrade the ozone layer.

Conjugate acid The species formed from a Bronsted base after it has accepted a proton.

Conjugate base The species formed from a Bronsted acid after it has donated a proton.

Electrophile A species (ion or compound) which can act as a Lewis acid or electron-pair acceptor. An "electron-seeker." Carbocations are one type of electrophile.

Epoxide A molecule in which a three-membered oxygen-containing ring is a structural unit.

$$R_2C \overset{\displaystyle }{\underset{\displaystyle O}{\diagup\!\!\!\!\diagdown}} CR_2$$

Ether A molecule that contains a C-O-C unit such as ROR', ROAr, or ArOAr

Free radical A neutral species in which one of the electrons in the valence shell of carbon is unpaired. An example is methyl radical $\cdot CH_3$.

Functional group An atom or group of atoms in a molecule responsible for its reactivity under a given set of conditions.

Hydrogen bonding A type of dipole-dipole attractive force in which a positively polarized hydrogen of one molecule is weakly bonded to a negatively polarized atom of an adjacent molecule. Hydrogen bonds typically involve the hydrogen of one -OH group and the oxygen of another.

Hydronium ion The species H_3O^+.

Hydroxyl group The group –OH.

Initiation step A process which causes a reaction, usually a free-radical reaction, to begin but which by itself is not the principal source of products.

The initiation step in the halogenation of an alkane is the dissociation of a halogen molecule to two halogen atoms.

Intermediate A transient species formed during a chemical reaction. Typically, an intermediate is not stable under the conditions of its formation and proceeds further to form the product.

Lewis acid (See "Acid.")

Lewis base (See "Base.")

Mechanism The sequence of steps which describes how a chemical reaction occurs. A description of the intermediates and transition states which are involved during the transformation of reactants to products.

Nitroalkane A compound of the type RNO_2.

Nucleophile A species (ion or compound) which can act as a Lewis base or electron-pair donor. Halide ions, for example, are nucleophiles.

Oxonium ion Specific name for the species H_3O^+ (also called *hydronium ion*). General name for species such as alkyloxonium ions ROH_2^+ derived from H_3O^+.

Photochemical reaction A chemical reaction which occurs when light is absorbed by a substance.

pK_a A measure of acid strength defined as -log K_a. The stronger the acid, the smaller the value of pK_a.

Primary alkyl group An alkyl group in which the carbon that bears the potential point of attachment is itself directly bonded to only one other carbon. A structural unit of the type RCH_2-.

Propagation steps The fundamental steps which repeat over and over again in a chain reaction. Almost all of the products in the free-radical halogenation of an alkane arise from the propagation steps.

R Symbol for an alkyl group.

Secondary alkyl group An alkyl group in which the carbon that bears the potential point of attachment is itself directly bonded to two other carbons. A structural unit of the type R_2CH-.

Substitution reaction A chemical reaction in which an atom or group of a molecule is replaced by a different atom or group.

Termination steps   Reactions that halt a chain reaction. In a free-radical chain reaction, termination steps consume free radicals without generating new radicals to continue the chain.

Tertiary alkyl group   An alkyl group in which the carbon that bears the potential point of attachment is itself directly bonded to three other carbons. A structural unit of the type R_3C-.

Thiol   A compound of the type RSH or ArSH.

SOLUTIONS TO TEXT PROBLEMS

3.1 (b)   First write the structure of the starting material. Its structural formula reveals that all of the hydrogens of 2,2,3,3-tetramethylbutane are equivalent. Substitution of any one of them with chlorine gives the same product as substitution of any other.

| 2,2,3,3-Tetramethylbutane | Chlorine | 1-Chloro-2,2,3,3-tetramethylbutane | Hydrogen chloride |

(c)   All ten hydrogens of cyclopentane are equivalent, so replacement of any one of them gives the same product.

| Cyclopentane | Chlorine | Chlorocyclopentane | Hydrogen chloride |

3.2   The $C_4H_{10}O$ alcohols contain a hydroxyl group attached to one of the four isomeric C_4H_9 alkyl groups. The longest chain of carbon atoms is numbered in the direction that gives the –OH group the lowest number. The suffix -*ol* is used in place of -*e* in the alkane name corresponding to the number of carbons in the longest chain. The $C_4H_{10}O$ alcohols and their systematic names are:

| 1-Butanol | 2-Butanol | 2-Methyl-1-propanol | 2-Methyl-2-propanol |

3.3   The four C_4H_9 alkyl groups give rise to four isomeric alkyl chlorides having the molecular formula C_4H_9Cl. Halogens and alkyl groups have the same precedence in a systematic name, and are listed alphabetically. The isomeric chlorides are:

| 1-Chlorobutane | 2-Chlorobutane | 1-Chloro-2-methylpropane | 2-Chloro-2-methylpropane |

41

3.4   A carbon bonded to four others accounts for all five carbons present. The C_5H_{12} alkane containing a quaternary carbon is **2,2-dimethylpropane**.

$$CH_3-\overset{\displaystyle CH_3}{\underset{\displaystyle CH_3}{C}}-CH_3 \quad \text{quaternary carbon}$$

3.5   Alcohols are classified as primary, secondary, or tertiary according to the number of carbon atoms directly attached to the carbon that bears the hydroxyl group. A carbon atom that is directly attached to one other carbon is a primary carbon.

$$CH_3-\overset{\displaystyle H}{\underset{\displaystyle CH_3}{C}}-\overset{\displaystyle H}{\underset{\displaystyle H}{C}}-OH \quad \text{a primary carbon}$$

2-Methyl-1-propanol is a **primary** alcohol.

3.6   The name indicates that the compound is a methyl-substituted alcohol and that the hydroxyl group is attached to C-3 of a seven carbon chain (3-heptanol). The structure reveals that the carbon bearing the hydroxyl group is directly attached to two other carbon atoms.

4-Methyl-3-heptanol is a **secondary** alcohol.

$$\text{or} \quad CH_3CH_2\overset{\displaystyle CH_3}{\underset{\displaystyle OH}{CHCHCH}}CH_2CH_2CH_3$$

3.7   Ammonia is a base and abstracts (accepts) a proton from the acid (proton donor) hydrogen chloride.

$$H_3N: \quad + \quad H-\ddot{C}\ddot{l}: \quad \rightleftharpoons \quad \overset{+}{N}H_4 \quad + \quad :\ddot{\underset{\cdot\cdot}{C}l}:^-$$

Base   Acid   Conjugate acid   Conjugate base

3.8   (b)   As described in the sample solution, $K_a = 10^{-pK_a}$; therefore K_a for vitamin C is given by the expression:

$$K_a = 10^{-4.17} = 10^{(-5 + 0.83)} = 10^{-5} \times 10^{0.83}$$

$$K_a = 6.7 \times 10^{-5}$$

(c)   Similarly, $K_a = 1.8 \times 10^{-4}$ for formic acid (pK_a 3.75).

(d) $K_a = 6.5 \times 10^{-2}$ for oxalic acid (pK_a 1.19).

In ranking the acids in order of decreasing acidity, remember that the larger the equilibrium constant K_a, the stronger the acid and the lower the pK_a value.

Acid	K_a	pK_a
Oxalic (strongest)	6.5×10^{-2}	1.19
Aspirin	3.3×10^{-4}	3.48
Formic acid	1.8×10^{-4}	3.75
Vitamin C (weakest)	6.7×10^{-5}	4.17

3.9 The reaction that takes place is:

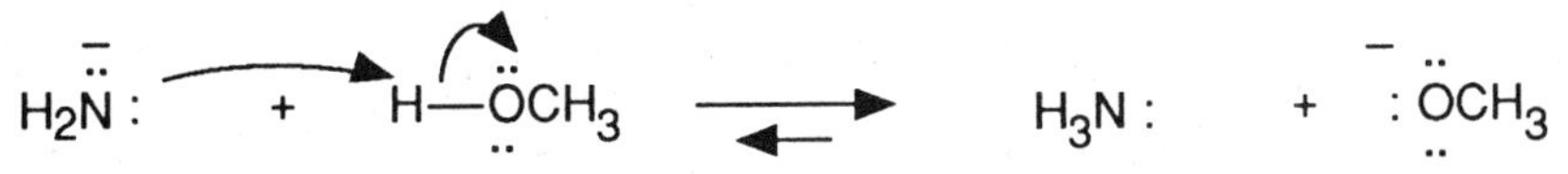

Amide ion	Methanol	Ammonia	Methoxide ion
(stronger base)	(stronger acid)	(weaker acid)	(weaker base)
	$K_a = 10^{-16}$	$K_a = 10^{-36}$	

The position of equilibrium lies to the right, to favor formation of the weaker acid and the weaker base.

3.10 Sulfuric acid has the structure shown in the equation below. When it transfers a proton to a base, the hydrogen sulfate anion is formed. In this case the base is methanol. It accepts a proton from sulfuric acid to give the corresponding oxonium ion.

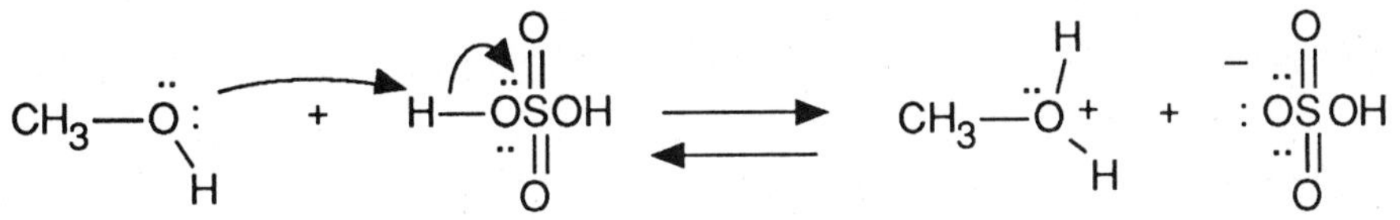

Methanol	Sulfuric acid	Methyloxonium ion	Hydrogen sulfate ion
(base)	(acid)	(conjugate acid)	(conjugate base)

3.11 (b) Hydrogen chloride converts tertiary alcohols to tertiary alkyl chlorides.

$$(CH_3CH_2)_3COH \quad + \quad HCl \quad \longrightarrow \quad (CH_3CH_2)_3CCl \quad + \quad H_2O$$

3-Ethyl-3-pentanol	Hydrogen chloride	3-Chloro-3-ethylpentane	Water

(c) 1-Tetradecanol is a primary alcohol having an unbranched 14-carbon chain. Hydrogen bromide reacts with a primary alcohol to give the corresponding primary alkyl bromide.

$$CH_3(CH_2)_{12}CH_2OH \quad + \quad HBr \quad \longrightarrow \quad CH_3(CH_2)_{12}CH_2Br \quad + \quad H_2O$$

1-Tetradecanol	Hydrogen bromide	1-Bromotetradecane	Water

3.12 A carbocation can exist for each of the four isomeric C_4H_9 alkyl groups (see Problems 3.2 and 3.3). The number of carbon atoms directly attached to the carbon bearing the positive charge determines if the carbocation is primary, secondary, or tertiary.

$$CH_3CH_2CH_2\overset{+}{C}H_2 \qquad CH_3CH_2\overset{+}{C}HCH_3 \qquad CH_3\overset{+}{C}HCH_2 \qquad CH_3\overset{+}{C}CH_3$$
$$\underset{CH_3}{|} \qquad\qquad \underset{CH_3}{|}$$

Primary   Secondary   Primary   Tertiary

3.13 The order of carbocation stability is tertiary > secondary > primary. There is only one $C_5H_{11}^+$ carbocation that is tertiary, so that is the most stable one.

$$CH_3CH_2-\overset{\overset{\displaystyle CH_3}{\diagup}}{\underset{\diagdown}{C}}+$$
$$CH_3$$

3.14 Tertiary alcohols react fastest with hydrogen halides. The only tertiary alcohol having the molecular formula $C_5H_{12}O$ is **2-methyl-2-butanol**.

$$\overset{\displaystyle OH}{\underset{\displaystyle |}{}}$$
$$CH_3CH_2CCH_3$$
$$\underset{CH_3}{|}$$

3.15 The most stable alkyl free radicals are tertiary. The tertiary free radical having the formula C_5H_{11} has the same skeleton as the carbocation in Problem 3.13.

$$CH_3CH_2-\overset{\overset{\displaystyle CH_3}{\diagup}}{\underset{\diagdown}{C}}\cdot$$
$$CH_3$$

3.16 The free radical intermediate in alkane chlorination has the same carbon-hydrogen skeleton as the alkyl chloride product. Thus the propagation steps leading to 1-chlorobutane and 2-chlorobutane are:

$$CH_3CH_2CH_2\overset{\cdot}{C}H_2 \quad + \quad Cl_2 \quad \longrightarrow \quad CH_3CH_2CH_2CH_2Cl \quad + \quad \cdot\overset{..}{\underset{..}{Cl}}:$$

Primary radical   Chlorine   1-Chlorobutane   Chlorine atom

$$CH_3CH_2\overset{\cdot}{C}HCH_3 \quad + \quad Cl_2 \quad \longrightarrow \quad CH_3CH_2CHCH_3 \quad + \quad \cdot\overset{..}{\underset{..}{Cl}}:$$
$$\underset{Cl}{|}$$

Secondary radical   Chlorine   2-Chlorobutane   Chlorine atom

3.17 (b) Write a structural formula of pentane to determine the number of nonequivalent sets of hydrogen substituents. There are three different hydrogen environments in pentane designated as a, b, and c in the structural formula. Chlorination of pentane will yield three monochloro derivatives corresponding to substitution at each of those positions.

$$\overset{a}{CH_3}-\overset{b}{CH_2}-\overset{c}{CH_2}-\overset{b}{CH_2}-\overset{a}{CH_3} \qquad \text{(Pentane)}$$

Cl₂

a → b → c

$CH_3CH_2CH_2CH_2CH_2Cl$

1-Chloropentane

$CH_3CH_2CH_2\underset{Cl}{CH}CH_3$

2-Chloropentane

$CH_3CH_2\underset{Cl}{CH}CH_2CH_3$

3-Chloropentane

(c)   The structure of 2-methylpentane reveals five different hydrogen environments. The six methyl hydrogens designated as (a) are equivalent to one another, but they are not equivalent to the methyl protons designated as (e). Proton (b) is unique, and the protons (c) of one CH_2 group are different from those (d) of the other. There are **five** isomeric monochloro derivatives.

$$\overset{a}{CH_3}-\overset{b}{CH}-\overset{c}{CH_2}-\overset{d}{CH_2}-\overset{e}{CH_3} \qquad \text{(2-Methylpentane)}$$
$$\underset{\underset{a}{CH_3}}{|}$$

Cl₂

a → b → c → d → e

$ClCH_2\underset{CH_3}{CH}CH_2CH_2CH_3$

1-Chloro-2-methylbutane

$CH_3\overset{Cl}{\underset{CH_3}{C}}CH_2CH_2CH_3$

2-Chloro-2-methylbutane

$CH_3\overset{Cl}{\underset{CH_3}{CH}}CHCH_2CH_3$

3-Chloro-2-methylbutane

$CH_3\underset{CH_3}{CH}CH_2\underset{Cl}{CH}CH_3$

2-Chloro-4-methylbutane

$CH_3\underset{CH_3}{CH}CH_2CH_2CH_2Cl$

1-Chloro-4-methylbutane

(d) 2,4-Dimethylpentane has three different hydrogen environments so forms three different monochlorides. The methyl hydrogens (a) are all equivalent, as are the two methine (C–H) hydrogens (b), and there are two methylene hydrogens (c).

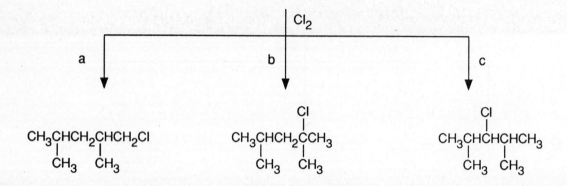

(2,4-Dimethylpentane)

1-Chloro-2,4-dimethylpentane   2-Chloro-2,4-dimethylpentane   3-Chloro-2,4-dimethylpentane

3.18 (a) Cyclobutanol has a hydroxyl group attached to a four-membered ring.

Cyclobutanol

(b) The hydroxyl group is at C-3 of an unbranched seven-carbon chain in 3-heptanol.

$$\overset{1\quad 2\quad\; 3\quad\; 4\quad\;\; 5\quad\;\;\; 6\quad\; 7}{CH_3CH_2CHCH_2CH_2CH_2CH_3}$$
$$|$$
$$OH$$

3-Heptanol

(c) A chlorine at C-2 is on the opposite side of the ring from the hydroxyl group in *trans*-2-chlorocyclopentanol.   Note that it is not necessary to assign a number to the carbon that bears the hydroxyl group. Naming the compound as a derivative of cyclopentanol automatically requires the hydroxyl group to be located at C-1.

trans-2-Chlorocyclopentanol

(d) The longest continuous chain has six carbon atoms in 4-methyl-2-hexanol. The hydroxyl group is at C-2, the methyl is at C-4.

$$\overset{1\quad 2\quad\;\; 3\quad\; 4\quad\; 5\quad\; 6}{CH_3CHCH_2CHCH_2CH_3}$$
$$|\qquad\quad |$$
$$OH\qquad CH_3$$

4-Methyl-2-hexanol

46

(e) A bromine and an iodine are substituents on carbons 1 and 3, respectively, of a four-carbon chain in 1-bromo-3-iodobutane.

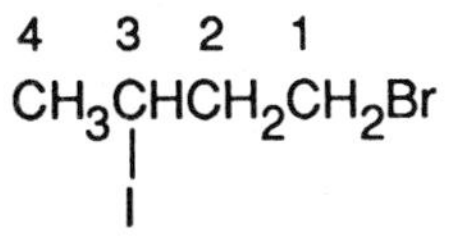

$$\overset{4}{C}H_3\overset{3}{C}H\overset{2}{C}H_2\overset{1}{C}H_2Br \qquad \text{1-Bromo-3-iodobutane}$$

(f) This compound is an alcohol in which the longest continuous chain that incorporates the hydroxyl function has eight carbons. It bears chlorine substituents at C-2 and C-6 and methyl and hydroxyl groups at C-4.

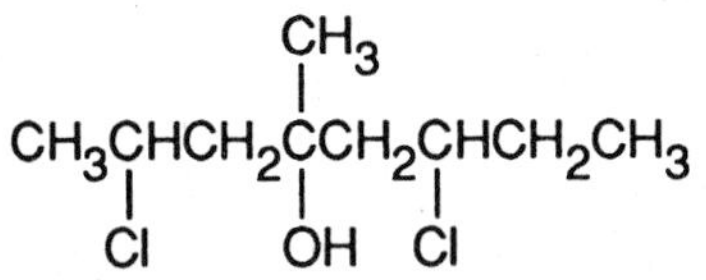

$$
\begin{array}{c}
\hspace{3.5cm}CH_3 \\
\hspace{3.5cm}| \\
CH_3CHCH_2CCH_2CHCH_2CH_3 \\
\hspace{1.0cm}| \hspace{2.3cm}| \hspace{0.6cm}| \\
\hspace{1.0cm}Cl \hspace{1.5cm}OH \hspace{0.3cm}Cl
\end{array}
\qquad \text{2,6-Dichloro-4-methyl-4-octanol}
$$

3.19   The classification of a functional group as primary, secondary, or tertiary is determined by the classification of the carbon atom to which it is attached. A primary carbon is directly attached to one other carbon, a secondary carbon to two other carbons, and a tertiary carbon to three.   The classifications of the molecules in the previous problem are:

(a)   Cyclobutanol is a **secondary** alcohol; the carbon that bears the hydroxyl group is directly attached to two other carbons (marked in the structure shown below).

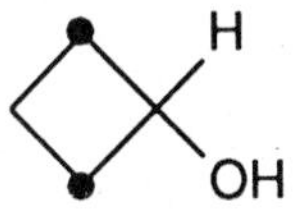

(b)   3-Heptanol is a **secondary** alcohol. There are two carbons (C-2 and C-4) directly attached to the carbon that bears the hydroxyl group (C-3).

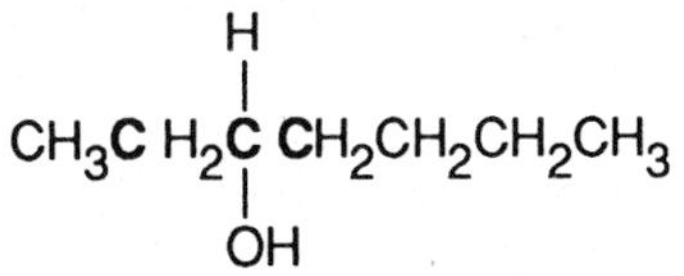

$$
\begin{array}{c}
\hspace{1.5cm}H \\
\hspace{1.5cm}| \\
CH_3CH_2C\ CH_2CH_2CH_2CH_3 \\
\hspace{1.6cm}| \\
\hspace{1.6cm}OH
\end{array}
$$

(c)   Both the carbon that bears the chlorine substituent and the one that bears the hydroxyl group in *trans*-2-chlorocyclopentanol are **secondary** carbons. Each is directly attached to two other carbons.

(d)   4-Methyl-2-hexanol is a **secondary** alcohol. As with 3-heptanol (part b of this problem) the alcohol function is part of a CHOH unit.

(e)   Bromine is attached to a **primary** carbon and iodine to a **secondary** carbon.

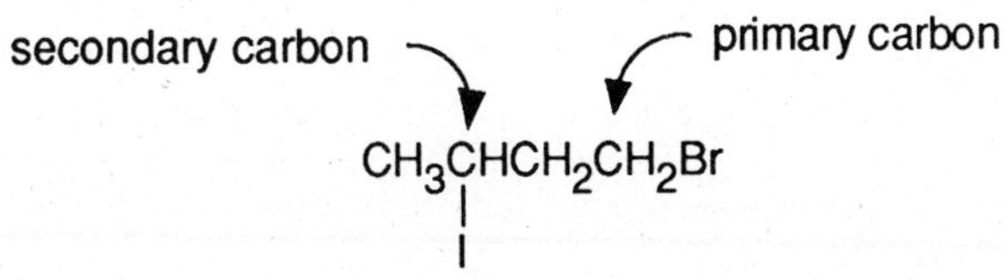

secondary carbon ⟍   ⟋ primary carbon

$$
\begin{array}{c}
CH_3CHCH_2CH_2Br \\
\hspace{0.4cm}| \\
\hspace{0.4cm}I
\end{array}
$$

(f)   Both halogens in 2,6-dichloro-4-methyl-4-octanol are attached to **secondary** carbon atoms and the hydroxyl group is attached to **tertiary** carbon.

tertiary carbon

secondary carbon   CH₃   secondary carbon

CH₃CHCH₂CCH₂CHCH₂CH₃
        | | |
        Cl   OH   Cl

3.20   (a) This compound has a five-carbon chain that bears a hydroxyl group at C-1 and a methyl substituent at C-4.

CH₃CHCH₂CH₂CH₂OH   4-Methyl-1-pentanol
        |
        CH₃

(b) This compound is a derivative of ethane bearing three chlorines and one bromine. The direction of numbering gives bromine, which precedes chlorine in the alphabet, the lower number.

Cl₂CHCHBr   1-Bromo-1,2,2-trichloroethane
        |
        Cl

(c)   This is a trifluoro derivative of ethanol. The direction of numbering is dictated by the hydroxyl group, which takes precedence over the halogens.

CF₃CH₂OH   2,2,2-Trifluoroethanol

(d)   Italicized prefixes are not considered when ranking substituents alphabetically. Therefore, *tert*-butyl precedes fluoro. Both substituents are on the same side of the ring, so they are cis.

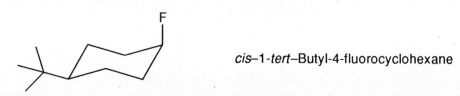

cis–1-*tert*–Butyl-4-fluorocyclohexane

(e)   The longest continuous chain that contains the hydroxyl group has six carbons. Number in the direction that assigns the lower number to the hydroxyl group.

5,5-Dimethyl-2-hexanol

48

(f)   The six carbons that form the longest continuous chain have substituents at C-2, C-3, and C-5 when numbering proceeds in the direction that gives the lowest locants to substituents at the first point of difference. Cite the substituents in alphabetical order.

5-Bromo-2,3-dimethylhexane

3.21   (a) This compound is named as a derivative of ethane, and the halogen substituents are listed alphabetically.

F₃CCHBrCl   2-Bromo-2-chloro-1,1,1-trifluoroethane

(b)   Freon 114 is also named as a derivative of ethane.

1,2-Dichloro-1,1,2,2-tetrafluoroethane

(c)   A cyclobutane ring has eight fluorine substituents. since all the possible locations bear fluorines, no numerical locants are necessary.

Octafluorocyclobutane

3.22   Primary alcohols are alcohols in which the hydroxyl group is attached to a carbon atom that has one alkyl substituent and two hydrogens. There are four primary alcohols which have the molecular formula $C_5H_{12}O$.

CH₃CH₂CH₂CH₂CH₂OH   CH₃CH₂CHCH₂OH   CH₃CHCH₂CH₂OH   CH₃CCH₂OH
                                    | | (with CH₃ above and CH₃ below)
                                   CH₃   CH₃

1-Pentanol   2-Methyl-1-butanol   3-Methyl-1-butanol   2,2-Dimethyl-1-propanol

Secondary alcohols are alcohols in which the hydroxyl group is attached to a carbon atom that has two alkyl substituents and one hydrogen. There are three secondary alcohols of molecular formula $C_5H_{12}O$.

CH₃CHCH₂CH₂CH₃   CH₃CH₂CHCH₂CH₃   CH₃CHCHCH₃
      | | (with CH₃ above and OH below)
      OH   OH

2-Pentanol   3-Pentanol   3-Methyl-2-butanol

49

Only 2-methyl-2-butanol is a tertiary alcohol (three alkyl substituents on the carbon that bears the hydroxyl group).

$$
\begin{array}{c}
CH_3 \\
| \\
CH_3CCH_2CH_3 \\
| \\
OH
\end{array}
$$

2-Methyl-2-butanol

3.23 The higher boiling points of alcohols relative to alkyl chlorides are due primarily to intermolecular hydrogen-bonding forces present in the liquid state of the alcohols. The intermolecular forces present in alkyl chlorides (dipole-dipole association) are much weaker. 1-Propanol has a higher boiling point (97°C) than ethyl chloride (12°C).

3.24 Hydrogen fluoride (bp 19°C) has a higher boiling point than neon (bp -246°C) because of association due to hydrogen bonding.

$$
\overset{\delta+}{H}\!-\!\overset{\delta-}{\ddot{F}}\!:\,-\,-\,-\,\overset{\delta+}{H}\!-\!\overset{\delta-}{\ddot{F}}\!:\,-\,-\,-\,\overset{\delta+}{H}\!-\!\overset{\delta-}{\ddot{F}}\!:\,-\,-\,-\,\overset{\delta+}{H}\!-\!\overset{\delta-}{\ddot{F}}\!:
$$

3.25 The conjugate acid of a substance or species is formed when a proton is gained. The conjugate base is formed when a proton is lost.

	Compound	Conjugate Acid	Conjugate Base
(a)	H_2O	H_3O^+	HO^-
(b)	CH_3OH	$CH_3OH_2^+$	CH_3O^-
(c)	NH_3	NH_4^+	NH_2^-

3.26 (a) The stronger base is the conjugate base of the weaker acid. From Table 3.3 (Section 3.8) it can be seen that ammonia (NH_3, $K_a \sim 10^{-36}$) is a weaker acid than water (H_2O, $K_a = 1.8 \times 10^{-16}$). Therefore the conjugate base of ammonia, amide ion (**NH_2^-**), is a stronger base than hydroxide ion, HO^-.

(b) Fluoride ion (**F^-**) is the conjugate base of the weak acid HF, and is a stronger base than chloride ion. Chloride ion is the conjugate base of HCl, a strong acid.

(c) The stronger base is ethoxide ion, **$CH_3CH_2O^-$**, the conjugate base of ethanol CH_3CH_2OH ($K_a \sim 10^{-16}$. Acetate ion, $CH_3CO_2^-$, is the conjugate base of acetic acid ($K_a = 1.8 \times 10^{-5}$).

(d) Both nitrate and hydrogen sulfate are derived from strong acids and are very weak bases. From Table 3.3 we find that nitric acid has the smaller ionization constant K_a (2.5×10^1) and is, therefore, a weaker acid than sulfuric acid ($K_a = 1.6 \times 10^5$). Therefore, nitrate ion (**$^-ONO_2$**) is a stronger base than hydrogen sulfate ($^-OSO_2OH$).

3.27 In all these reactions the negatively charged atom abstracts a proton from an acid.

(a)

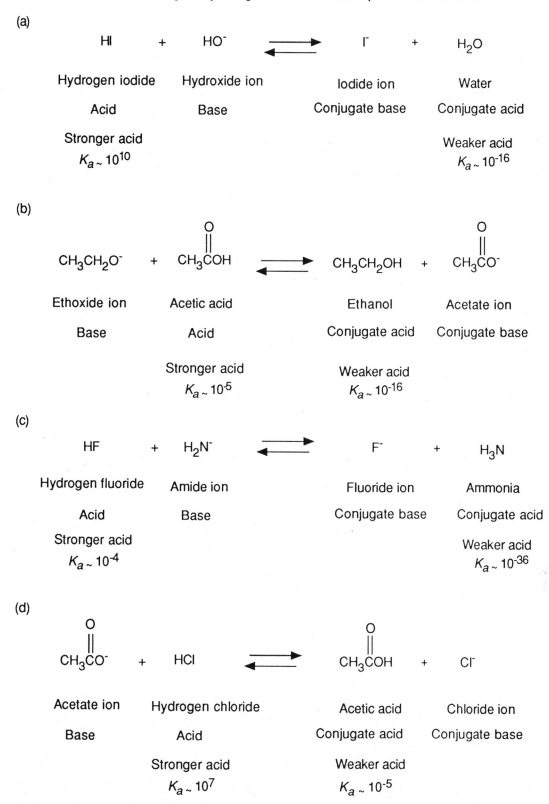

HI + HO⁻ ⇌ I⁻ + H_2O

Hydrogen iodide   Hydroxide ion   Iodide ion   Water

Acid   Base   Conjugate base   Conjugate acid

Stronger acid   Weaker acid
$K_a \sim 10^{10}$ $K_a \sim 10^{-16}$

(b)

$CH_3CH_2O^-$ + $CH_3\overset{\displaystyle O}{\overset{\|}{C}}OH$ ⇌ CH_3CH_2OH + $CH_3\overset{\displaystyle O}{\overset{\|}{C}}O^-$

Ethoxide ion   Acetic acid   Ethanol   Acetate ion

Base   Acid   Conjugate acid   Conjugate base

Stronger acid   Weaker acid
$K_a \sim 10^{-5}$ $K_a \sim 10^{-16}$

(c)

HF + H_2N^- ⇌ F⁻ + H_3N

Hydrogen fluoride   Amide ion   Fluoride ion   Ammonia

Acid   Base   Conjugate base   Conjugate acid

Stronger acid   Weaker acid
$K_a \sim 10^{-4}$ $K_a \sim 10^{-36}$

(d)

$CH_3\overset{\displaystyle O}{\overset{\|}{C}}O^-$ + HCl ⇌ $CH_3\overset{\displaystyle O}{\overset{\|}{C}}OH$ + Cl⁻

Acetate ion   Hydrogen chloride   Acetic acid   Chloride ion

Base   Acid   Conjugate acid   Conjugate base

Stronger acid   Weaker acid
$K_a \sim 10^7$ $K_a \sim 10^{-5}$

3.28 (a) An acid-base reaction will favor products ($K > 1$) when the stronger acid reacts with the stronger base. Use the acid dissociation constants in Table 3.3 to determine the relative acid strengths in the reaction.

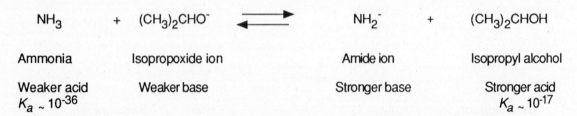

$(CH_3)_3CO^-$ + H_2O ⇌ $(CH_3)_3COH$ + HO^-

tert-Butoxide ion   Water   tert-Butyl alcohol   Hydroxide ion

Stronger base   Stronger acid   Weaker acid   Weaker base
                        $K_a \sim 10^{-16}$ $K_a \sim 10^{-18}$

The reaction meets the criterion for an equilibrium that favors products; K is greater than 1.

(b) Proceeding as in part (a), write the equilibrium and determine the relative acid strengths of the species involved.

NH_3 + $(CH_3)_2CHO^-$ ⇌ NH_2^- + $(CH_3)_2CHOH$

Ammonia   Isopropoxide ion   Amide ion   Isopropyl alcohol

Weaker acid   Weaker base   Stronger base   Stronger acid
$K_a \sim 10^{-36}$ $K_a \sim 10^{-17}$

The reaction in this part of the problem favors reactants (lies to the left); K is less than 1.

(c) Proceeding as before:

HF + HO^- ⇌ F^- + H_2O

Hydrogen fluoride   Hydroxide ion   Fluoride ion   Water

Stronger acid   Stronger base   Weaker base   Weaker acid
$K_a = 3.5 \times 10^{-4}$ $K_a = 1.8 \times 10^{-16}$

The reaction between hydrofluoric acid and hydroxide ion favors products; K is greater than 1.

3.29 (a) pK_a is calculated from the acid dissociation constant by:

$$pK_a = -\log K_a$$

Therefore the pK_a of benzoic acid is:

$$pK_a = -\log (6.3 \times 10-5)$$
$$pK_a = -\log 6.3 - \log (10^{-5})$$
$$pK_a = -0.80 - (-5) = \mathbf{4.20}$$

(b) The pK_a given for chloroacetic acid in the problem is 2.9. This is smaller than the pK_a of 4.2 just calculated for benzoic acid. The smaller the pK_a, the stronger the acid. **Chloroacetic acid** is a stronger acid than benzoic acid.

(c) Benzoic acid is a weaker acid than chloroacetic acid, so **benzoate ion** (the conjugate base of benzoic acid) is a stronger base than chloroacetate ion.

3.30 The larger pK_a value describes the weaker acid, thus methane (pK_a 60) is a weaker acid than methanol (pK_a 16).

$$CH_3O^- \quad + \quad CH_4 \quad \rightleftharpoons \quad CH_3^- \quad + \quad CH_3OH$$

Methoxide ion	Methane	Methide ion	Methanol
Weaker base	Weaker acid	Stronger base	Stronger acid

(b) The equilibrium constant favors reactants and is less than 1.

3.31 Each reagent in this problem substitutes a halogen for the hydroxyl group of 1-butanol.

(a) Primary alcohols are converted efficiently to alkyl bromides on being heated with hydrogen bromide.

$$CH_3CH_2CH_2CH_2OH \quad \xrightarrow[\text{heat}]{HBr} \quad CH_3CH_2CH_2CH_2Br$$

1-Butanol   1-Bromobutane

(b) Primary alcohols do not react very readily with hydrogen chloride. Therefore, when primary alkyl chlorides are desired the reaction of a primary alcohol with thionyl chloride is used.

$$CH_3CH_2CH_2CH_2OH \quad \xrightarrow{SOCl_2} \quad CH_3CH_2CH_2CH_2Cl$$

1-Butanol   1-Chlorobutane

(c) As an alternative to their reaction with hydrogen bromide, alcohols may be converted to alkyl bromides with phosphorus tribromide.

$$CH_3CH_2CH_2CH_2OH \quad \xrightarrow{PBr_3} \quad CH_3CH_2CH_2CH_2Br$$

1-Butanol   1-Bromobutane

3.32 Similar reactions occur with a secondary alcohol such as cyclohexanol.

(a)

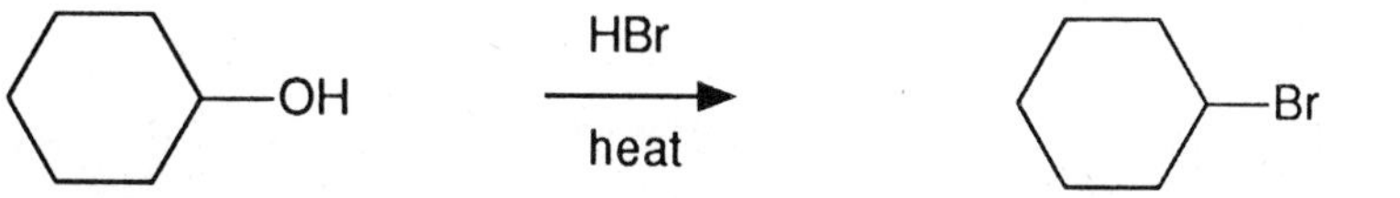

Cyclohexanol   Bromocyclohexane

(b)

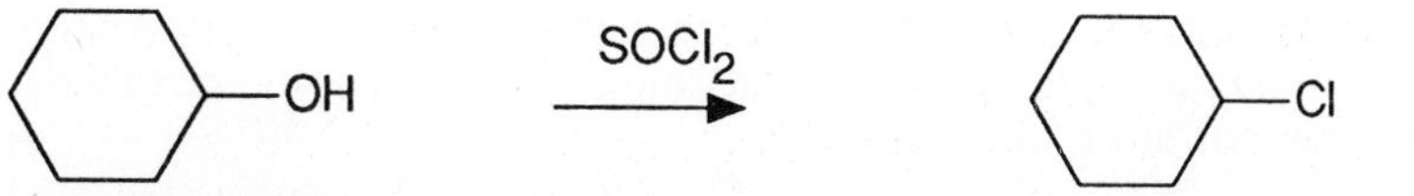

Cyclohexanol   Chlorocyclohexane

53

(c)

$$\text{Cyclohexanol} \xrightarrow{\ \text{PBr}_3\ } \text{Bromocyclohexane}$$

Cyclohexanol   Bromocyclohexane

3.33 Alcohols react with hydrogen bromide to give alkyl bromides.

$$\text{ROH} \quad + \quad \text{HBr} \quad \longrightarrow \quad \text{RBr} \quad + \quad \text{H}_2\text{O}$$

Alcohol   Hydrogen bromide   Alkyl bromide   Water

The order of reactivity of alcohols with is tertiary > secondary > primary. A primary alcohol is one in which the –OH group is attached to a primary carbon; in a secondary alcohol the –OH group is attached to a secondary carbon; the –OH group is attached to a tertiary carbon in a tertiary alcohol.

(a) **2-Butanol** (secondary) is more reactive than 1-butanol (primary).

2-Butanol   1-Butanol

(b) **2-Methyl-2-butanol** (tertiary) is more reactive than 2-butanol (secondary).

2-Methyl-2-butanol   2-Butanol

(c) **2-Butanol** is more reactive. 2-Methylbutane [$CH_3CH_2CH(CH_3)_2$] is an alkane and does not react with HBr.

3.34 The carbocation formed on reaction with hydrogen chloride has the same carbon skeleton as the starting alcohol. The order of carbocation stability is tertiary > secondary > primary.

$$(CH_3)_2\overset{+}{C}CH_2CH_3 \qquad (CH_3)_2CH\overset{+}{C}HCH_3 \qquad (CH_3)_2CHCH_2\overset{+}{C}H_2$$

Tertiary carbocation   Secondary carbocation   Primary carbocation
most stable   least stable

3.35 As in the preceding problem, the order of carbocation stability is tertiary > secondary > primary. *Write the structural formula for the carbocation and determine whether the positively charged carbon is directly bonded to one, two, or three other carbons.*

Secondary carbocation

Primary carbocation
least stable

Tertiary carbocation
most stable

3.36 While both cations are secondary, the positively charged sp^2-hybridized carbon of cyclopropyl cation is incorporated into a three-membered ring making it impossible for this carbon to possess the 120° bond angles which are optimal for sp^2-hybridization. Cyclopropyl cation is less stable than isopropyl cation.

$$CH_3 \overset{\overset{\overset{H}{|}}{\overset{C}{}}}{\underset{+}{}} CH_3$$

Isopropyl cation
more stable

Cyclopropyl cation
less stable

3.37 The free radicals having the formula C_4H_9 have the same carbon skeletons as the carbocations in Problem 3.12. The tertiary radical (*tert*-butyl) is the most stable.

$$CH_3CH_2CH_2\overset{\overset{H}{|}}{\underset{\underset{H}{|}}{C}}\cdot$$

Primary

$$CH_3CH_2\overset{\overset{H}{|}}{\underset{\underset{CH_3}{|}}{C}}\cdot$$

Secondary

$$CH_3\overset{\overset{H}{|}}{\underset{\underset{CH_3}{|}}{C}}\overset{H}{\underset{}{C}}\cdot$$

Primary

$$CH_3\overset{\overset{CH_3}{|}}{\underset{\underset{CH_3}{|}}{C}}\cdot$$

Tertiary

most stable

3.38 The compound $(CH_3)_3CCH_2CH_3$ contains three nonequivalent sets of hydrogens, and therefore gives three monochlorides on chlorination. They are:

$$ClCH_2\overset{\overset{CH_3}{|}}{\underset{\underset{CH_3}{|}}{C}}CH_2CH_3$$

1-Chloro-2,2-dimethylbutane

$$CH_3\overset{\overset{CH_3}{|}}{\underset{\underset{CH_3}{|}}{C}}\overset{}{\underset{\underset{Cl}{|}}{C}}HCH_3$$

3-Chloro-2,2-dimethylbutane

$$CH_3\overset{\overset{CH_3}{|}}{\underset{\underset{CH_3}{|}}{C}}CH_2CH_2Cl$$

1-Chloro-3,3-dimethylbutane

3.39 Free-radical chlorination leads to substitution at each distinct position that bears a hydrogen substituent. This problem essentially requires you to recognize structures that possess various numbers of nonequivalent hydrogens.

(a) The C_5H_{12} isomer that gives a single monochloride is **2,2-dimethylpropane**, since this is the isomer which has all of its hydrogens equivalent to one another.

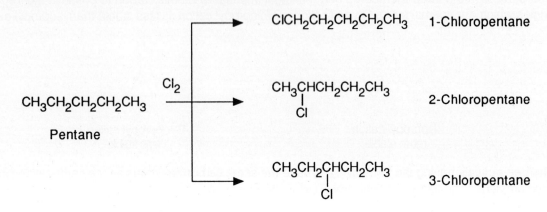

2,2-Dimethylpropane   1-Chloro-2,2-dimethylpropane

(b) The C_5H_{12} isomer that has three nonequivalent sets of hydrogens is **pentane**. It yields three isomeric monochlorides on free-radical chlorination.

$$CH_3CH_2CH_2CH_2CH_3 \xrightarrow{Cl_2}$$

Pentane

$ClCH_2CH_2CH_2CH_2CH_3$ 1-Chloropentane

$CH_3CHCH_2CH_2CH_3$ with Cl   2-Chloropentane

$CH_3CH_2CHCH_2CH_3$ with Cl   3-Chloropentane

(c) **2-Methylbutane** forms four different monochlorides.

$$CH_3CHCH_2CH_3 \text{ with } CH_3 \xrightarrow{Cl_2}$$

2-Methylbutane

$ClCH_2CHCH_2CH_3$ with CH_3 1-Chloro-2-methylbutane

$CH_3CCH_2CH_3$ with Cl and CH_3 2-Chloro-2-methylbutane

$CH_3CHCHCH_3$ with Cl and CH_3 2-Chloro-3-methylbutane

$CH_3CHCH_2CH_2Cl$ with CH_3 1-Chloro-3-methylbutane

(d) In order that only two dichlorides be formed, the starting alkane must have a structure that in which most (or all) of the hydrogens are equivalent. **2,2-Dimethylpropane** satisfies this requirement.

$$
\underset{\text{2,2-Dimethylpropane}}{\underset{\displaystyle |}{\overset{\displaystyle CH_3}{\underset{CH_3}{\overset{|}{CH_3CCH_3}}}}}
\quad \xrightarrow[\text{light}]{2\ Cl_2} \quad
\underset{\text{1,1-Dichloro-2,2-dimethylpropane}}{\underset{\displaystyle CH_3}{\overset{\displaystyle CH_3}{CH_3CCHCl_2}}}
\quad + \quad
\underset{\text{1,3-Dichloro-2,2-dimethylpropane}}{\underset{\displaystyle CH_3}{\overset{\displaystyle CH_3}{ClCH_2CCH_2Cl}}}
$$

3.40 The steps in the chlorination of ethane are:

Initiation: $\qquad Cl_2 \xrightarrow[\text{heat}]{\text{light or}} 2Cl\cdot$

Propagation: $\qquad Cl\cdot \ + \ CH_3CH_3 \longrightarrow CH_3\overset{\displaystyle\cdot}{C}H_2 \ + \ HCl$

$\qquad\qquad\qquad CH_3\overset{\displaystyle\cdot}{C}H_2 \ + \ Cl_2 \longrightarrow CH_3CH_2Cl \ + \ \cdot Cl$

Termination: $\qquad Cl\cdot \ + \ Cl\cdot \longrightarrow Cl_2$

$\qquad\qquad\qquad CH_3\overset{\displaystyle\cdot}{C}H_2 \ + \ CH_3\overset{\displaystyle\cdot}{C}H_2 \longrightarrow CH_3CH_2CH_2CH_3$

$\qquad\qquad\qquad CH_3\overset{\displaystyle\cdot}{C}H_2 \ + \ Cl\cdot \longrightarrow CH_3CH_2Cl$

3.41 The product of monochlorination of ethane is ethyl chloride. Writing the structural formula for ethyl chloride reveals that there are two nonequivalent sets of hydrogen atoms which can be replaced by a second chlorine atom.

$$
\underset{\text{Chloroethane}}{CH_3CH_2Cl} \quad \xrightarrow[\substack{\text{light or}\\\text{heat}}]{Cl_2} \quad \underset{\text{1,1-Dichloroethane}}{CH_3CHCl_2} \quad + \quad \underset{\text{1,2-Dichloroethane}}{ClCH_2CH_2Cl}
$$

CHAPTER 4

ALKENES, ALKADIENES, AND ALKYNES I. STRUCTURE AND PREPARATION

GLOSSARY OF TERMS

Acetylene The simplest alkyne C_2H_2. The molecule is linear; both carbons are *sp* hybridized are are joined by a triple bond.

Alkadiene A hydrocarbon that contains two carbon-carbon double bonds; commonly referred to as "dienes."

Alkene A hydrocarbon that contains a carbon-carbon double bond; also known by older name "olefin."

Alkyne A hydrocarbon that contains a carbon-carbon triple bond.

Allyl group The group $CH_2=CHCH_2-$. (Do not confuse with vinyl group; see below.)

cis When applied to alkenes, the prefix *cis-* indicates that similar substituents are on the **same side** of the double bond. (Contrast with the prefix *trans-*.)

Concerted reaction A reaction involving a single transition state in which bond-making and bond-breaking take place simultaneously. The E2 mechanism is concerted; the E1 mechanism is a nonconcerted two-step process.

Conjugated diene A system of the type

$$C=C-C=C$$

in which two double bonds are joined by a single bond. The π electrons are delocalized over the unit of four consecutive sp^2 hybridized carbons.

Cumulated diene A system of the type $C=C=C$ in which a single carbon atom participates in double bonds with two others. Cumulated dienes are less stable than conjugated dienes and are infrequently encountered.

Cycloalkene A cyclic hydrocarbon which includes a double bond between two of the carbons in the ring.

Dehydration A reaction in which an alcohol, on being heated in the presence of an acid, is converted to an alkene by loss of its OH group along with an H from an adjacent carbon.

Dehydrohalogenation A reaction in which an alkyl halide, on being treated with a base such as sodium ethoxide, is converted to an alkene by loss of a proton from one carbon and the halogen from the adjacent carbon.

Disubstituted alkene An alkene of the type $R_2C=CH_2$ or $RCH=CHR$. The groups R may be the same or different, they may be any length and may be branched or unbranched. The significant point is that there are two carbons *directly* bonded to the carbons of the double bond.

Double dehydrohalogenation A reaction in which a geminal dihalide or vicinal dihalide, on being treated with a very strong base such as sodium amide, is converted to an alkyne by loss of two protons and the two halogen substituents.

E1 Mechanism *Elimination-unimolecular*. A mechanism for elimination characterized by the slow formation of a carbocation intermediate followed by rapid loss of a proton from the carbocation to form the alkene.

E2 Mechanism *Elimination-bimolecular*. A mechanism for elimination of alkyl halides characterized by a transition state in which the attacking base removes a proton at the same time that the bond to the halide leaving group is broken.

Elimination A reaction in which a double or triple bond is formed by loss of atoms or groups from adjacent atoms. (See *dehydration, dehydrohalogenation,* and *double dehydrohalogenation*.)

Ethene The IUPAC name for $CH_2=CH_2$. The common name ethylene, however, is used far more often and the IUPAC rules permit its use.

Ethylene Ethylene is the simplest alkene and the most important industrial organic chemical. Its structural formula is $CH_2=CH_2$. Ethylene is a planar molecule and both of its carbon atoms are sp^2 hybridized. The double bond is composed of a σ component and a π component.

Geminal dihalide A dihalide in which the two halogen substituents are located on the same carbon.

Internal triple bond A triple bond which appears at a position other than at the end of a carbon chain; the alkyne represented by the formula below has an internal triple bond.

$$RC\equiv CR$$

Isolated diene A compound of the type

$$C=C-(C)_x-C=C$$

in which the two double bonds are separated by one or more sp^3 hybridized carbons. Isolated dienes are slightly less stable than isomeric conjugated dienes.

Monosubstituted alkene An alkene of the type $RCH=CH_2$ in which there is only one carbon *directly* bonded to the carbons of the double bond.

π Bond In alkenes, a bond formed by overlap of p orbitals in a side-by-side manner. A π bond is weaker than a σ bond. The carbon-carbon double bond in alkenes consists of two sp^2-hybridized carbons joined by a σ bond and a π bond.

sp Hybridization The hybridization state adopted by carbon when it bonds to two other atoms as, for example, in alkynes. The s orbital and one of the $2p$ orbitals mix to form two equivalent sp-hybridized orbitals. A linear geometry is characteristic of sp hybridization.

Stereoisomeric alkenes Isomeric alkenes which possess the same constitution, but which differ in the arrangement of their atoms in space. *cis*- and *trans*-2-Butene are stereoisomers of one another. Both have the constitution $CH_3CH=CHCH_3$ but the methyl groups are on the same side of the double bond in the cis isomer while they are on opposite sides in the trans isomer.

Terminal alkyne An alkyne of the type shown below in which the triple bond appears at the end of the chain.

$$RC\equiv CH$$

Tetrasubstituted alkene An alkene of the type $R_2C=CR_2$ in which there are four carbons *directly* bonded to the carbons of the double bond. (The R groups may be the same or different.)

trans When applied to alkenes, the prefix *trans*- indicates that analogous substituents are on **opposite sides** of the double bond. (Contrast with the prefix *cis*-.)

Transition state The point of maximum energy on the passage of starting materials to products. The species which exists at the transition state is called the *activated complex*.

Trisubstituted alkene An alkene of the type $R_2C=CHR$ in which there are three carbons *di*-

rectly bonded to the carbons of the double bond. (The R groups may be the same or different.)

Vicinal dihalide A dihalide in which the two halogen substituents are located on adjacent carbons.

Vinyl group The group $CH_2=CH-$. (Do not confuse with allyl group; see above.)

Zaitsev's rule When two or more alkenes are capable of being formed by an elimination reaction, the one with the more highly substituted double bond (the more stable alkene) is the major product.

SOLUTIONS TO TEXT PROBLEMS

4.1 (b) Writing the structure in more detail, we see that the longest continuous chain contains four carbon atoms.

$$\overset{CH_3}{\underset{CH_3}{\overset{|}{\underset{|}{\overset{4}{CH_3}-\overset{3}{C}-\overset{2}{CH}=\overset{1}{CH_2}}}}}$$

The double bond is located at the end of the chain, so the alkene is named as a derivative of 1-butene. Two methyl groups are substituents at C-3. The correct IUPAC name is **3,3-dimethyl-1-butene**.

(c) Expanding the structural formula reveals the molecule to be a methyl-substituted derivative of hexene.

$$\overset{1}{CH_3}-\underset{\underset{CH_3}{|}}{\overset{2}{C}}=\overset{3}{CH}\overset{4}{CH_2}\overset{5}{CH_2}\overset{6}{CH_3}$$

The chain is numbered so as to give the lowest number to the doubly bonded carbons, which makes the alkene a derivative of 2-hexene (not 4-hexene). Its IUPAC name is **2-methyl-2-hexene**.

(d) The IUPAC rules require that the chain be numbered in the direction that gives lower numbers to the doubly bonded carbons rather than the direction that gives the lower numbers to alkyl substituents.

$$\overset{1}{CH_3}-\underset{\underset{CH_3}{|}}{\overset{2}{C}}=\overset{3}{CH}-\underset{\underset{CH_3}{|}}{\overset{4}{\overset{\overset{CH_3}{|}}{C}}}-\overset{5}{CH_3}$$

2,4,4-Trimethyl-2-pentene

(*not* 2,2,4-trimethyl-3-pentene)

(e) First identify the longest continuous chain that includes the double bond, then number it in the direction that gives the lowest numbers to the doubly bonded carbons.

$$\overset{1}{CH_3}\underset{\underset{CH_3-CH}{\underset{\underset{CH_3}{|}}{|}}}{\overset{2}{CH}}\overset{3}{C}=\overset{4}{CH}\overset{5}{CH_2}\underset{\underset{CH_3}{|}}{\overset{6}{CH}}\overset{7}{CH_3}$$

60

This compound is named as a derivative of 3-heptene. It bears methyl substituents at C-2 and C-6, along with an isopropyl group at C-3. The substituents are listed in alphabetical order in the IUPAC name. This alkene is **3-isopropyl-2,6-dimethyl-3-heptene**.

4.2   There are three sets of nonequivalent positions on a cyclopentene ring, identified as *a*, *b*, and *c* on the cyclopentene structure shown at the top of the next page.

Thus, there are three different monochloro-substituted derivatives of cyclopentene. The carbons that bear the double bond are numbered C-1 and C-2 in each isomer, and the other positions are numbered in sequence in the direction that gives the chlorine-bearing carbon its lower locant.

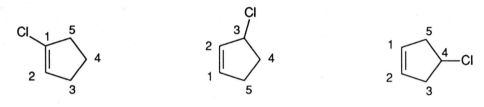

| 1-Chlorocyclopentene | 3-Chlorocyclopentene | 4-Chlorocyclopentene |

4.3   (b)   The longest chain of carbon atoms containing the double bond has six carbon atoms, so this compound is named as a derivative of hexene.

$$\underset{H}{\overset{\overset{6}{CH_3}\overset{5}{CH_2}}{\diagdown}} \underset{4}{C} = \underset{3}{C} \underset{CH_2CH_3}{\overset{\overset{2}{CH_2}\overset{1}{CH_3}}{\diagup}}$$

3-Ethyl-3-hexene

(c)   The two carbons connected by the double bond (C-3 and C-4) are sp^2 hybridized. The remaining six carbon atoms are sp^3 hybridized.

(d)   The three alkyl groups attached directly to the double bond are linked by sp^2-sp^3 σ bonds. The remaining three carbon-carbon bonds are of the sp^3-sp^3 type.

4.4   (b)   Stereoisomerism is not possible in 2-chloropropene. One of the carbons of the double bond (C-1) bears two identical substituents (hydrogens).

$$\underset{H}{\overset{H}{\diagdown}} C = C \underset{Cl}{\overset{CH_3}{\diagup}}$$

(c)   This compound can exist as a pair of stereoisomers. In one, the two chlorines are on the same side of the double bond; in the other, they are on opposite sides.

cis-1,2-Dichloropropene   *trans*-1,2-Dichloropropene

4.5   Apply the two general rules for alkene stability to rank these compounds. First, more highly substituted double bonds are more stable than less substituted ones. Second, when two double bonds are similarly constituted, the trans stereoisomer is more stable than the cis. Therefore, the predicted order of decreasing stability is:

2-Methyl-2-butene	*trans*-2-Pentene	*cis*-2-Pentene	1-Pentene
(trisubstituted)	(disubstituted)	(disubstituted)	(monosubstituted)
most stable			least stable

4.6   Write out the structure of the alcohol, recognizing that the alkene is formed by loss of a hydrogen and a hydroxyl group from adjacent carbons.

(b), (c) Both 1-propanol and 2-propanol give propene on acid-catalyzed dehydration.

1-Propanol   Propene   2-Propanol

(d)   Carbon-3 has no hydrogen substituents in 2,3,3-trimethyl-2-butanol. Elimination can involve only the hydroxyl group at C-2 and a hydrogen at C-1.

2,3,3-Trimethyl-2-butanol   2,3,3-Trimethyl-1-butene

62

4.7 **(b)** Elimination can involve loss of a hydrogen from the methyl group or from C-2 of the ring in 1-methylcyclohexanol.

1-Methylcyclohexanol   Methylenecyclohexane   1-Methylcyclohexene
                                     (disubstituted alkene;   (trisubstituted alkene;
                                     minor product)   major product)

According to the Zaitsev rule, the major alkene is the one corresponding to loss of a hydrogen from the alkyl group that has the fewest number of hydrogens. Thus hydrogen is removed from the methylene group rather than from the methyl group, and 1-methylcyclohexene is formed in greater amounts than methylenecyclohexane.

(c) The two alkenes are as shown in the equation.

Compound has a   Compound has a
trisubstituted   tetrasubstituted
double bond   double bond
(minor product)   (major product)

The more highly substituted alkene is formed in greater amounts, according to the Zaitsev rule.

4.8 **(b)** The site of positive charge in the carbocation is the carbon atom that bears the hydroxyl group in the starting alcohol.

1-Methylcyclohexanol   1-Methylcyclohexyl cation   Water

The curved arrow notation for proton abstraction by hydrogen sulfate anion from the methyl group is demonstrated in the following equation.

1-Methylcyclohexyl cation   Methylenecyclohexane

Abstraction of a proton from the ring gives 1-methylcyclohexene.

1-Methylcyclohexyl cation   1-Methylcyclohexene

(c)   Loss of the hydroxyl group under conditions of acid catalysis yields a tertiary carbocation.

Hydrogen sulfate ion may abstract a proton from an adjacent methylene group to give a trisubstituted alkene.

Abstraction of the methine proton affords a tetrasubstituted alkene.

4.9   (b)   The dehydrohalogenation of 2-bromohexane is similar to that of 2-bromobutane described in (a) of this problem. Elimination can occur to give 1-hexene, *cis*-2-hexene, or *trans*-2-hexene.

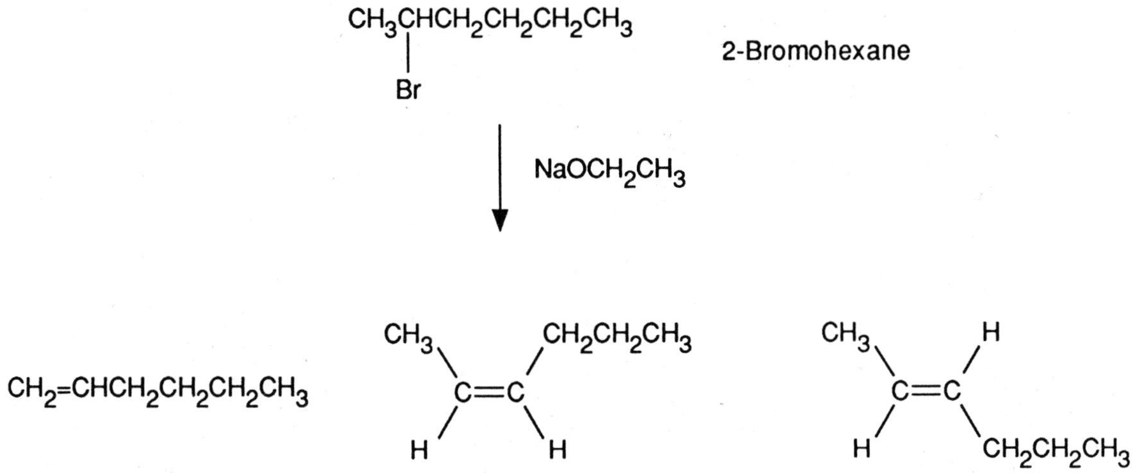

$$CH_3CHCH_2CH_2CH_2CH_3$$
$$|$$
$$Br$$

2-Bromohexane

NaOCH_2CH_3

$$CH_2=CHCH_2CH_2CH_2CH_3$$

1-Hexene

cis-2-Hexene

trans-2-Hexene

(c)   In 2-iodo-4-methylpentane elimination can involve a proton at C-1 or C-3. Three alkenes are capable of being formed.

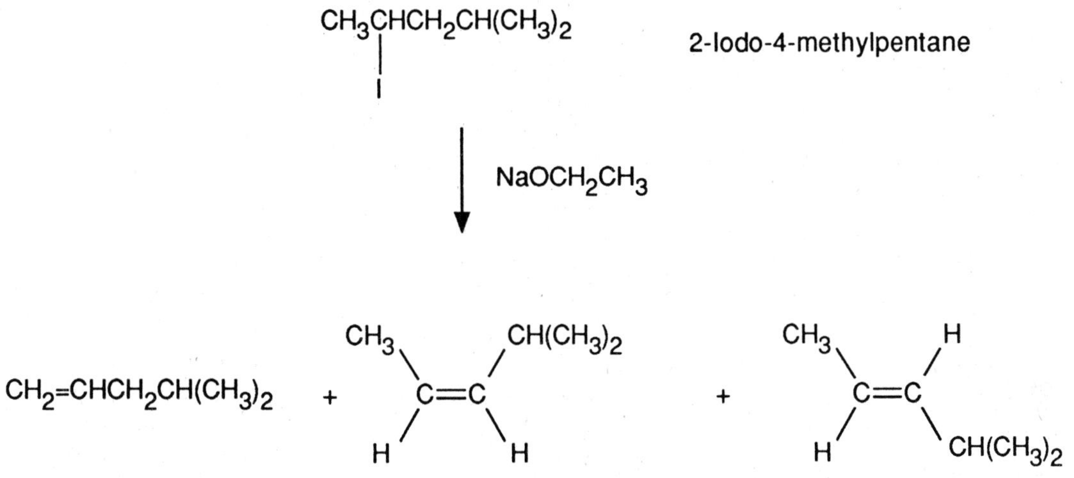

$$CH_3CHCH_2CH(CH_3)_2$$
$$|$$
$$I$$

2-Iodo-4-methylpentane

NaOCH_2CH_3

$$CH_2=CHCH_2CH(CH_3)_2$$ +

4-Methyl-1-pentene

+

cis-4-Methyl-2-pentene

trans-4-Methyl-2-pentene

(d)   Elimination can proceed in only one direction in 3-iodo-2,2-dimethylpentane. A mixture of *cis* and *trans*-4,4-dimethyl-2-pentene is formed.

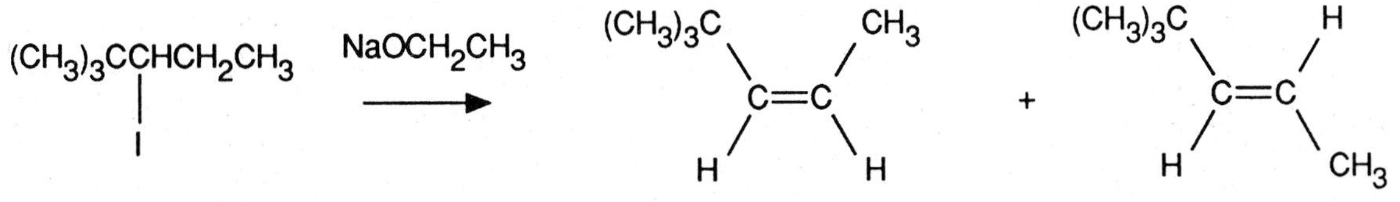

$$(CH_3)_3CCHCH_2CH_3$$
$$|$$
$$I$$

3-Iodo-2,2-dimethylpentane

NaOCH_2CH_3

cis-4,4-Dimethyl-2-pentene   +

trans-4,4-Dimethyl-2-pentene

65

4.10 **(b)** Proceeding as in (a), this compound is named **1,3-cyclooctadiene**. The double bonds are separated by one single bond; the compound is a conjugated diene.

(c) This compound is **1,4-cyclooctadiene**, an isolated diene. The two double bonds are separated from one another by an sp^3-hybridized carbon (C-3).

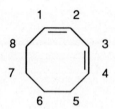

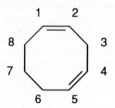

4.11 A triple bond may connect C-1 and C-2 or C-2 and C-3 in an unbranched chain of five carbons.

$$CH_3CH_2CH_2C{\equiv}CH \qquad\qquad CH_3CH_2C{\equiv}CCH_3$$

1-Pentyne   2-Pentyne

The remaining C_5H_8 isomer has a branched carbon chain.

$$CH_3CHC{\equiv}CH$$
$$\vert$$
$$CH_3$$

3-Methyl-1-butyne

4.12 Each of the dibromides shown yields 3,3-dimethyl-1-butyne when subjected to double dehydrohalogenation with strong base.

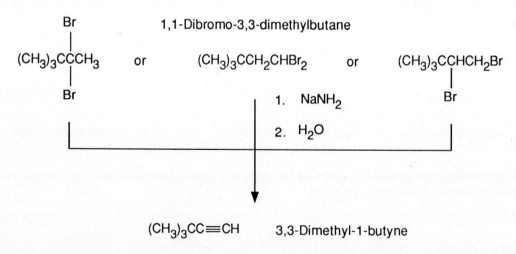

2,2-Dibromo-3,3-dimethylbutane

1,2-Dibromo-3,3-dimethylbutane

1,1-Dibromo-3,3-dimethylbutane

$(CH_3)_3CCCH_3$ or $(CH_3)_3CCH_2CHBr_2$ or $(CH_3)_3CCHCH_2Br$

1. $NaNH_2$
2. H_2O

$(CH_3)_3CC{\equiv}CH$ 3,3-Dimethyl-1-butyne

4.13 **(a)** 1-Heptene is $CH_2{=}CH(CH_2)_4CH_3$

(b) 3-Ethyl-1-pentene is $CH_2{=}CHCH(CH_2CH_3)_2$

(c) 3-Isopropyl-2-methyl-2-heptene is

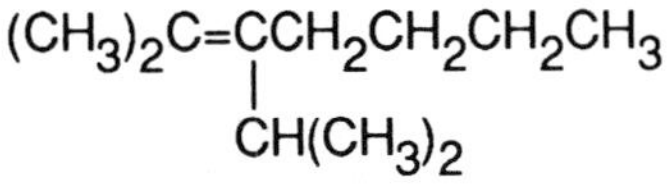
(CH3)2C=CCH2CH2CH2CH3
CH(CH3)2

(d) *cis*-3-Octene is

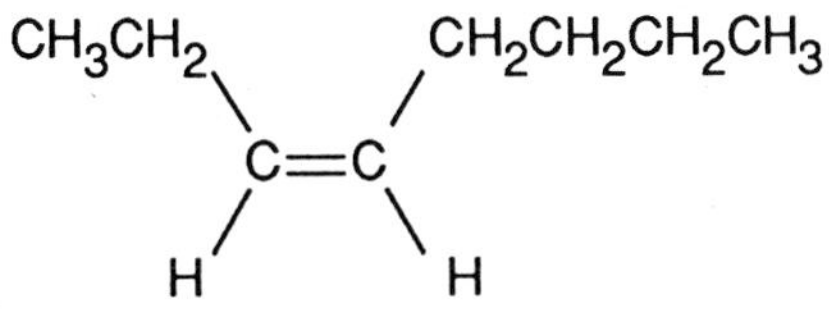
CH3CH2
CH2CH2CH2CH3
C=C
H
H

(e) *trans*-2-Hexene is

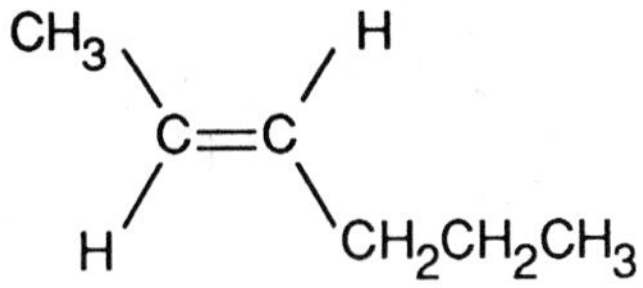
CH3
H
C=C
H
CH2CH2CH3

(f) The groups that are the same (methyl groups) are on opposite sides of the double bond in *trans*-3-methyl-2-pentene.

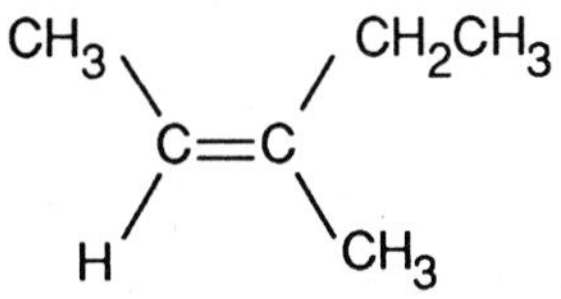
CH3
CH2CH3
C=C
H
CH3

(g) 1-Bromocyclohexene is

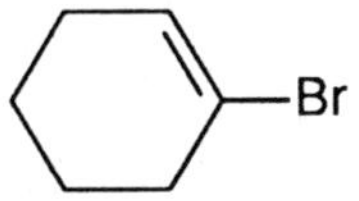
Br

(h) Vinylcycloheptane is

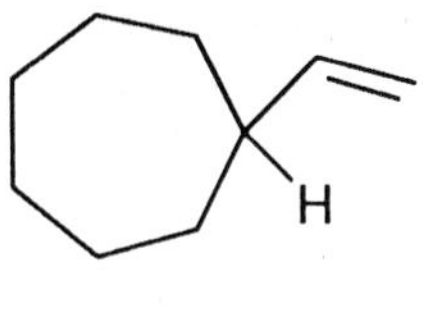
H

(i) 1,1-Diallylcyclopropane is

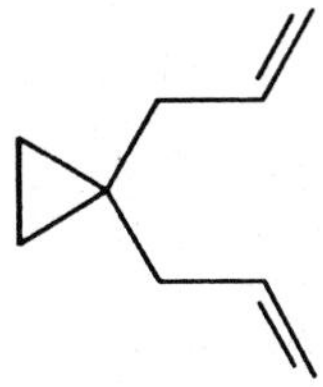

4.14 (a) The longest chain that includes the double bond contains five carbon atoms, so the compound is named as a derivative of pentene. The chain is numbered to give the double bond the lowest number.

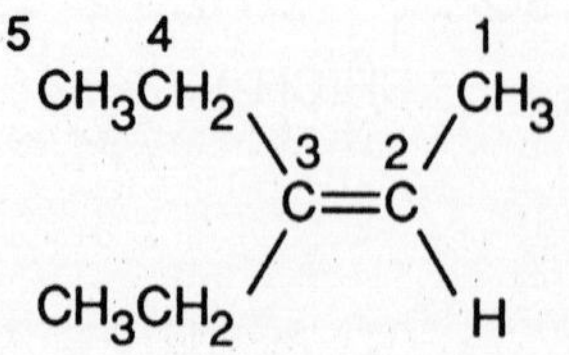

$$\underset{\text{5}}{CH_3}\underset{\text{4}}{CH_2}\diagdown\overset{\text{3}}{\underset{\diagup}{C}}=\overset{\text{2}}{\underset{\diagdown}{C}}\underset{\text{1}}{\diagup}CH_3$$

3-Ethyl-2-pentene

(b) Writing out the structure in detail reveals the longest continuous chain that includes the double bond to contain six carbon atoms. The double bond is between C-3 and C-4, and so the compound is named as a derivative of 3-hexene.

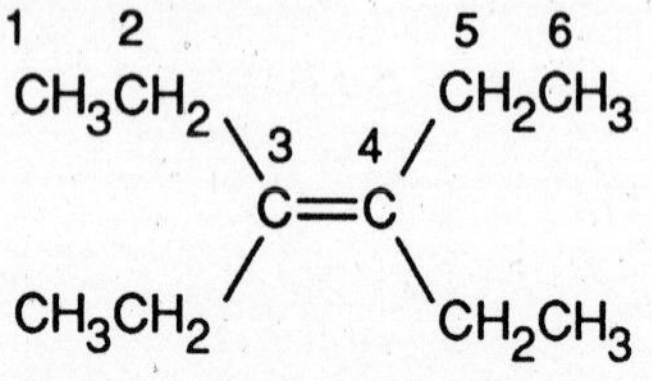

3,4-Diethyl-3-hexene

(c) The longest chain has five carbon atoms, the double bond is at C-1 and there are two methyl substituents at C-4.

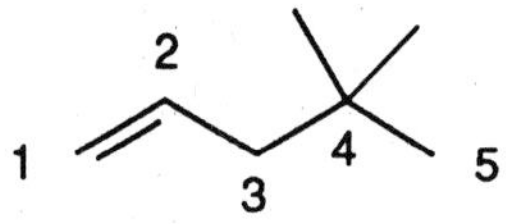

4,4-Dimethyl-1-pentene

(d) This compound contains a trans double bond between C-2 and C-3 of a methyl-substituted six carbon chain.

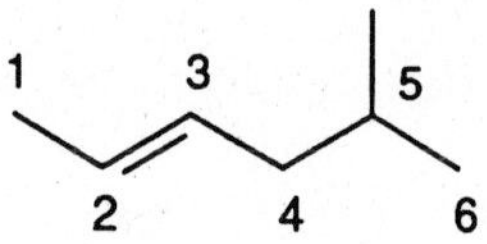

trans-5-Methyl-2-hexene

(e) This compound is a dimethyl derivative of cyclopentene. The double bond of a cycloalkene is between C-1 and C-2, so the methyl substituents are located at C-1 and C-3.

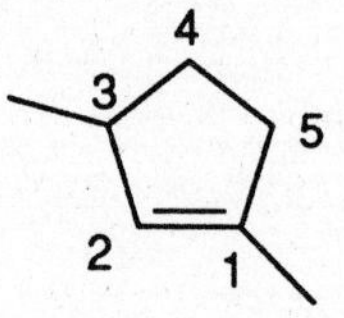

1,3-Dimethylcyclopentene

4.15 (a) This compound is a four-carbon alcohol with a double bond at C-3 (3-buten-) and the hydroxyl group at C-2 (2-ol).

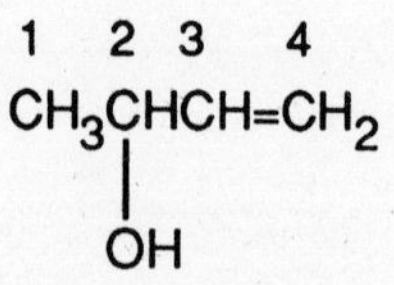

$$\underset{\text{1}}{CH_3}\underset{\underset{\displaystyle OH}{|}}{\overset{\text{2}}{CH}}\underset{\text{3}}{CH}=\underset{\text{4}}{CH_2}$$

3-Buten-2-ol

(b)   The name 3-chloro-1-butene describes a four-carbon alkene with a chlorine substituent at C-3.

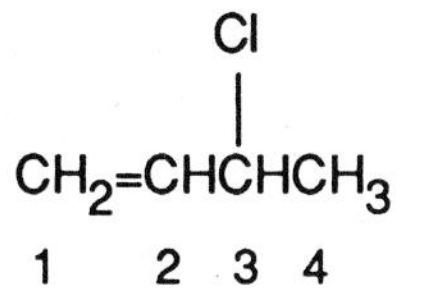

$$\underset{1 \quad\; 2 \; 3 \; 4}{CH_2{=}CHCHCH_3} \overset{\displaystyle Cl}{\underset{\displaystyle |}{}}$$

3-Chloro-1-butene

(c)   This seven-membered cycloalkene derivative has a hydroxyl group at C-1, the double bond between C-2 and C-3, and a chlorine at C-4.

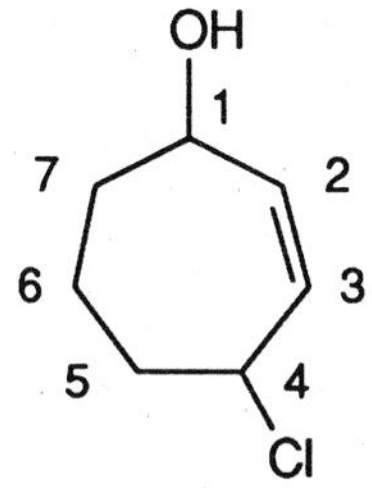

4-Chloro-2-cyclohepten-1-ol

4.16   The carbon atoms of each double bond are sp^2 hybridized. The double bonds themselves consist of one σ component and one π component. All carbon-carbon single bonds are σ bonds.

(a)   The compound has four sp^3 hybridized carbon atoms and two sp^2 hybridized carbons. There are five C-C σ bonds and one π bond.

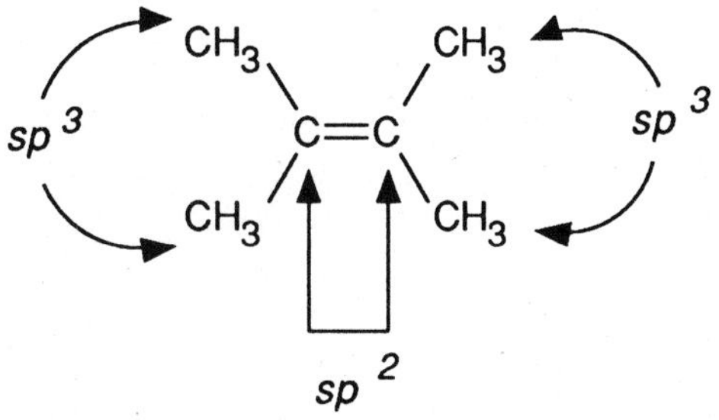

(b)   1,5-Hexadiene has four sp^2 carbon atoms and two sp^3 carbons. Carbons 1,2,5, and 6 are sp^2 hybridized while carbons 3 and 5 are sp^3 hybridized. There are five C-C σ bonds and two π bonds.

$$CH_2{=}CH-CH_2-CH_2-CH{=}CH_2$$

(c)   1,2-Dimethylcyclohexene has two sp^2 carbon atoms and six sp^3 carbons. There are eight C-C σ bonds and one π bond. The only carbons that are sp^2 hybridized are the ones which are doubly bonded.

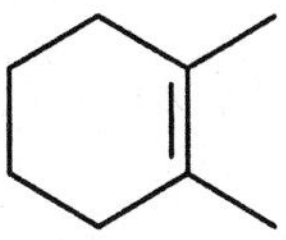

4.17   The degree of substitution of an alkene is determined by the number of carbon atoms *directly* attached to the doubly bonded carbons. In the structural formulas shown in this problem, carbons directly attached to the doubly bonded carbons are indicated by heavy dots.

69

(a)   There are three carbons directly attached to the doubly bonded carbons in 3-ethyl-2-pentene. C-2 bears one carbon substituent and C-3 bears two. The double bond is **trisubstituted**.

3-Ethyl-2-pentene

(b)   There are four ethyl groups attached to the double bond in 3,4-diethyl-3-hexene. The alkene is **tetrasubstituted**.

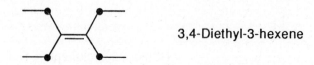

3,4-Diethyl-3-hexene

(c)   Three of the atoms directly attached to the doubly bonded carbons in 4,4-dimethyl-1-pentene are hydrogens. There is only one carbon substituent. This alkene is **monosubstituted**.

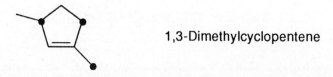

4,4-Dimethyl-1-pentene

(d)   This alkene has two carbons directly attached to the double bond; it is **disubstituted**.

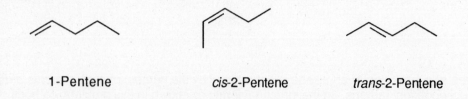

trans-5-Methyl-2-hexene

(e)   Remember to count the ring carbons in determining the substitution pattern. The double bond in 1,3-dimethylcyclopentene is **trisubstituted**.

1,3-Dimethylcyclopentene

4.18   Consider first the C_5H_{10} alkenes that have an unbranched carbon chain. Two of these, *cis*- and *trans*-2-pentene, are a pair of stereoisomers.

1-Pentene   *cis*-2-Pentene   *trans*-2-Pentene

There are three additional isomers. These have a four-carbon chain with a methyl substituent.

| 2-Methyl-1-butene | 2-Methyl-2-butene | 3-Methyl-1-butene |

4.19 The most stable alkene is the most substituted. The trisubstituted alkene, 2-methyl-2-butene is the most stable. Among the disubstituted alkenes, the trans stereoisomer is more stable than cis.

The most stable isomer has a trisubstituted double bond

The three disubstituted alkenes are of similar stability, with the trans isomer of 2-pentene being more stable than the cis

The two monosubstituted alkenes are the least stable of all of the isomers

4.20 The most stable C_6H_{12} alkene is tetrasubstituted, and is 2,3-dimethyl-2-butene.

$$CH_3 \diagdown \diagup CH_3$$
$$C=C$$
$$CH_3 \diagup \diagdown CH_3$$

4.21 Alkenes dehydrate by loss of a hydroxyl group (as water) and a hydrogen from an adjacent carbon. Dehydration of 2-pentanol gives a mixture of 1-pentene and the stereoisomers of 2-pentene. The trans stereoisomer of the more substituted alkene 2-pentene will predominate.

| 1-Pentanol | | 1-Pentene | | cis-2-Pentene | | trans-2-Pentene (major product) |

4.22 (a) The isomeric $C_5H_{12}O$ alcohols are:

$$CH_3CH_2CH_2CH_2CH_2OH$$

1-Pentanol

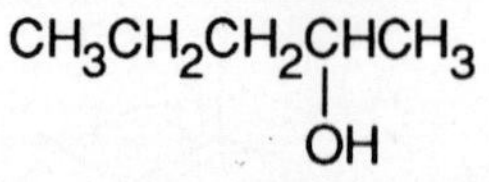

$$CH_3CH_2CH_2\underset{\underset{OH}{|}}{C}HCH_3$$

2-Pentanol

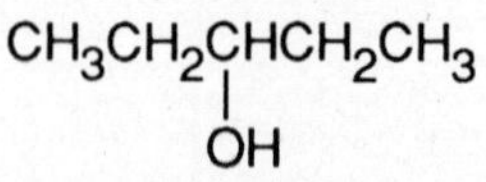

$$CH_3CH_2\underset{\underset{OH}{|}}{C}HCH_2CH_3$$

3-Pentanol

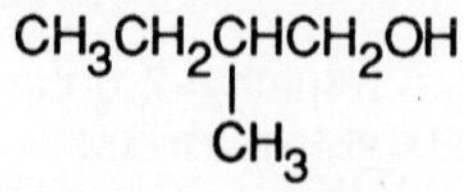

$$CH_3CH_2\underset{\underset{CH_3}{|}}{C}HCH_2OH$$

2-Methyl-1-butanol

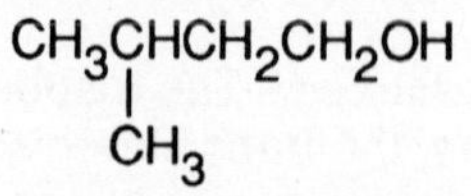

$$CH_3\underset{\underset{CH_3}{|}}{C}HCH_2CH_2OH$$

3-Methyl-1-butanol

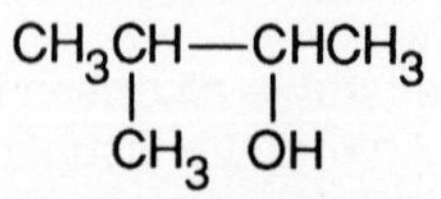

$$CH_3\underset{\underset{CH_3}{|}}{C}H\!-\!\underset{\underset{OH}{|}}{C}HCH_3$$

3-Methyl-2-butanol

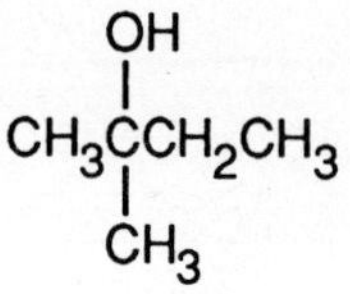

$$CH_3\overset{\overset{OH}{|}}{\underset{\underset{CH_3}{|}}{C}}CH_2CH_3$$

2-Methyl-2-butanol

$$CH_3\overset{\overset{CH_3}{|}}{\underset{\underset{CH_3}{|}}{C}}CH_2OH$$

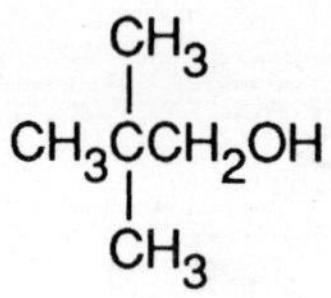

2,2-Dimethyl-1-propanol

(b) The order of reactivity in alcohol dehydration is tertiary > secondary > primary. The only tertiary alcohol in the group is 2-methyl-2-butanol. It will dehydrate fastest.

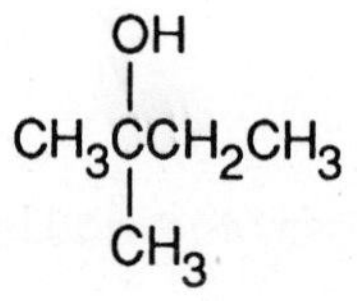

$$CH_3\overset{\overset{OH}{|}}{\underset{\underset{CH_3}{|}}{C}}CH_2CH_3$$

2-Methyl-2-butanol

(c) The most stable C_5H_{11} carbocation is the tertiary carbocation.

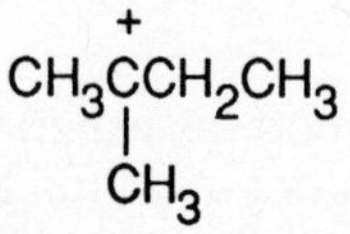

$$CH_3\overset{+}{\underset{\underset{CH_3}{|}}{C}}CH_2CH_3$$

(d) A proton may be lost from C-1 or C-3:

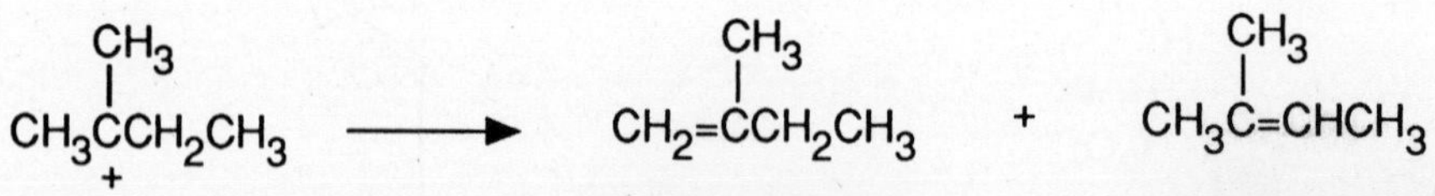

$$CH_3\underset{\underset{+}{}}{\overset{\overset{CH_3}{|}}{C}}CH_2CH_3 \longrightarrow CH_2{=}\overset{\overset{CH_3}{|}}{C}CH_2CH_3 \ + \ CH_3\overset{\overset{CH_3}{|}}{C}{=}CHCH_3$$

2-Methyl-1-butene
(minor alkene)

2-Methyl-2-butene
(major alkene)

(e)   The only alkenes formed by dehydration of 3-pentanol are the cis and trans stereoisomers of 2-pentene.

3-Pentanol   cis-2-Pentene   trans-2-Pentene

4.23 The first step is protonation of the alcohol oxygen by the acid.

2-Butanol   Phosphoric acid   Oxonium ion   Dihydrogen
                                                                                   phosphate ion

The oxonium ion dissociates, forming a carbocation.

Oxonium ion   Carbocation   Water

In the last step, a proton is lost from the carbon adjacent to the carbocation. The base which removes the proton is the dihydrogen phosphate ion, $H_2PO_4^-$. According to Zaitsev's rule, loss of a proton from C-3 gives a mixture of 2-butene stereoisomers as the major products.

4.24 In all parts of this problem, identify the carbon that bears the halogen and the adjacent carbons that bear at least one hydrogen substituent. These are the carbons that become doubly bonded in the alkene product.

(a)   1-Bromohexane gives only 1-hexene upon dehydrohalogenation.

1-Bromohexane   1-Hexene

(b)   2-Bromohexane can give both 1-hexene and 2-hexene on dehydrobromination. The 2-hexene fraction is a mixture of cis and trans stereoisomers.

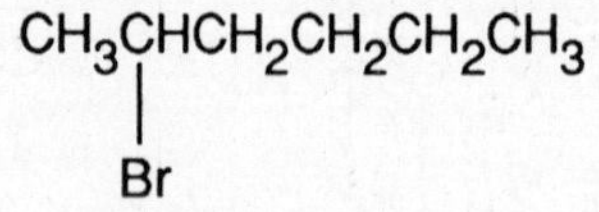

$$CH_3CHCH_2CH_2CH_2CH_3$$
$$|$$
$$Br$$

2-Bromohexane

base

E2

$$CH_2=CHCH_2CH_2CH_2CH_3$$ +

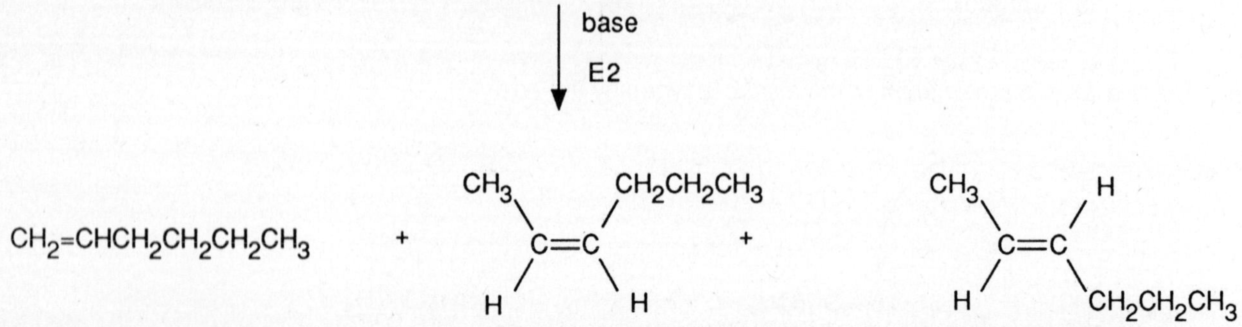

1-Hexene

cis-2-Hexene

trans-2-Hexene

(c)   Proton abstraction from the C-3 methyl group of 3-bromo-3-methylpentane yields 2-ethyl-1-butene.

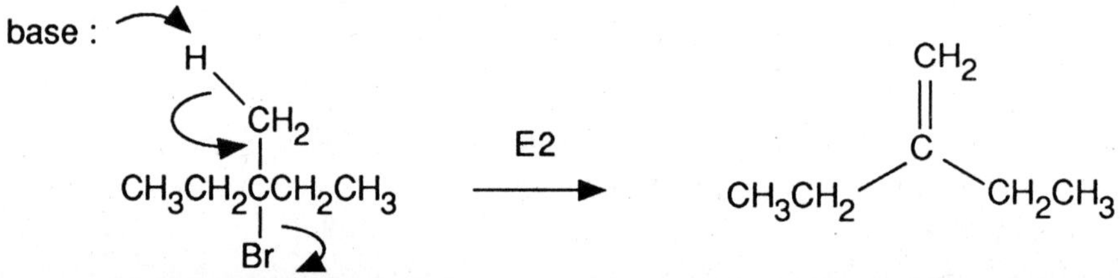

base :

E2

3-Bromo-3-methylpentane

2-Ethyl-1-butene

Stereoisomeric 3-methyl-2-pentenes are formed by proton abstraction from C-2.

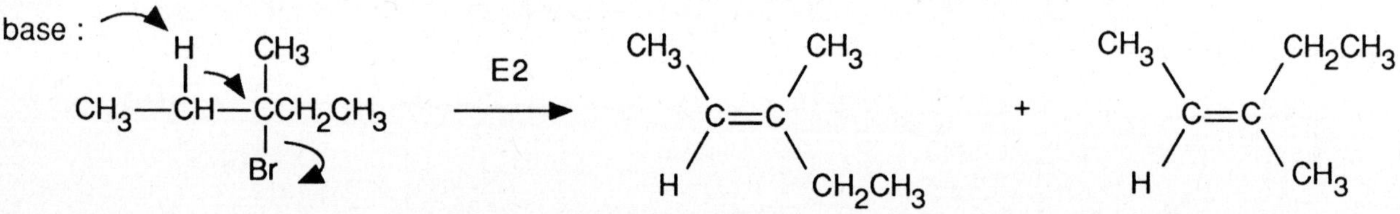

base :

E2

+

(d)   Only 3,3-dimethyl-1-butene may be formed by dehydrohalogenation of 2-bromo-3,3-dimethylbutane.

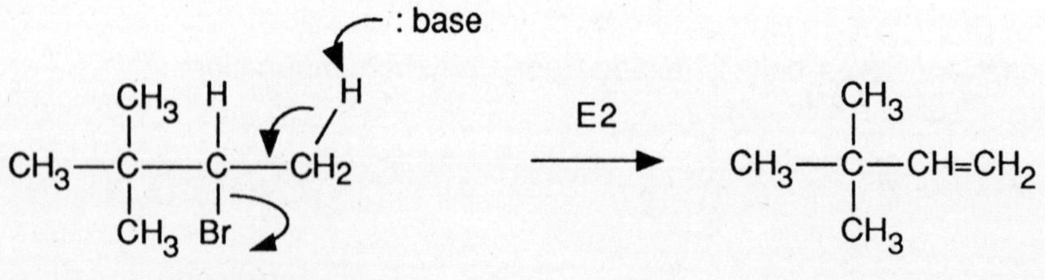

: base

E2

$$CH_3-C-CH=CH_2$$

2-Bromo-3,3-dimethylbutane

3,3-Dimethyl-1-butene

74

4.25 (a)   Write out the structure of the alkyl bromide to reveal the number of different hydrogen environments on carbons adjacent to the carbon bearing the bromine.

$$\overset{\underset{\displaystyle Br}{\displaystyle |}}{\underset{}{\overset{1}{CH_3}-\overset{2}{\underset{}{C}}}}\overset{\displaystyle CH_3}{}-\overset{\underset{\displaystyle CH_3}{}}{\overset{\displaystyle H}{\overset{3}{C}}}-\overset{4}{CH_3}$$

Bromine is lost from C-2; hydrogen may be lost from C-1 or C-3

Of the two possible alkenes shown below, the major product predicted on the basis of Zaitsev's rule is the one with the tetrasubstituted double bond.

$$CH_2{=}\overset{\overset{\displaystyle CH_3}{\diagup}}{\underset{\underset{\displaystyle CH(CH_3)_2}{\diagdown}}{C}}\qquad\text{and}\qquad \overset{CH_3}{\underset{CH_3}{}}{>}C{=}C\overset{CH_3}{\underset{CH_3}{}}$$

      2,3-Dimethyl-1-butene   2,3-Dimethyl-2-butene

          (minor product)   (major product)

(b)   All the hydrogens on carbons adjacent to the one bearing the bromine are equivalent, so elimination of any one gives the same product.

$$CH_3CH_2{-}\overset{\overset{\displaystyle CH_2CH_3}{|}}{\underset{\underset{\displaystyle CH_2CH_3}{|}}{C}}{-}Br\quad\overset{-\,HBr}{\longrightarrow}\quad CH_3CH{=}C\overset{\diagup CH_2CH_3}{\diagdown CH_2CH_3}$$

        3-Bromo-3-ethylpentane   3-Ethyl-2-pentene

(c)   Only one alkene can be formed by dehydrohalogenation of 1-bromo-3-methylbutane.

$$BrCH_2CH_2CH(CH_3)_2\quad\overset{-\,HBr}{\longrightarrow}\quad CH_2{=}CHCH(CH_3)_2$$

    1-Bromo-3-methylbutane   3-Methyl-1-butene

(d)   Two alkenes may be formed here. The more highly substituted one is 1-methylcyclohexene, and this is predicted to be the major product in accordance with Zaitsev's rule.

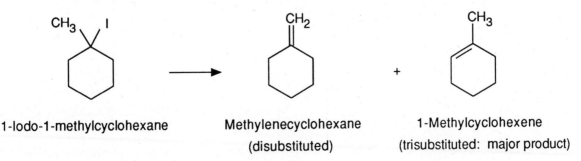

  1-Iodo-1-methylcyclohexane   Methylenecyclohexane   1-Methylcyclohexene

                        (disubstituted)   (trisubstituted: major product)

4.26 You need to reason backward from an alkene to a bromide of molecular formula $C_7H_{13}Br$. Recall that the carbon-carbon double bond is formed by loss of a proton from one of the doubly bonded carbons and a bromide from the other.

(a) Cycloheptene is the only alkene formed by an E2 elimination reaction of cycloheptyl bromide.

Cycloheptyl bromide   Cycloheptene

(b) (Bromomethyl)cyclohexane is the correct answer. It gives methylenecyclohexane as the *only* alkene under E2 conditions.

(Bromomethyl)cyclohexane   Methylenecyclohexane

1-Bromo-1-methylcyclohexane is not correct. It gives a mixture of 1-methylcyclohexene and methylenecyclohexane on elimination.

(Bromomethyl)cyclohexane   Methylenecyclohexane   1-Methylcyclohexene

(c) In order for 4-methylcyclohexene to be the only alkene, the starting alkyl bromide must be 1-bromo-4-methylcyclohexane.

1-Bromo-4-methylcyclohexane   4-Methylcyclohexene

1-Bromo-3-methylcyclohexene is incorrect; its dehydrobromination yields a mixture of 3-methylcyclohexene and 4-methylcyclohexene.

1-Bromo-3-methylcyclohexane   4-Methylcyclohexene   3-Methylcyclohexene

76

4.27 (a) Both 1-bromopropane and 2-bromopropane yield propene as the exclusive product of E2 elimination.

$$CH_3CH_2CH_2Br \quad \text{or} \quad CH_3CHCH_3 \xrightarrow[\text{base}]{E2} CH_3CH=CH_2$$
$$\underset{Br}{|}$$

| 1-Bromopropane | 2-Bromopropane | Propene |

(b) Isobutene is formed on dehydrobromination of either *tert*-butyl bromide or isobutyl bromide.

$$(CH_3)_3CBr \quad \text{or} \quad (CH_3)_2CHCH_2Br \xrightarrow[\text{base}]{E2} (CH_3)_2C=CH_2$$

tert-Butyl bromide   Isobutyl bromide   Isobutene

(c) The bromine substituent may be at either C-2 or C-3.

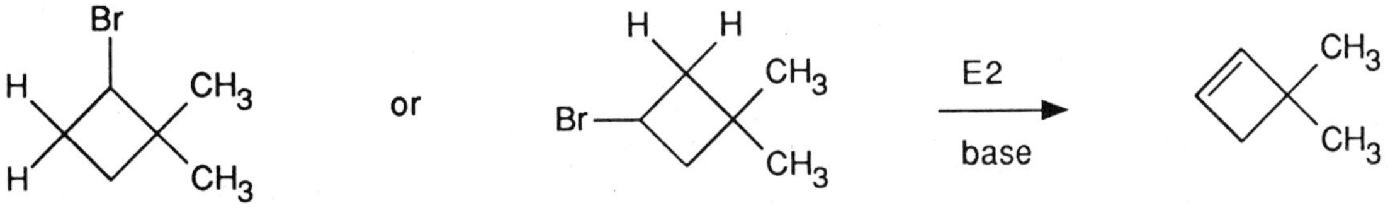

2-Bromo-1,1-dimethylcyclobutane   3-Bromo-1,1-dimethylcyclobutane   3,3-Dimethylcyclobutene

4.28 (a) β-Springene has three isolated double bonds and a pair of conjugated double bonds.

conjugated double bonds

isolated double bonds

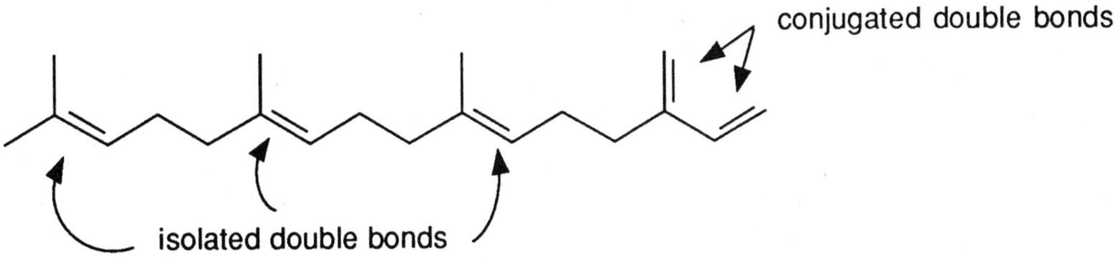

(b) All the double bonds in humulene are isolated from each other.

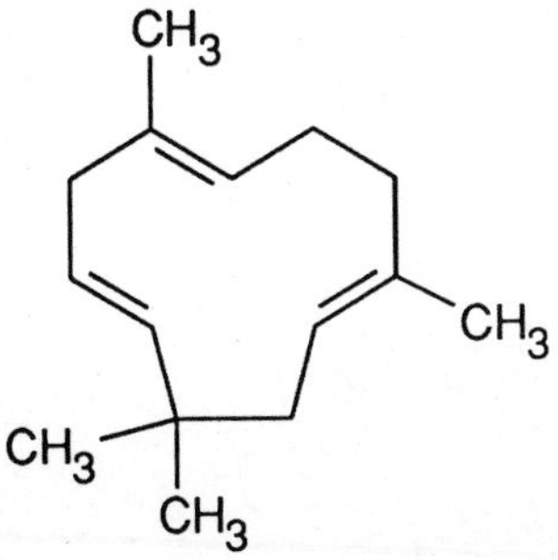

Humulene

77

(c)   The C-1 and C-3 double bonds of cembrene are conjugated to each other.

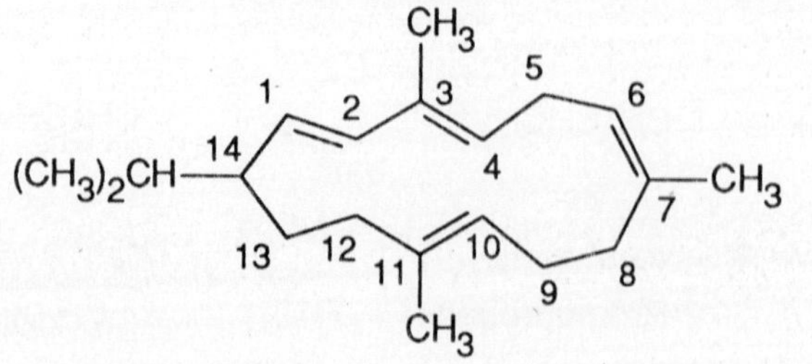

Cembrene

The double bonds at C-6 and C-10 are isolated from each other and from the conjugated diene system.

(d)   The sex attractant of the dried-bean beetle has a cumulated diene system involving C-4, C-5, and C-6. This allenic system is conjugated with the C-2 double bond.

$$\overset{6\quad 5\quad 4\quad 3\quad 2\quad 1}{CH_3(CH_2)_6CH_2CH=C=CHCH=CHCO_2CH_3}$$

4.29 (a)

$CH_2=CH(CH_2)_5CH=CH_2$ 1,8-Nonadiene

(b)

$(CH_3)_2C=\overset{\overset{\displaystyle CH_3}{|}}{\underset{\underset{\displaystyle CH_3}{|}}{C}}C=C(CH_3)_2$ 2,3,4,5-Tetramethyl-2,4-hexadiene

(c)

$CH_2=CH\overset{\underset{\underset{\displaystyle CH=CH_2}{|}}{}}{C}HCH=CH_2$ 3-Vinyl-1,4-pentadiene

4.30 Elimination of two moles of water gives a conjugated diene. The compound is named as a dimethyl derivative of butadiene.

$(CH_3)_2\underset{\underset{\displaystyle HO}{|}}{C}-\underset{\underset{\displaystyle OH}{|}}{C}(CH_3)_2$ $\xrightarrow[\text{heat}]{\text{HBr}}$ $CH_2=\underset{\underset{\displaystyle H_3C}{|}}{C}-\underset{\underset{\displaystyle CH_3}{|}}{C}=CH_2$

2,3-Dimethyl-2,3-butanediol   2,3-Dimethyl-1,3-butadiene

4.31 There are three isomers that have unbranched carbon chains.

$CH_3CH_2CH_2CH_2C\equiv CH$ $CH_3CH_2CH_2C\equiv CCH_3$ $CH_3CH_2C\equiv CCH_2CH_3$

1-Hexyne   2-Hexyne   3-Hexyne

78

Next consider all the alkynes with a single methyl branch.

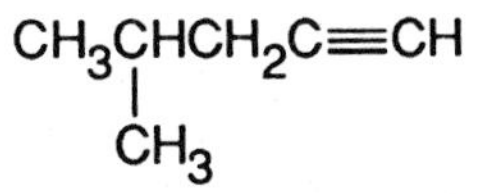

CH₃CHCH₂C≡CH
|
CH₃

CH₃CH₂CHC≡CH
|
CH₃

CH₃CHC≡CCH₃
|
CH₃

4-Methyl-1-pentyne 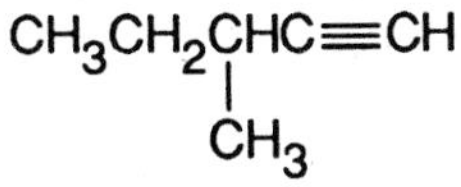   3-Methyl-1-pentyne 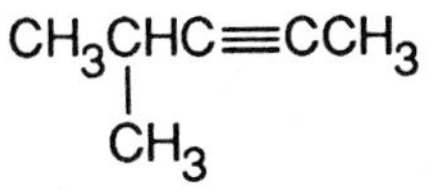   4-Methyl-2-pentyne

There is one isomer with two methyl branches. None is possible with an ethyl branch.

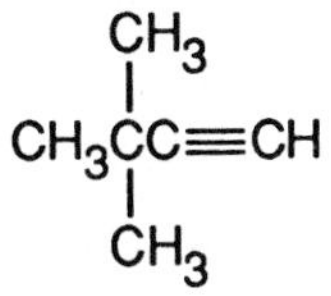

CH₃
|
CH₃CC≡CH   3,3-Dimethyl-1-butyne
|
CH₃

4.32 (a)

5   4   3   2   1
CH₃CH₂CH₂C≡CH   is 1-pentyne

(b)

5   4   3   2 1
CH₃CH₂C≡CCH₃   is 2-pentyne

(c)

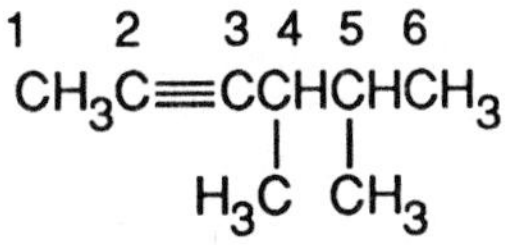

1   2   3 4 5 6
CH₃C≡CCHCHCH₃   is 4,5-dimethyl-2-hexyne
| |
H₃C CH₃

4.33 (a)   The structure of 1-octyne is

HC≡CCH₂CH₂CH₂CH₂CH₂CH₃

(b)   2-Octyne is

CH₃C≡CCH₂CH₂CH₂CH₂CH₃

(c)   2,5-Dimethyl-3-hexyne is

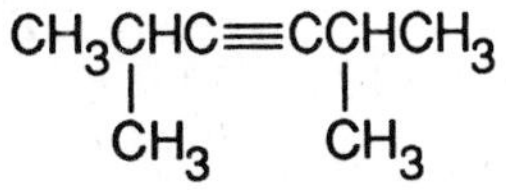

CH₃CHC≡CCHCH₃
| |
CH₃   CH₃

(d)   4-Ethyl-1-hexyne is

CH₃CH₂CHCH₂C≡CH
|
CH₂CH₃

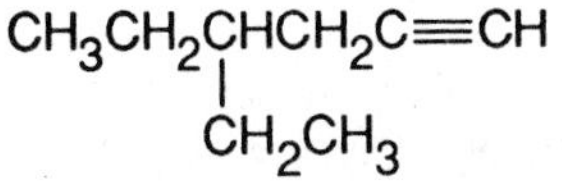

(e)   3-Ethyl-3-methyl-1-pentyne is

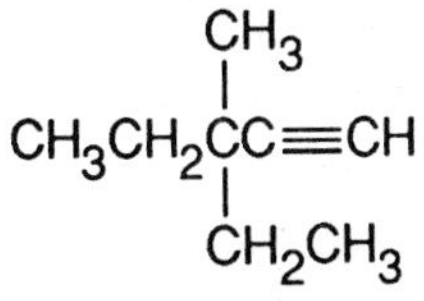

CH₃
|
CH₃CH₂CC≡CH
|
CH₂CH₃

4.34 A terminal alkyne has one hydrogen directly bonded to a triply bonded carbon. The alkynes in parts a, d, and e of the previous problem have **terminal** triple bonds. The alkynes in parts b and d have **internal** triple bonds.

4.35 Alkynes can be prepared by double dehydrohalogenation of geminal dihalides (both halogens on the same carbon) or vicinal dihalides (the halides on adjacent carbons). Treatment of either 1,1-dichlorohexane or 1,2-dichlorohexane with a strong base such as sodium amide yields 1-hexyne.

1,1-Dichlorohexane

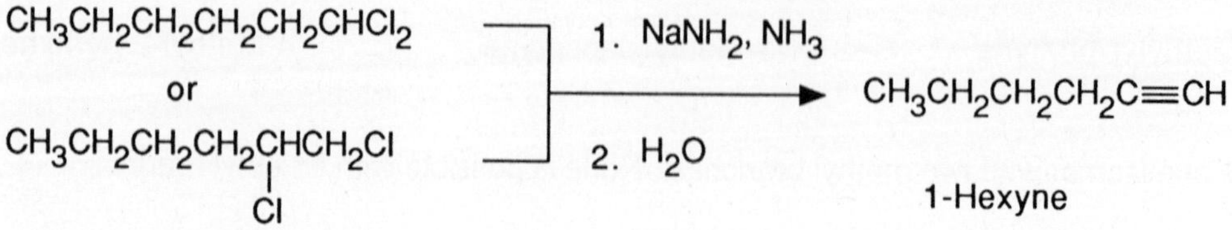

$CH_3CH_2CH_2CH_2CH_2CHCl_2$
or
$CH_3CH_2CH_2CH_2CHCH_2Cl$
Cl
1. $NaNH_2$, NH_3
2. H_2O
$CH_3CH_2CH_2CH_2C{\equiv}CH$

1-Hexyne

1,2-Dichlorohexane

4.36 (a) Alcohols dehydrate on heating in strong acid. The more substituted alkene is the major product.

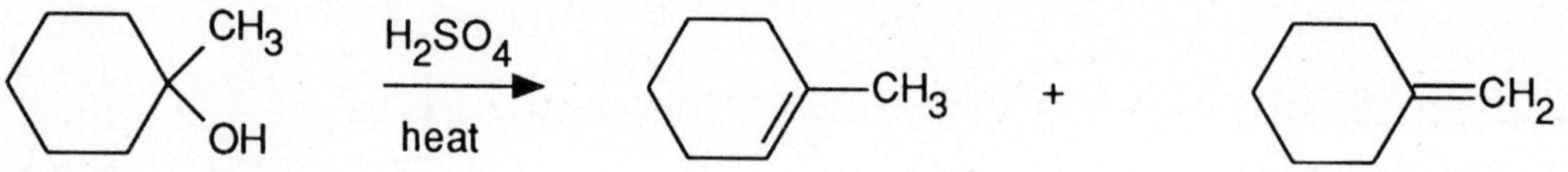

CH_3
OH
H_2SO_4
heat
CH_3
+
CH_2

1-Methylcyclohexanol | 1-Methylcyclohexene (major product) | Methylenecyclohexane

(b) Alcohols undergo substitution, not elimination, on treatment with hydrochloric acid at room 25°C.

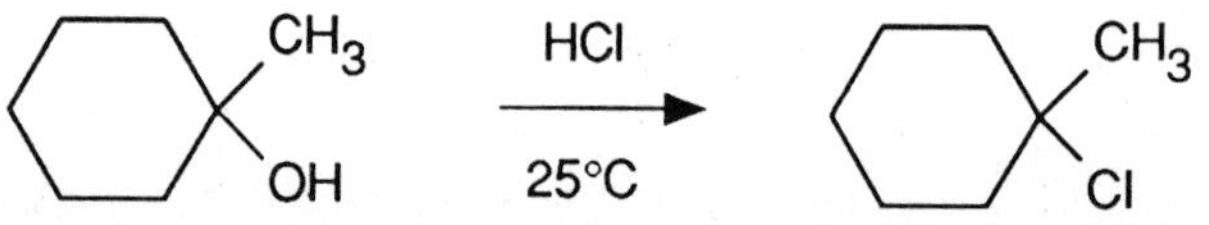

CH_3
OH
HCl
25°C
CH_3
Cl

1-Methylcyclohexanol | 1-Chloro-1-methylcyclohexane

(c) This reaction is a dehydrohalogenation; the products are the same as those in part (a). The major product is 1-methylcyclohexene which has a trisubstituted double bond.

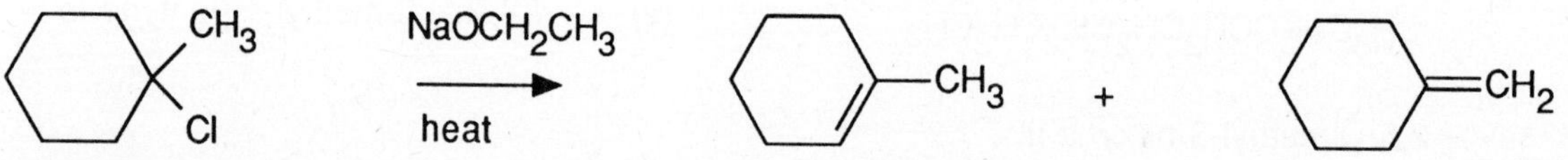

CH_3
Cl
$NaOCH_2CH_3$
heat
CH_3
+
CH_2

1-Chloro-1-methylcyclohexane | 1-Methylcyclohexene (major product) | Methylenecyclohexane

(d) Phosphorus tribromide (PBr_3) converts an alcohol to the corresponding alkyl bromide (Section 3.14).

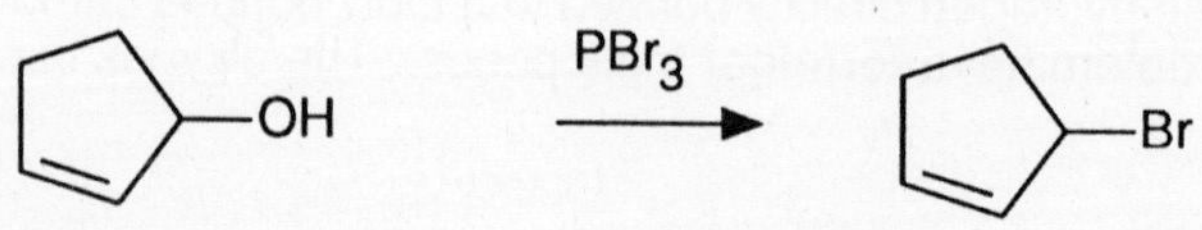

OH
PBr_3
Br

2-Cyclopenten-1-ol | 3-Bromocyclopentene

(e) Dehydrobromination of the product from (d) with sodium ethoxide gives a conjugated diene.

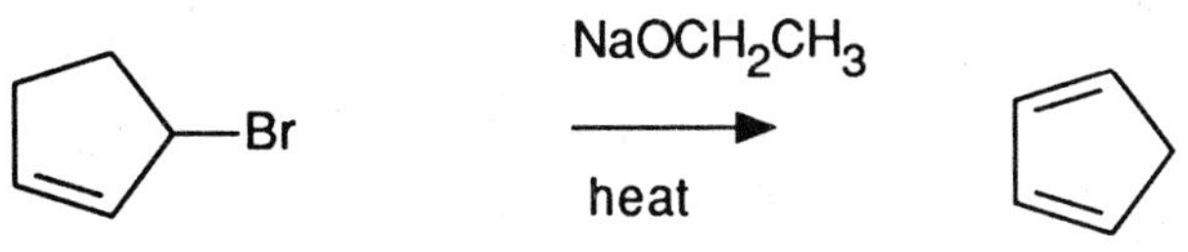
Br
NaOCH2CH3
heat

3-Bromocyclopentene　　　　1,3-Cyclopentadiene

(f) Dehydrobromination occurs with loss of a hydrogen from either ring (both give the same product).

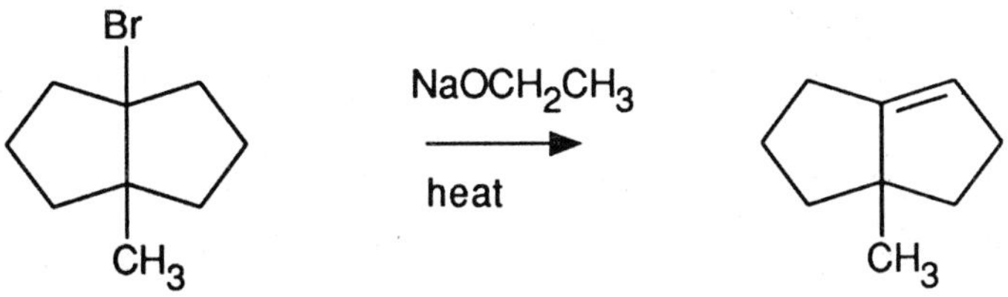
Br
CH3
NaOCH2CH3
heat
CH3

(g) Double dehydrohalogenation of a geminal dihalide gives an alkyne.

$$(CH_3)_3CCH_2CHClBr \xrightarrow[\text{2. } H_2O]{\text{1. } NaNH_2,\ NH_3} (CH_3)_3CC{\equiv}CH$$

1-Bromo-1-chloro-3,3-dimethylbutane　　　　3,3-Dimethyl-1-butyne

CHAPTER 5

ALKENES, ALKADIENES, AND ALKYNES II. REACTIONS

GLOSSARY OF TERMS

1,2-Addition and 1,4-Addition When reagents of the type X-Y add to conjugated dienes, X and Y may add to adjacent doubly-bonded carbons (1,2-addition)

$$R_2C=CH-CH=CR_2 \xrightarrow{X-Y} R_2C-CH-CH=CR_2$$
$$\qquad\qquad\qquad\qquad\qquad\quad | \quad |$$
$$\qquad\qquad\qquad\qquad\qquad\quad X \quad Y$$

or to the termini of the diene system (1,4-addition).

$$R_2C=CH-CH=CR_2 \xrightarrow{X-Y} R_2C-CH=CH-CR_2$$
$$\qquad\qquad\qquad\qquad\qquad\quad | \qquad\qquad\quad |$$
$$\qquad\qquad\qquad\qquad\qquad\quad X \qquad\qquad\quad Y$$

Addition reaction A reaction in which a reagent adds to a multiple bond; one portion of the reactant becomes attached to one of the carbons of the multiple bond and the remaining portion is attached to the other carbon of the double bond.

Aldehyde A compound of the type

$$\begin{array}{c} O \\ || \\ RCH \end{array}$$

Alkylation reaction A reaction in which an alkyl group is attached to some structural unit in a molecule.

Allyl cation The carbocation $CH_2=CHCH_2^+$. The carbocation is stabilized by delocalization of the p electrons of the double bond, and the positive charge is shared by the two CH_2 groups. Substituted analogs of ally cation are called *allylic carbocations*.

Anti addition An addition reaction in which the two portions of the attacking reagent X—Y add to opposite faces of the double bond.

anti-Markovnikov addition A addition reaction for which the regioselectivity is opposite to that predicted on the basis of Markovnikov's rule.

Carbanion An anion in which the negative charge is borne by carbon. An example is acetylide ion.

Carboxylic Acid A compound of the type

$$\begin{array}{c} O \\ || \\ RCOH \end{array}$$

Conjugate addition When applied to conjugated dienes, the term refers to an addition reaction in which the reagent adds to the termini of the conjugated system with migration of the double bond. Synonymous with 1,4-addition.

Diels-Alder reaction The conjugate addition of an alkene to a conjugated diene to give a cyclohexene derivative. Diels-Alder reactions are extremely useful in synthesis.

Dienophile   The alkene which adds to the diene in a Diels-Alder reaction.

Dimer   A molecule formed by the combination of two identical molecules.

Electrophilic addition   A mechanism of addition reaction in which the species which first attacks the multiple bond is an electrophile ("electron seeker").

Enol   A type of alcohol in which the hydroxyl group is bonded to the sp^2 hybridized carbon of an alkene. Enols are intermediates in the hydration of alkynes.

Halonium ion   A species which incorporates a positively charged halogen. Bridged halonium ions are intermediates in the addition of halogens to the double bond of an alkene.

Hydration   The addition of the elements of water (H, OH) to a multiple bond.

Hydroboration-oxidation   A reaction sequence involving a separate hydroboration stage and oxidation stage. In the hydroboration stage, diborane adds to an alkene to give an alkylborane. In the oxidation stage, the alkylborane is oxidized with hydrogen peroxide to give an alcohol. The reaction product is an alcohol corresponding to the anti-Markovnikov, syn-hydration of an alkene.

Hydrogenation   The addition of H_2 to a multiple bond.

Ketone   A compound of the type

LeChatelier's principle   A reaction at equilibrium responds to any stress imposed upon by shifting the equilibrium in the direction which minimizes the stress.

Lindlar palladium   A palladium-on-barium sulfate catalyst which has been deactivated by treatment with lead acetate and quinoline so that it is suffi-ciently reactive to catalyze the hydrogenation of an alkyne to an alkene, but no further. Hydrogenation over Lindlar palladium is an effective method for the preparation of *cis*-alkenes.

Markovnikov's rule   An unsymmetrical reagent adds to an unsymmetrical double bond in the direction that places the positive part of the reagent on the carbon of the double bond that has the greater number of hydrogen substituents.

Monomer   The simplest stable molecule from which a particular polymer may be prepared.

Ozonolysis   The ozone-induced cleavage of a carbon-carbon double or triple bond.

Peroxide   A compound of the type ROOR.

Polyethylene   A polymer of ethylene.

Polymer   A large molecule formed by the combination of many smaller identical molecules (monomers).

Regioselective   A reaction is regioselective if it can produce two (or more) constitutional isomers but gives one of them in greater amounts than the other. A reaction which is 100% regioselective is termed **regiospecific**.

Regiospecific reaction   A reaction which is 100% regioselective.

Saturated hydrocarbon   A hydrocarbon in which there are no multiple bonds.

Syn addition   An addition reaction in which the two portions of the reagent which add to a multiple bond add from the same side.

Terminal alkyne   An alkyne in which the carbon-carbon triple bond is at the end of the chain.

Unsaturated hydrocarbon   A hydrocarbon that can undergo addition reactions; i.e. one which contains multiple bonds.

SOLUTIONS TO TEXT PROBLEMS

5.1   Catalytic hydrogenation converts an alkene to an alkane having the same carbon skeleton. Since 2-methylbutane is the product of hydrogenation, all three alkenes must have a four-carbon chain with a one-carbon branch. Therefore, the three alkenes are:

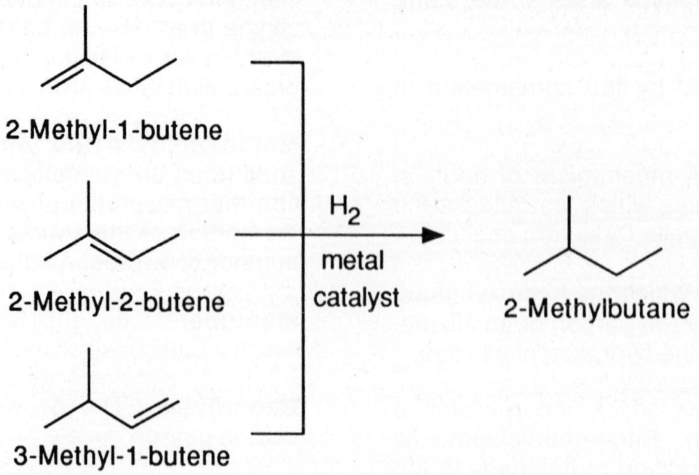

2-Methyl-1-butene

2-Methyl-2-butene

$\xrightarrow[\text{metal catalyst}]{H_2}$

2-Methylbutane

3-Methyl-1-butene

5.2 (b) Begin by writing out the structure of the starting alkene. Identify the doubly bonded carbon that has the greater number of hydrogen substituents; this is the one to which the proton of hydrogen chloride adds. Chloride adds to the carbon atom of the double bond that has the fewer hydrogen substituents.

chlorine adds to this carbon →

$$\begin{array}{c} CH_3 \\ CH_3CH_2 \end{array} C=C \begin{array}{c} H \\ H \end{array}$$

← hydrogen adds to this carbon

2-Methyl-1-butene

↓ HCl

$$CH_3CH_2\underset{\underset{Cl}{|}}{\overset{\overset{CH_3}{|}}{C}}CH_3$$

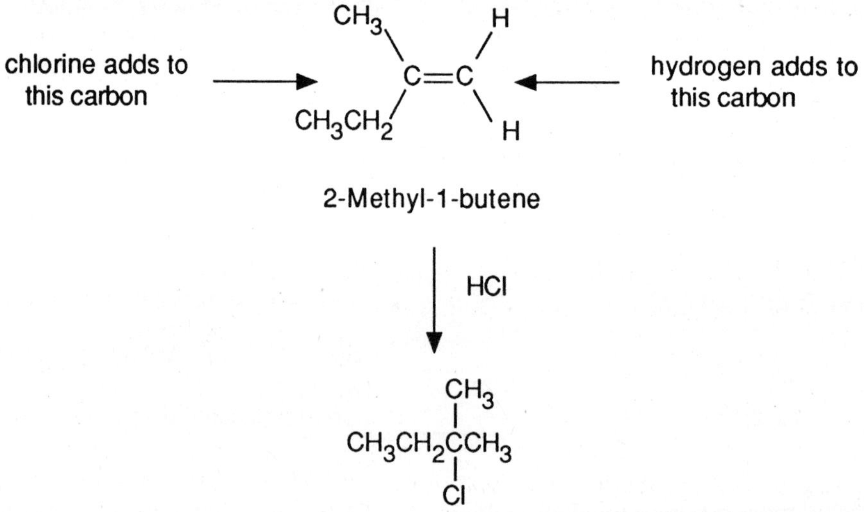

By applying Markovnikov's rule, we see that the major product is 2-chloro-2-methylbutane.

(c) Regioselectivity of addition is not an issue here because the two carbons of the double bond are equivalent in *cis*-2-butene. Hydrogen chloride adds to *cis*-2-butene to give 2-chlorobutane.

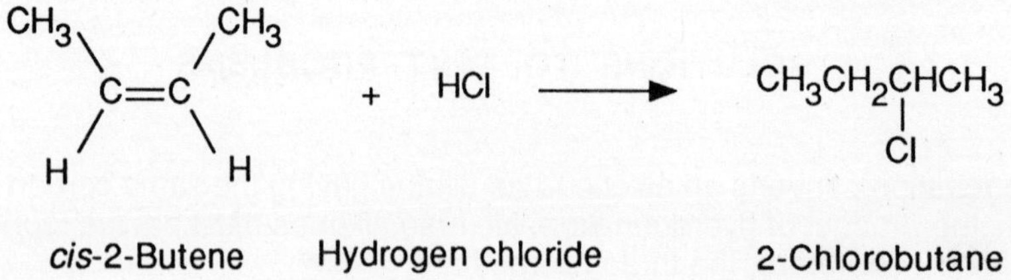

$$\begin{array}{c} CH_3 \\ H \end{array} C=C \begin{array}{c} CH_3 \\ H \end{array}$$ +   HCl   ⟶ $CH_3CH_2\underset{\underset{Cl}{|}}{C}HCH_3$

cis-2-Butene   Hydrogen chloride   2-Chlorobutane

(d)   One end of the double bond has no hydrogen substituents while the other end has one. In accordance with Markovnikov's rule, the proton adds to the carbon that already has one hydrogen. The product is 1-chloro-1-ethylcyclohexane.

Ethylidenecyclohexane   Hydrogen chloride   1-Chloro-1-ethylcyclohexane

5.3   (b)   A proton is transferred to the terminal carbon atom of 2-methyl-1-butene so as to produce a tertiary carbocation.

2-Methyl-1-butene   Hydrogen chloride   *tert*-Pentyl cation   Chloride ion

This is the carbocation that leads to the observed product, 2-chloro-2-methylbutane.

(c)   A secondary carbocation is an intermediate in the reaction of *cis*-2-butene with hydrogen chloride.

cis-2-Butene   Hydrogen chloride   *sec*-Butyl cation   Chloride ion

Capture of this carbocation by chloride gives *sec*-butyl chloride (2-chlorobutane).

(d)   A tertiary carbocation is formed by proton transfer from hydrogen chloride to the indicated alkene.

1-Ethylcyclohexyl cation   Chloride

5.4   In the hydroboration of an unsymmetrical alkene, boron becomes attached to the less-substituted carbon of the double bond.

$$6 \ (CH_3)_2C=CH_2 \quad + \quad B_2H_6 \quad \longrightarrow \quad 2 \ [(CH_3)_2CH{-}CH_2]_3B$$

85

5.5   Oxidation of a trialkylborane replaces the boron atom with a hydroxyl group. The overall reaction is formation of the less-substituted alcohol from an unsymmetrical alkene.

$$[(CH_3)_2CH-CH_2]_3B \xrightarrow[HO^-]{H_2O_2} 3\ (CH_3)_2CH-CH_2OH$$

5.6   (b)   Hydroboration-oxidation adds water (H and OH) to an alkene so the hydroxyl group is bonded to the less substituted end of the double bond.

$$(CH_3)_3CCH=CH_2 \xrightarrow[\text{2. } H_2O_2,\ HO^-]{\text{1. } B_2H_6} (CH_3)_3CCH_2CH_2OH$$

3,3-Dimethyl-1-butene   3,3-Dimethyl-1-butanol

(c)   The carbon-carbon double bond is symmetrically substituted in cis-2-butene, so the regioselectivity of hydroboration-oxidation is not an issue. Hydration of the double bond gives 2-butanol.

cis-2-Butene   2-Butanol

5.7   Reasoning backwards from the product is the best way to solve this problem. 1,2-Dibromo-2-methylpropane is the product of addition of bromine to 2-methylpropene. 2-Methylpropene is prepared by dehydrohalogenation of 2-bromo-2-methylpropane (Section 4.1.9). Therefore the correct synthesis is

2-Bromo-2-methylpropane   2-Methylpropene   1,2-Dibromo-2-methylpropane

5.8   Reason backwards as in the previous problem. 1-Bromo-2-methylpropane is the product of HBr addition, in the presence of peroxides, to 2-methylpropene. The alkene is prepared as in the previous problem.

$$(CH_3)_2C=CH_2 \quad + \quad HBr \xrightarrow{\text{peroxides}} (CH_3)_2CHCH_2Br$$

2-Methylpropene (from 5.7)   Hydrogen bromide   1-Bromo-2-methylpropane

5.9   The products of ozonolysis are formaldehyde and 4,4-dimethyl-2-pentanone.

Formaldehyde   4,4-Dimethyl-2-pentanone

86

The two carbons that were doubly bonded to each other in the alkene become the carbons that are doubly bonded to oxygen in the products of ozonolysis. Therefore, mentally remove the oxygens and connect these two carbons by a double bond to reveal the structure of the starting alkene.

$$\underset{H}{\overset{H}{\diagdown}}C=C\underset{CH_2C(CH_3)_3}{\overset{CH_3}{\diagup}}$$

2,4,4-Trimethyl-1-pentene

5.10 (b) The desired alkyne has a methyl group and a butyl group as substituents on a pair of triply bonded carbons. Therefore, two alkylations of acetylene are required--one with a methyl halide, the other with a butyl halide.

$$HC\equiv CH \quad \xrightarrow[\text{2. } CH_3Br]{\text{1. } NaNH_2, NH_3} \quad CH_3C\equiv CH \quad \xrightarrow[\text{2. } CH_3CH_2CH_2CH_2Br]{\text{1. } NaNH_2, NH_3} \quad CH_3C\equiv CCH_2CH_2CH_2CH_3$$

Acetylene   Propyne   2-Heptyne

It does not matter whether the methyl group or the butyl group is introduced first; the order of steps shown in the above synthetic scheme may be inverted.

(c) An ethyl group and a propyl group need to be introduced as substituents on a pair of triply bonded carbons. As in (b), it does not matter which of the two is introduced first.

$$HC\equiv CH \quad \xrightarrow[\text{2. } CH_3CH_2CH_2Br]{\text{1. } NaNH_2, NH_3} \quad CH_3CH_2CH_2C\equiv CH \quad \xrightarrow[\text{2. } CH_3CH_2Br]{\text{1. } NaNH_2, NH_3} \quad CH_3CH_2CH_2C\equiv CCH_2CH_3$$

Acetylene   1-Pentyne   3-Heptyne

5.11 The desired alkene can best be prepared by hydrogenation of 2-pentyne in the presence of Lindlar palladium. The necessary alkyne is prepared by alkylation of 1-butyne with methyl iodide.

$$CH_3CH_2C\equiv CH \quad \xrightarrow[\text{2. } CH_3I]{\text{1. } NaNH_2, NH_3} \quad CH_3CH_2C\equiv CCH_3 \quad \xrightarrow[\text{Lindlar Pd}]{H_2} \quad \underset{H}{\overset{CH_3CH_2}{\diagdown}}C=C\underset{H}{\overset{CH_3}{\diagup}}$$

1-Butyne   2-Pentyne   *cis*-2-Pentene

5.12 The preparation of a trans alkene is best accomplished by metal-ammonia reduction of the corresponding alkyne. 2-Pentyne, prepared in the previous problem, is treated with sodium in liquid ammonia to yield the desired product.

$$CH_3CH_2C{\equiv}CCH_3 \xrightarrow{\text{Na, NH}_3} \begin{array}{c} CH_3CH_2 \\ \diagdown \\ C=C \\ \diagup \quad \diagdown \\ H \qquad CH_3 \end{array}$$

2-Pentyne   *trans*-2-Pentene

5.13 (b) Addition of hydrogen chloride to vinyl chloride gives the geminal dichloride 1,1-dichloroethane.

$$CH_2{=}CHCl \xrightarrow{\text{HCl}} CH_3CHCl_2$$

Vinyl chloride   1,1-Dichloroethane

(c) Since 1,1-dichloroethane can be prepared by adding two moles of hydrogen chloride to acetylene, first convert vinyl bromide to acetylene by dehydrohalogenation.

$$CH_2{=}CHBr \xrightarrow[\text{2. H}_2\text{O}]{\text{1. NaNH}_2,\ \text{NH}_3} HC{\equiv}CH \xrightarrow{\text{2 HCl}} CH_3CHCl_2$$

Vinyl bromide   Acetylene   1,1-Dichloroethane

(d) As in (c), first convert the designated starting material to acetylene.

$$CH_3CHBr_2 \xrightarrow[\text{2. H}_2\text{O}]{\text{1. NaNH}_2,\ \text{NH}_3} HC{\equiv}CH \xrightarrow{\text{2 HCl}} CH_3CHCl_2$$

1,1-Dibromoethane   Acetylene   1,1-Dichloroethane

5.14 The enol arises by addition of water across the triple bond.

$$CH_3C{\equiv}CCH_3 \xrightarrow[\text{H}_2\text{SO}_4]{\text{H}_2\text{O, Hg}^{2+}} \underset{\underset{OH}{|}}{CH_3C}{=}CHCH_3 \longrightarrow \overset{\overset{O}{\|}}{CH_3C}CH_2CH_3$$

2-Butyne   2-Buten-2-ol (enol form)   2-Butanone

5.15 Each of the carbons that are part of $-CO_2H$ groups was once a carbon of a triple bond. The two fragments $CH_3(CH_2)_4CO_2H$ and $HO_2CCH_2CH_2CO_2H$ and account for only 10 of the original 16 carbons. The full complement of carbons can be accommodated by assuming that two molecules of $CH_3(CH_2)_{14}CH_3$ are formed, along with one molecule $HO_2CCH_2CH_2CO_2H$. The starting alkyne is therefore deduced from the ozonolysis data to be as shown:

$$CH_3(CH_2)_4C{\equiv}CCH_2CH_2C{\equiv}C(CH_2)_4CH_3$$

$CH_3(CH_2)_4CO_2H$ $HO_2CCH_2CH_2CO_2H$ $HO_2C(CH_2)_4CH_3$

5.16 (b) The 1,2 and 1,4 addition products are the constitutional isomers shown:

2,3-Dimethyl-1,3-butadiene   Product of 1,2-addition   Product of 1,4-addition

(c) The 1,2 and 1,4 addition of HBr to 1,3-cyclopentadiene give the same product.

1,3-Cyclopentadiene   3-Bromocyclopentene Product of 1,2-addition   3-Bromocyclopentene Product of 1,4-addition

5.17 The two double bonds of 2-methyl-1,3-butadiene are not equivalent, so two different products of 1,2-addition are possible, along with one 1,4-addition product.

2-Methyl-1,3-butadiene   3,4-Dibromo-3-methyl-1-butene (1,2-addition)   3,4-Dibromo-2-methyl-1-butene (1,2-addition)   1,4-Dibromo-2-methyl-2-butene (1,4-addition)

5.18 This problem illustrates the reactions of alkenes with various reagents and requires application of the Markovnikov rule in the addition of unsymmetrical electrophiles.

(a) Markovnikov addition of hydrogen chloride to 1-pentene will give 2-chloropentane.

$$CH_2=CHCH_2CH_2CH_3 \quad + \quad HCl \longrightarrow CH_3CHCH_2CH_2CH_3$$
$$\underset{Cl}{|}$$

1-Pentene   2-Chloropentane

(b) Ionic addition of hydrogen bromide will give 2-bromopentane.

$$CH_2=CHCH_2CH_2CH_3 \quad + \quad HBr \longrightarrow CH_3CHCH_2CH_2CH_3$$
$$\underset{Br}{|}$$

1-Pentene   2-Bromopentane

(c) The presence of peroxides in the reaction medium will cause free-radical addition of hydrogen bromide, and anti-Markovnikov regioselectivity will be observed.

$$CH_2=CHCH_2CH_2CH_3 \quad + \quad HBr \xrightarrow{\text{peroxides}} BrCH_2CH_2CH_2CH_2CH_2$$

1-Pentene   1-Bromopentane

(d) Hydrogen iodide will add according to Markovnikov's rule.

$$CH_2=CHCH_2CH_2CH_3 \quad + \quad HI \longrightarrow CH_3CHCH_2CH_2CH_3$$
$$\underset{I}{|}$$

1-Pentene   2-Iodopentane

(e) Dilute sulfuric acid will cause hydration of the double bond with regioselectivity in accord with Markovnikov's rule.

$$CH_2=CHCH_2CH_2CH_3 \quad + \quad H_2O \xrightarrow{H_2SO_4} CH_3CHCH_2CH_2CH_3$$
$$\underset{OH}{|}$$

1-Pentene   2-Pentanol

(f) Hydroboration-oxidation of an alkene brings about anti-Markovnikov hydration of the double bond; 1-pentanol will be the product.

$$CH_2=CHCH_2CH_2CH_3 \xrightarrow[\text{2. } H_2O_2, HO^-]{\text{1. } B_2H_6} HOCH_2CH_2CH_2CH_2CH_3$$

1-Pentene   1-Pentanol

(g) Bromine adds across the double bond to give the vicinal dibromide.

$$CH_2=CHCH_2CH_2CH_3 \quad + \quad Br_2 \quad \longrightarrow \quad BrCH_2\overset{\underset{|}{Br}}{C}HCH_2CH_2CH_3$$

1-Pentene   1,2-Dibromopentane

5.19 When we compare the reactions of 2-methyl-2-butene with the analogous reactions of 1-pentene, we find that the reactions proceed in a similar manner.

(a)

$$(CH_3)_2C=CHCH_3 \quad + \quad HCl \quad \longrightarrow \quad (CH_3)_2\overset{\underset{|}{Cl}}{C}CH_2CH_3$$

2-Methyl-2-butene   2-Chloro-2-methylbutane

(b)

$$(CH_3)_2C=CHCH_3 \quad + \quad HBr \quad \longrightarrow \quad (CH_3)_2\overset{\underset{|}{Br}}{C}CH_2CH_3$$

2-Methyl-2-butene   2-Bromo-2-methylbutane

(c)

$$(CH_3)_2C=CHCH_3 \quad + \quad HBr \quad \xrightarrow{\text{peroxides}} \quad (CH_3)_2\overset{\underset{|}{Br}}{C}HCHCH_3$$

2-Methyl-2-butene   2-Bromo-3-methylbutane

(d)

$$(CH_3)_2C=CHCH_3 \quad + \quad HI \quad \longrightarrow \quad (CH_3)_2\overset{\underset{|}{I}}{C}CH_2CH_3$$

2-Methyl-2-butene   2-Iodo-2-methylbutane

(e)

$$(CH_3)_2C=CHCH_3 \quad + \quad H_2O \quad \xrightarrow{H_2SO_4} \quad (CH_3)_2\overset{\underset{|}{OH}}{C}CH_2CH_3$$

2-Methyl-2-butene   2-Methyl-2-butanol

(f)

$$(CH_3)_2C=CHCH_3 \quad \xrightarrow[\text{2. } H_2O_2, HO^-]{\text{1. } B_2H_6} \quad (CH_3)_2\overset{\underset{|}{OH}}{C}HCHCH_3$$

2-Methyl-2-butene   3-Methyl-2-butanol

(g)

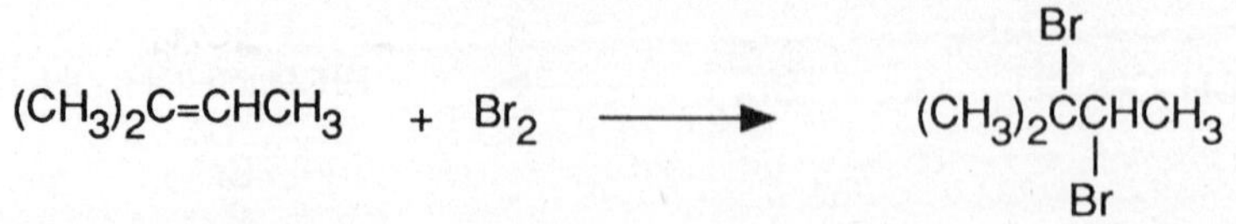

$$(CH_3)_2C=CHCH_3 \quad + \quad Br_2 \quad \longrightarrow$$

2-Methyl-2-butene

$$(CH_3)_2\overset{\displaystyle Br}{\underset{\displaystyle Br}{\overset{|}{\underset{|}{C}}}}CHCH_3$$

2,3-Dibromo-2-methylbutane

5.20 Cycloalkenes undergo the same kinds of reactions as do noncyclic ones.

(a)

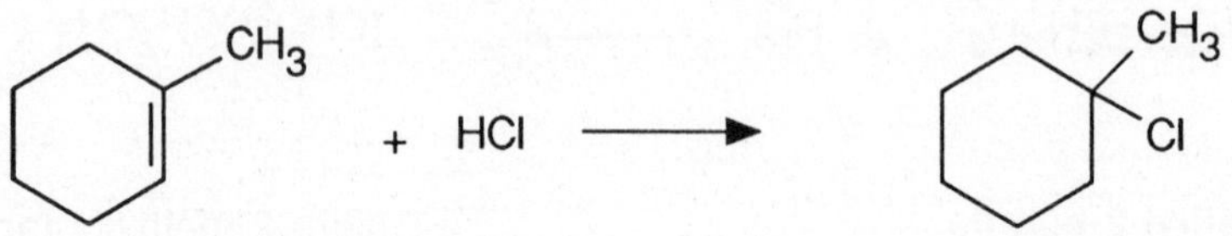

1-Methylcyclohexene + HCl ⟶ 1-Chloro-1-methylcyclohexane

(b)

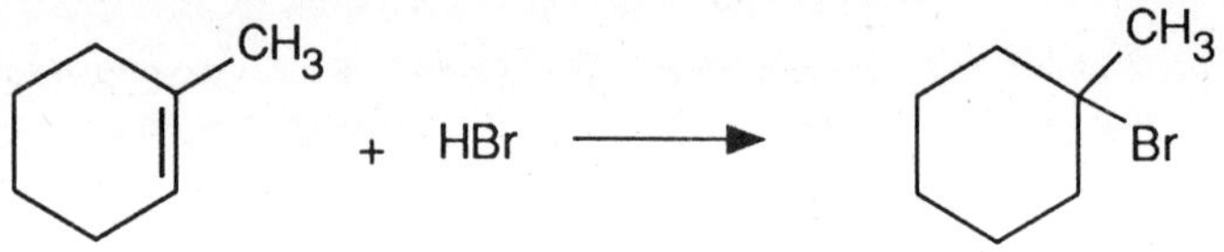

1-Methylcyclohexene + HBr ⟶ 1-Bromo-1-methylcyclohexane

(c)

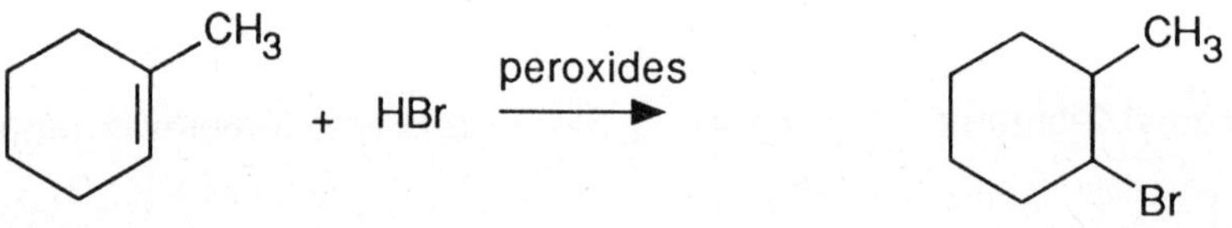

1-Methylcyclohexene + HBr --peroxides--> 1-Bromo-2-methylcyclohexane
(mixture of cis and trans)

(d)

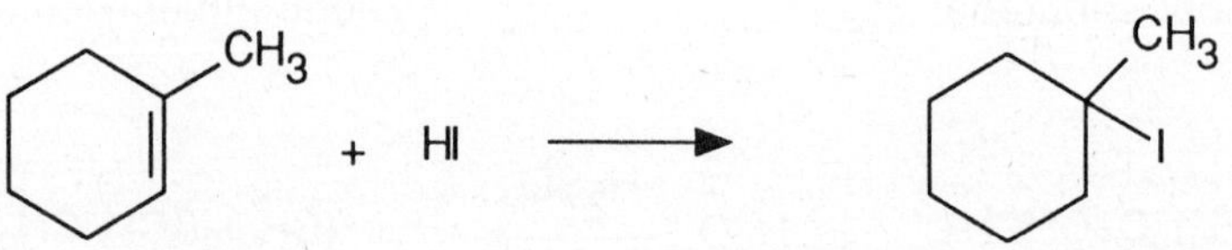

1-Methylcyclohexene + HI ⟶ 1-Iodo-1-methylcyclohexane

(e)

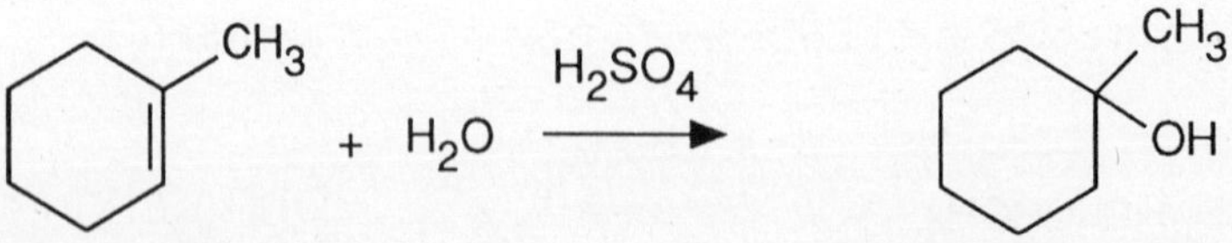

1-Methylcyclohexene + H_2O --H_2SO_4--> 1-Methylcyclohexanol

92

(f)

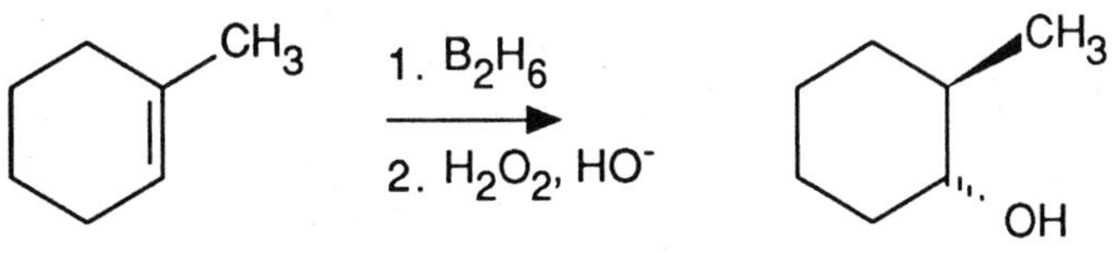

1-Methylcyclohexene   *trans*-2-Methylcyclohexanol

(g)

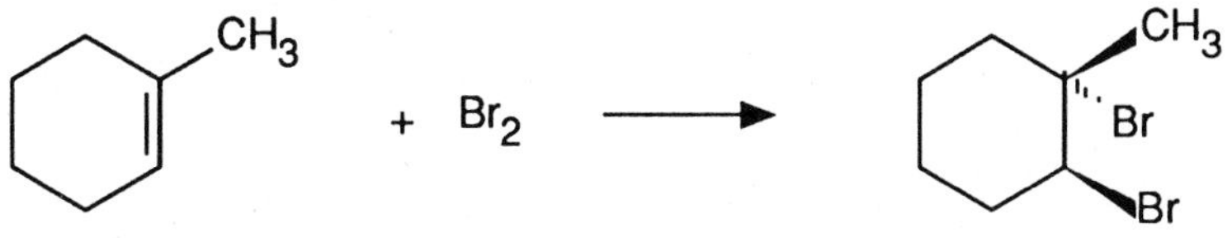

1-Methylcyclohexene   *trans*-1,2-Dibromo-1-methylcyclohexane

5.21 (a) Ozonolysis cleaves the double bond of alkenes, forming aldehydes and ketones.

$CH_2=CHCH_2CH_2CH_3$ $\xrightarrow[\text{2. Zn, H}_2\text{O}]{\text{1. O}_3}$ HCH + HCCH₂CH₂CH₃

1-Pentene   Formaldehyde   Butanal

(b) Potassium permanganate also cleaves alkenes, however aldehydes undergo further oxidation to carboxylic acids. Thus, cleavage of 1-pentene proceeds as follows:

$CH_2=CHCH_2CH_2CH_3$ $\xrightarrow[\text{H}^+]{\text{KMnO}_4}$ HOCOH + HOCCH₂CH₂CH₃

1-Pentene   Carbonic acid   Butanoic acid

Carbonic acid, however, is unstable and decomposes to carbon dioxide and water.

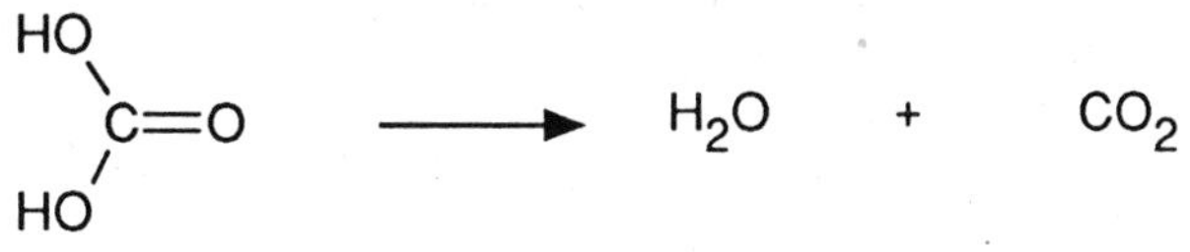

Carbonic acid   Water   Carbon dioxide

Thus, oxidation of 1-pentene by potassium permanganate gives butanoic acid, carbon dioxide, and water.

5.22 (a) Ozonolysis of 2-methyl-2-butene yields a ketone and an aldehyde.

$(CH_3)_2C=CHCH_3$ $\xrightarrow[\text{2. Zn, H}_2\text{O}]{\text{1. O}_3}$ $(CH_3)_2C=O$ + HCCH₃

2-Methyl-2-butene   Acetone   Acetaldehyde

(b) The products of oxidation of 2-methyl-2-butene by potassium permanganate are acetic acid and acetone.

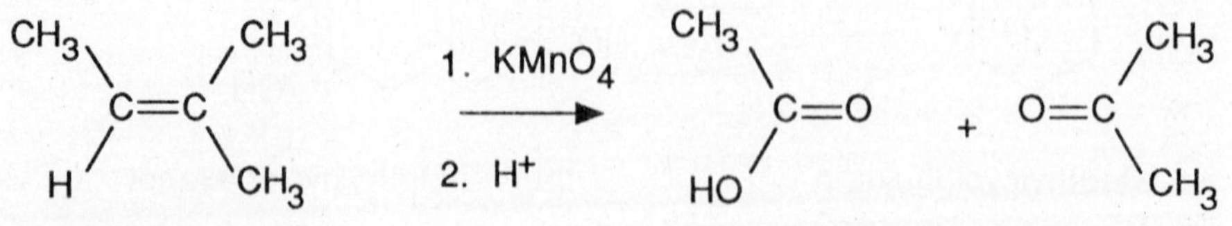

2-Methyl-2-butene   Acetic acid   Acetone

5.23 (a) Cleavage of the double bond by ozonolysis gives a noncyclic compound containing all seven carbon atoms of 1-methylcyclohexene.

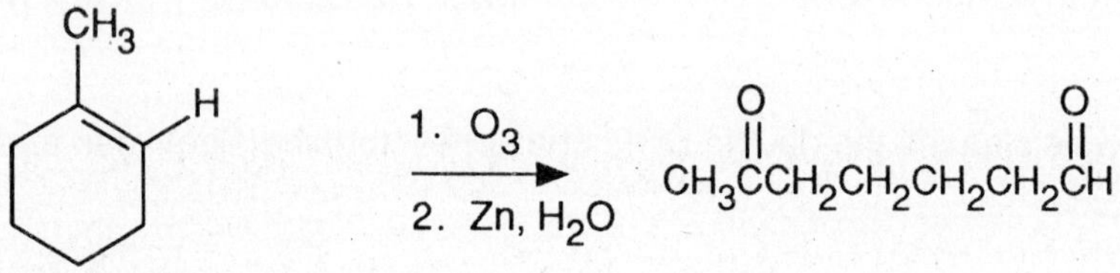

1-Methylcyclohexene   6-Oxoheptanal

(b) Likewise, permanganate cleavage of 1-methylcyclohexene gives a single noncyclic product.

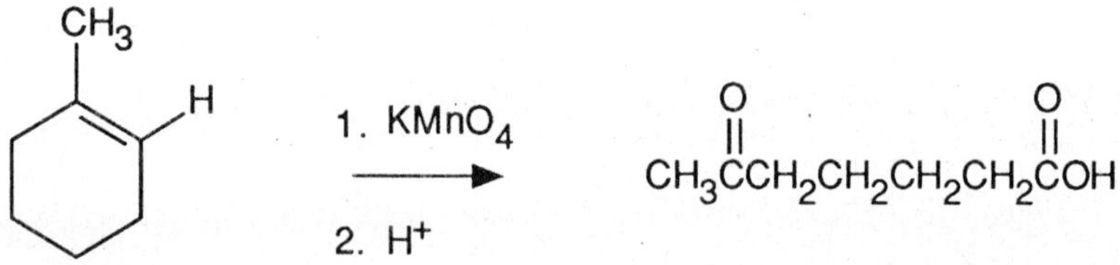

1-Methylcyclohexene   6-Oxoheptanoic acid

5.24 In all parts of this exercise we deduce the carbon skeleton on the basis of the alkane formed on hydrogenation of an alkene, then determine what carbon atoms may be connected by a double bond in that skeleton. Problems of this type are best done by using carbon skeleton formulas.

(a)

Product is 2,3-dimethylbutane   May be formed by hydrogenation of

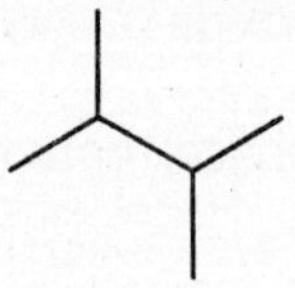

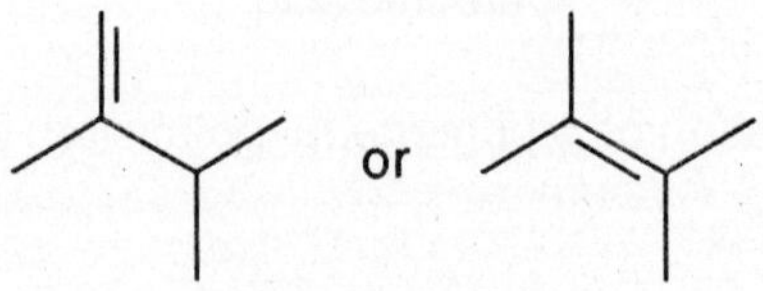

(b)

Product is methylcyclobutane   May be formed by hydrogenation of

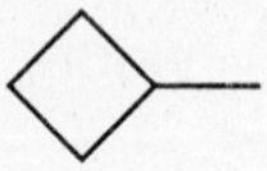

94

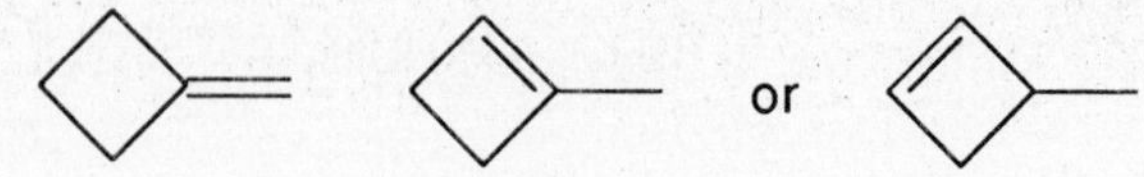

(c)

Product is cis-1,4-dimethylcyclohexane

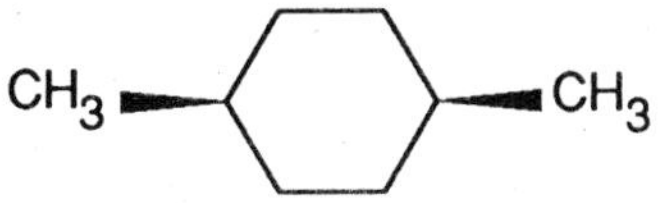

The alkene that gives only the cis isomer on hydrogenation is

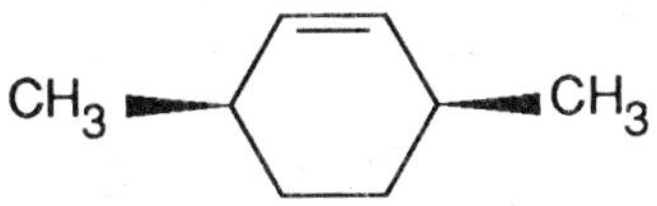

CH_3 ▰—⟨ ⟩—▰ CH_3

CH_3 ▰—⟨ ⟩—▰ CH_3

5.25 (a) The desired transformation is the conversion of an alkene to a vicinal dibromide.

$$CH_3CH=C(CH_2CH_3)_2 \xrightarrow{Br_2} CH_3CHC(CH_2CH_3)_2$$
$$\underset{Br \ \ Br}{| \ \ |}$$

3-Ethyl-2-pentene   2,3-Dibromo-3-ethylpentane

(b) Acid-catalyzed hydration will occur in accordance with Markovnikov's rule to yield the desired tertiary alcohol.

$$CH_3CH=C(CH_2CH_3)_2 \xrightarrow[H_2SO_4]{H_2O} CH_3CH_2C(CH_2CH_3)_2$$
$$\underset{OH}{|}$$

3-Ethyl-2-pentene   3-Ethyl-3-pentanol

(c) Markovnikov addition of hydrogen chloride is indicated.

$$CH_3CH=C(CH_2CH_3)_2 \xrightarrow{HCl} CH_3CH_2C(CH_2CH_3)_2$$
$$\underset{Cl}{|}$$

3-Ethyl-2-pentene   3-Chloro-3-ethylpentane

(d) Hydroboration-oxidation results in hydration of alkenes with a regioselectivity contrary to that of Markovnikov's rule.

$$CH_3CH=C(CH_2CH_3)_2 \xrightarrow[2. \ H_2O_2, \ HO^-]{1. \ B_2H_6} CH_3CHCH(CH_2CH_3)_2$$
$$\underset{OH}{|}$$

3-Ethyl-2-pentene   3-Ethyl-2-pentanol

(e) Hydrogenation of alkenes converts them to alkanes.

$$CH_3CH=C(CH_2CH_3)_2 \xrightarrow{H_2, \ Pt} CH_3CH_2CH(CH_2CH_3)_2$$

3-Ethyl-2-pentene   3-Ethylpentane

(f)   Free-radical addition of hydrogen bromide will produce the required anti-Markovnikov orientation.

$$CH_3CH=C(CH_2CH_3)_2 \xrightarrow[\text{peroxides}]{HBr} CH_3CHCH(CH_2CH_3)_2$$
$$\overset{|}{Br}$$

3-Ethyl-2-pentene   2-Bromo-3-ethylpentane

5.26 (a)   The addition of HBr to an alkene in the absence of peroxides proceeds by formation of the most stable *carbocation* intermediate. In this case protonation of the alkene leads to a tertiary carbocation.

$$(CH_3)_2C=CHCH_3 \xrightarrow{HBr} (CH_3)_2\overset{+}{C}—CH_2CH_3 \; + \; Br^- \longrightarrow (CH_3)_2C—CH_2CH_3$$
$$\overset{|}{Br}$$

2-Methyl-2-butene   Tertiary carbocation   2-Bromo-2-methylbutane

(b)   Addition of HBr to an alkene in the presence of peroxides proceeds by a *free radical* mechanism. In the free radical process, a bromine atom adds to the alkene in the first propagation step to form a tertiary radical.

Initiation: $ROOR \longrightarrow 2RO\bullet$

                    $RO\bullet + HBr \longrightarrow ROH + Br\bullet$

Propagation: $Br\bullet + (CH_3)_2C=CHCH_3 \longrightarrow (CH_3)_2\overset{\bullet}{C}—CHCH_3$
$$\overset{|}{Br}$$

$(CH_3)_2C—CHCH_3 + HBr \longrightarrow (CH_3)_2CH—CHCH_3 + Br\bullet$
$$\overset{|}{Br}$$

5.27 (a)

$$CH_3CH_2CH_2CH_2C\equiv CH + 2H_2 \xrightarrow{Pt} CH_3CH_2CH_2CH_2CH_2CH_3$$

1-Hexyne   Hexane

(b)

$$CH_3CH_2CH_2CH_2C\equiv CH + H_2 \xrightarrow{Lindlar\ Pd} CH_3CH_2CH_2CH_2CH=CH_2$$

1-Hexyne   1-Hexene

(c)

$$CH_3CH_2CH_2CH_2C\equiv CH \xrightarrow{Na,\ NH_3} CH_3CH_2CH_2CH_2CH=CH_2$$

1-Hexyne   1-Hexene

(d)

$$CH_3CH_2CH_2CH_2C{\equiv}CH \xrightarrow{\text{NaNH}_2,\ \text{NH}_3} CH_3CH_2CH_2CH_2C{\equiv}\overset{-}{C}{:}\ \ Na^+$$

1-Hexyne   Sodium 1-hexynide

(e)

$$CH_3CH_2CH_2CH_2C{\equiv}\overset{-}{C}{:}\ Na^+ \ + \ CH_3CH_2CH_2CH_2Br \longrightarrow CH_3CH_2CH_2CH_2C{\equiv}CCH_2CH_2CH_2CH_3$$

Sodium 1-hexynide   1-Bromobutane   5-Decyne

(f)

$$CH_3CH_2CH_2CH_2C{\equiv}CH \xrightarrow[\text{one mole}]{\text{HCl}} CH_3CH_2CH_2CH_2\underset{\underset{Cl}{|}}{C}{=}CH_2$$

1-Hexyne   2-Chloro-1-hexene

(g)

$$CH_3CH_2CH_2CH_2C{\equiv}CH \xrightarrow[\text{two moles}]{\text{HCl}} CH_3CH_2CH_2CH_2\overset{\overset{Cl}{|}}{\underset{\underset{Cl}{|}}{C}}CH_3$$

1-Hexyne   2,2-Dichlorohexane

(h)

$$CH_3CH_2CH_2CH_2C{\equiv}CH \xrightarrow[\text{two moles}]{\text{Cl}_2} CH_3CH_2CH_2CH_2\overset{\overset{Cl}{|}}{\underset{\underset{Cl}{|}}{C}}CHCl_2$$

1-Hexyne   1,1,2,2-Tetrachlorohexane

(i)

$$CH_3CH_2CH_2CH_2C{\equiv}CH \xrightarrow[\text{HgSO}_4]{\text{H}_2\text{O, H}_2\text{SO}_4} CH_3CH_2CH_2CH_2\overset{\overset{O}{||}}{C}CH_3$$

1-Hexyne   2-Hexanone

(j)

$$CH_3CH_2CH_2CH_2C{\equiv}CH \xrightarrow[\text{2. H}_2\text{O}]{\text{1. O}_3} CH_3CH_2CH_2CH_2\overset{\overset{O}{||}}{C}OH \ + \ HO\overset{\overset{O}{||}}{C}H$$

1-Hexyne   Pentanoic acid   Formic acid

5.28 (a)

$$CH_3CH_2C \equiv CCH_2CH_3 \quad + \quad 2\,H_2 \quad \xrightarrow{\text{Pt}} \quad CH_3CH_2CH_2CH_2CH_2CH_3$$

3-Hexyne   Hexane

(b)

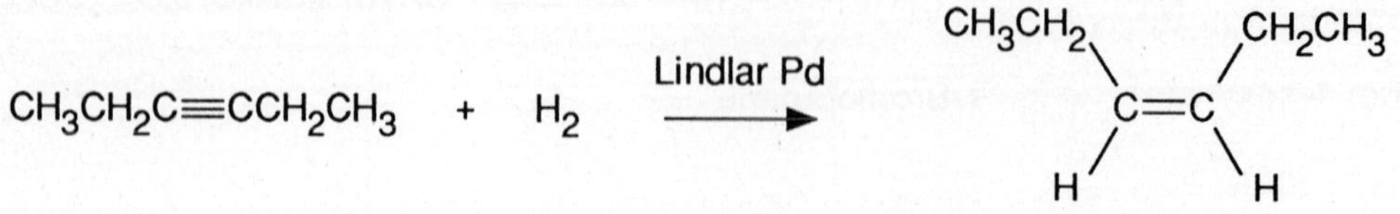

$$CH_3CH_2C \equiv CCH_2CH_3 \quad + \quad H_2 \quad \xrightarrow{\text{Lindlar Pd}}$$

3-Hexyne

$$
\begin{array}{ccc}
CH_3CH_2 & & CH_2CH_3 \\
 & C = C & \\
H & & H
\end{array}
$$

cis- 3-Hexene

(c)

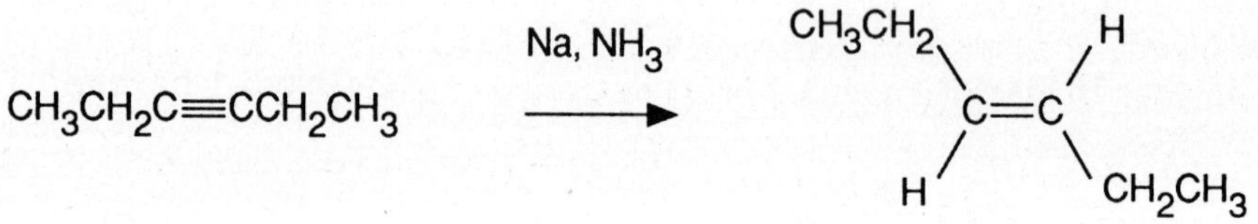

$$CH_3CH_2C \equiv CCH_2CH_3 \quad \xrightarrow{\text{Na, NH}_3}$$

3-Hexyne

$$
\begin{array}{ccc}
CH_3CH_2 & & H \\
 & C = C & \\
H & & CH_2CH_3
\end{array}
$$

trans- 3-Hexene

(d)

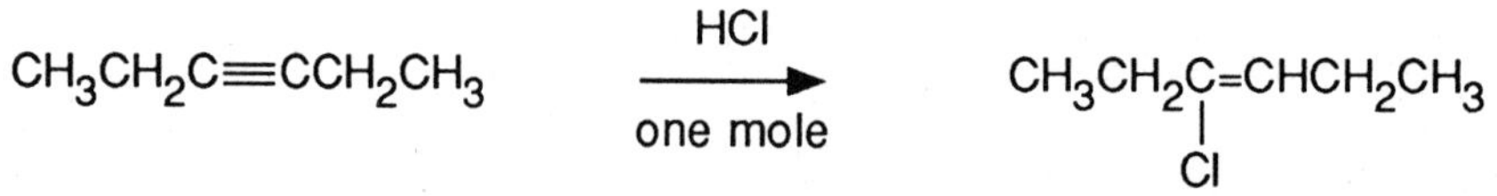

$$CH_3CH_2C \equiv CCH_2CH_3 \quad \xrightarrow[\text{one mole}]{\text{HCl}}$$

3-Hexyne

$$
\begin{array}{c}
CH_3CH_2C=CHCH_2CH_3 \\
\quad\quad | \\
\quad\quad Cl
\end{array}
$$

3-Chloro-3-hexene

(e)

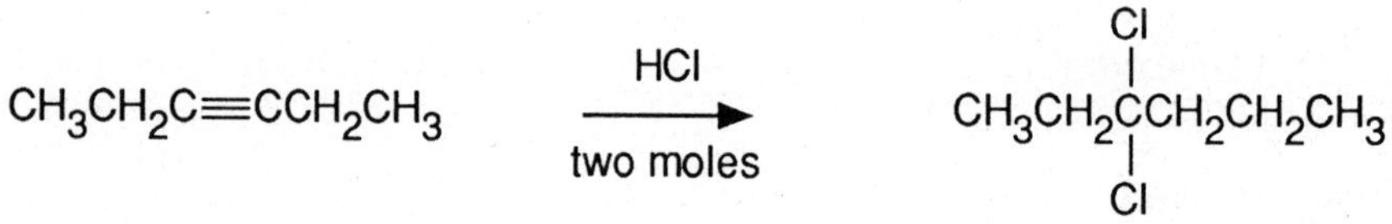

$$CH_3CH_2C \equiv CCH_2CH_3 \quad \xrightarrow[\text{two moles}]{\text{HCl}}$$

3-Hexyne

$$
\begin{array}{c}
\quad\quad Cl \\
\quad\quad | \\
CH_3CH_2CCH_2CH_2CH_3 \\
\quad\quad | \\
\quad\quad Cl
\end{array}
$$

3,3-Dichlorohexane

(f)

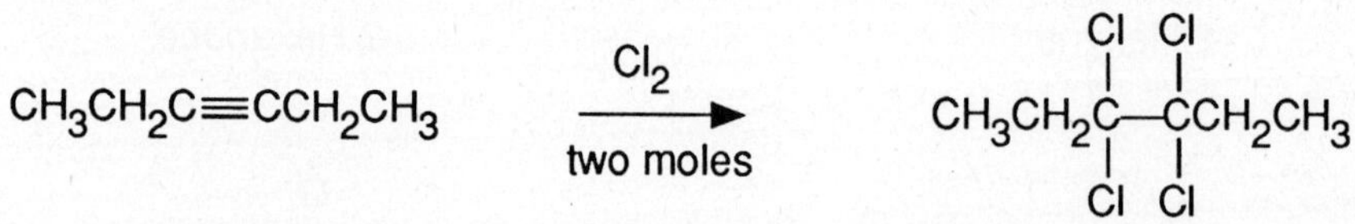

$$CH_3CH_2C \equiv CCH_2CH_3 \quad \xrightarrow[\text{two moles}]{\text{Cl}_2}$$

3-Hexyne

$$
\begin{array}{c}
\quad Cl \;\; Cl \\
\quad | \;\;\; | \\
CH_3CH_2C-CCH_2CH_3 \\
\quad | \;\;\; | \\
\quad Cl \;\; Cl
\end{array}
$$

3,3,4,4-Tetrachlorohexane

(g)

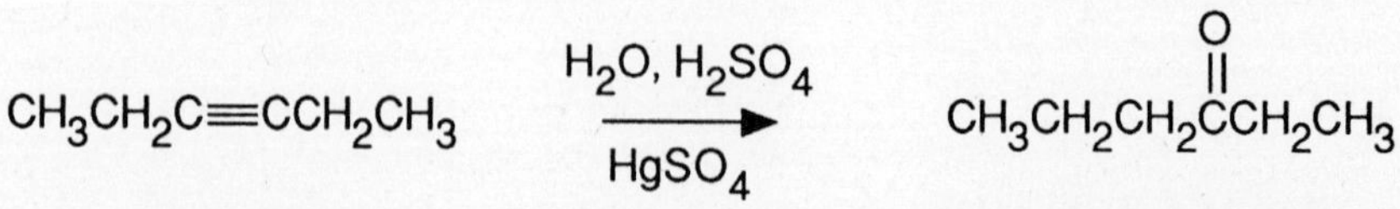

$$CH_3CH_2C \equiv CCH_2CH_3 \quad \xrightarrow[\text{HgSO}_4]{\text{H}_2\text{O, H}_2\text{SO}_4}$$

3-Hexyne

$$
\begin{array}{c}
\quad\quad\quad O \\
\quad\quad\quad \| \\
CH_3CH_2CH_2CCH_2CH_3
\end{array}
$$

3-Hexanone

98

(h)

$$CH_3CH_2C\equiv CCH_2CH_3 \xrightarrow[\text{2. }H_2O]{\text{1. }O_3} 2\ CH_3CH_2\overset{\overset{\displaystyle O}{||}}{C}OH$$

3-Hexyne   Propanoic acid

5.29 The two carbons of the triple bond are not identically substituted in 2-heptyne. Reaction of the alkyne with aqueous sulfuric acid containing mercuric sulfate gives two isomeric ketones.

$$CH_3C\equiv C(CH_2)_3CH_3 \xrightarrow[\text{HgSO}_4]{\text{H}_2\text{O, H}_2\text{SO}_4} CH_3\overset{\overset{\displaystyle O}{||}}{C}CH_2(CH_2)_3CH_3 \ + \ CH_3CH_2\overset{\overset{\displaystyle O}{||}}{C}(CH_2)_3CH_3$$

2-Heptyne   2-Heptanone   3-Heptanone

5.30 The enols leading to the ketones in the previous problem are formed by addition of water to the triple bond of 2-heptyne. Two regioisomeric enols are formed in the reaction.

$$CH_3C\equiv C(CH_2)_3CH_3 \xrightarrow[\text{HgSO}_4]{\text{H}_2\text{O, H}_2\text{SO}_4} \overset{\overset{\displaystyle OH}{|}}{CH_3C}=CH(CH_2)_3CH_3 \ + \ CH_3CH=\overset{\overset{\displaystyle OH}{|}}{C}(CH_2)_3CH_3$$

2-Heptyne   (gives 2-heptanone)   (gives 3-heptanone)

5.31 The starting material in all cases is the conjugated diene 2,3-dimethyl-1,3-butadiene.

(a) Hydrogenation of both double bonds will occur to yield 2,3-dimethylbutane.

$$\xrightarrow[\text{Pt}]{\text{H}_2\text{ (2 moles)}} (CH_3)_2CHCH(CH_3)_2$$

2,3-Dimethyl-1,3-butadiene   2,3-Dimethylbutane

(b) The problem specifies that 1,2-addition of one mole of hydrogen chloride is to be considered. The product therefore is the one formed by Markovnikov addition of HCl to one of the double bonds.

$$\xrightarrow{\text{HCl (1 mole)}}$$

2,3-Dimethyl-1,3-butadiene   3-Chloro-2,3-dimethyl-1-butene

(c) In 1,4-addition of hydrogen chloride, a proton adds to one end of the diene system and chloride to the other. Addition is accompanied by double bond migration.

2,3-Dimethyl-1,3-butadiene   1-Chloro-2,3-dimethyl-2-butene

(d) In 1,2-addition of Br_2 a vicinal dibromide is formed (bromines on adjacent carbons). The second double bond is not affected.

2,3-Dimethyl-1,3-butadiene   3,4-Dibromo-2,3-dimethyl-1-butene

(e) In 1,4-addition, bromines become attached to the end of the diene system and is accompanied by migration of the double bond.

2,3-Dimethyl-1,3-butadiene   1,4-Dibromo-2,3-dimethyl-2-butene

(f) Addition of two moles of bromine will lead to the same tetrabromide irrespective of whether the first addition step occurs by 1,2 or 1,4-addition.

2,3-Dimethyl-1,3-butadiene   1,2,3,4-Tetrabromo-2,3-dimethylbutane

(g) The reaction of a diene with maleic anhydride is a Diels-Alder reaction.

100

5.32 (a) Cyclohexane will be the product of hydrogenation of 1,3-cyclohexadiene:

1,3-Cyclohexadiene   Cyclohexane

(b)   1,2-Addition will occur to give 3-chlorocyclohexene. The proton adds to the terminus of the conjugated diene system.

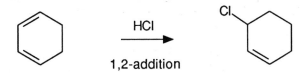

1,3-Cyclohexadiene   3-Chlorocyclohexene

(c)   The product of 1,4-addition is 3-chlorocyclohexene also. 1,2-Addition and 1,4-addition of hydrogen chloride to 1,3-cyclohexadiene give the same product.

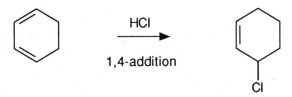

1,3-Cyclohexadiene   3-Chlorocyclohexene

(d)   Bromine can add to one of the double bonds to give 3,4-dibromocyclohexene:

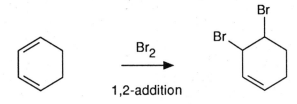

1,3-Cyclohexadiene   3,4-Dibromocyclohexene

(e)   1,4-Addition of bromine will give 3,6-dibromocyclohexene:

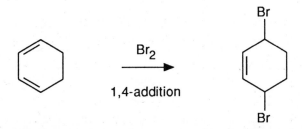

1,3-Cyclohexadiene   3,6-Dibromocyclohexene

(f)   Addition of two moles of bromine will yield 1,2,3,4-tetrabromocyclohexane.

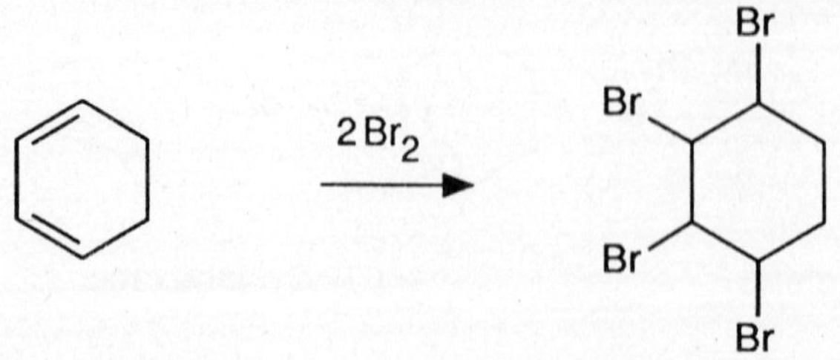

1,3-Cyclohexadiene   1,2,3,4-Tetrabromocyclohexane

(g)   The Diels-Alder adduct of 1,3-cyclohexadiene and maleic anhydride will have a bicyclic carbon skeleton.

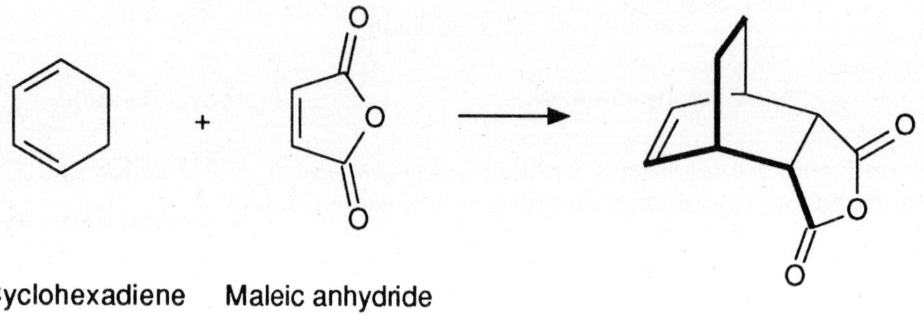

1,3-Cyclohexadiene   Maleic anhydride

5.33   1,3-Butadiene undergoes a Diels-Alder cycloaddition reaction with *cis*- and *trans*-cinnamic acid. The stereochemistry of substituent groups in the dienophile is preserved in the cyclohexene product.

(a)

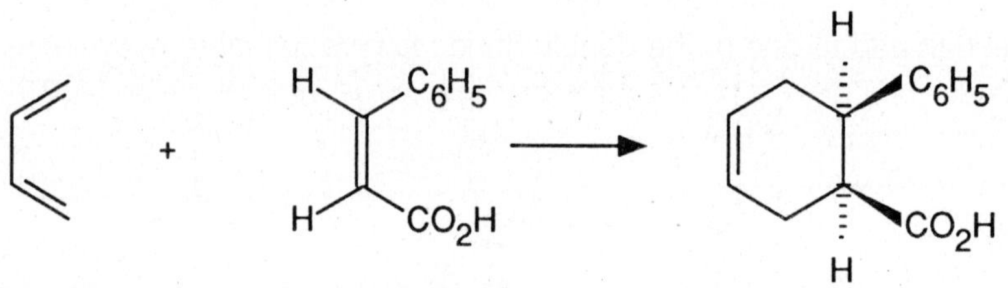

1,3-Butadiene   *cis*- Cinnamic acid

(b)

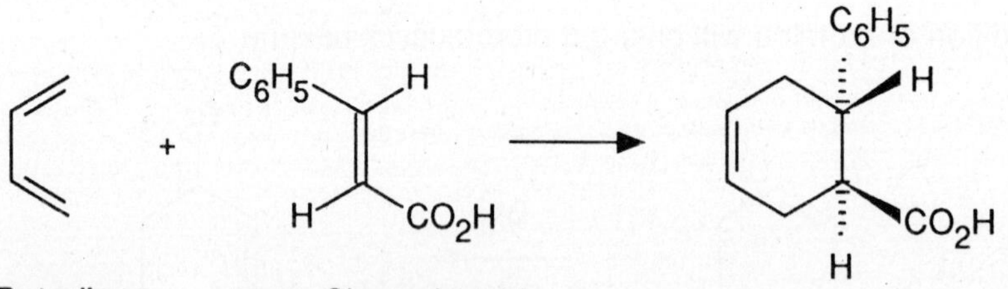

1,3-Butadiene   *trans*- Cinnamic acid

5.34 The only method that you have learned for preparing alkanes is hydrogenation of alkenes. Therefore, convert 2-methyl-2-hexanol to a mixture of alkenes by acid-catalyzed dehydration, then hydrogenate.

$$(CH_3)_2CCH_2CH_2CH_2CH_3 \xrightarrow[\text{heat}]{H_2SO_4} \begin{array}{c} (CH_3)_2C=CHCH_2CH_2CH_3 \\ + \\ CH_2=CCH_2CH_2CH_2CH_3 \end{array} \xrightarrow{H_2, Pt} (CH_3)_2CHCH_2CH_2CH_2CH_3$$

with OH on the starting material.

2-Methyl-2-hexanol   2-Methylhexane

(b) The product is a primary alcohol which can be prepared by the hydroboration - oxidation of the appropriate alkene, propene in this case. Propene is available by acid-catalyzed dehydration of 2-propanol.

$$CH_3CHCH_3 \xrightarrow[\text{heat}]{H_2SO_4} CH_3CH=CH_2 \xrightarrow[\text{2. } H_2O_2, HO^-]{\text{1. } B_2H_6} CH_3CH_2CH_2OH$$

with OH on the 2-propanol.

2-Propanol   Propene   1-Propanol

(c) The secondary alcohol product can be prepared by hydration of the appropriate alkene, propene. As in parts (a) and (b), propene is available by acid-catalyzed dehydration of the starting alcohol.

$$CH_3CH_2CH_2OH \xrightarrow[\text{heat}]{H_2SO_4} CH_3CH=CH_2 \xrightarrow[H_2SO_4]{H_2O} CH_3CHCH_3$$

with OH on the product.

1-Propanol   Propene   2-Propanol

(d) 1-Bromopropane can be prepared from propene by free-radical addition of HBr in the presence of peroxides. Propene can be prepared by dehydrohalogenation of 2-bromopropane.

$$CH_3CHCH_3 \xrightarrow[CH_3CH_2OH]{NaOCH_2CH_3} CH_3CH=CH_2 \xrightarrow[\text{peroxides}]{HBr} CH_3CH_2CH_2Br$$

with Br on the starting material, heat.

2-Bromopropane   heat   Propene   1-Bromopropane

(e) 1,2-Dibromopropane is the product of addition of Br_2 to propene. The alkene is prepared as in part (d).

$$CH_3CHCH_3 \xrightarrow[CH_3CH_2OH]{NaOCH_2CH_3} CH_3CH=CH_2 \xrightarrow{Br_2} CH_3CHCH_2Br$$

with Br on the starting material, heat; Br on the product.

2-Bromopropane   Propene   1,2-Dibromopropane

(f) The six-carbon dicarbonyl compound is available by ozonolysis of 1-methylcyclopentene. The cyclic alkene can be prepared by dehydrohalogenation of the starting material.

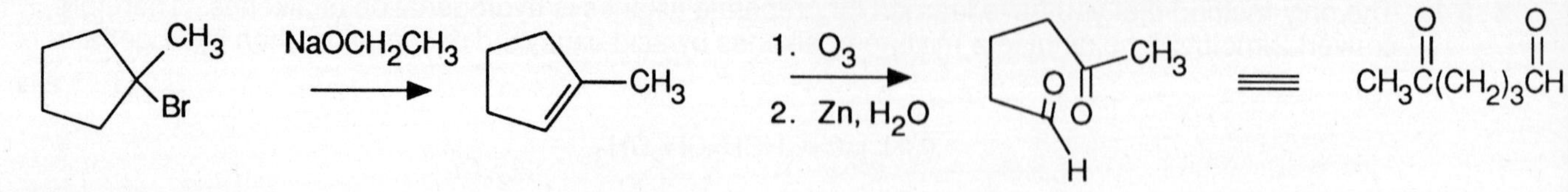

1-Bromo-1-methylcyclopentane — 1-Methylcyclopentene — 5-Oxohexanal

(g) 2,2-Dibromopropane is the product of addition of two moles of hydrogen bromide to propyne. The alkyne can be prepared by double dehydrohalogenation of 1,2-dibromopropane.

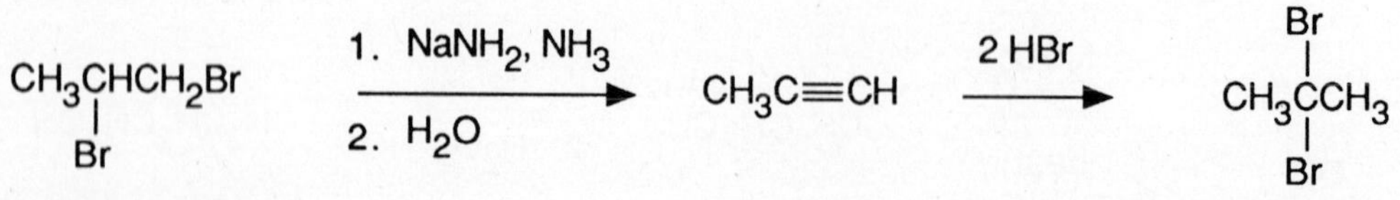

1,2-Dibromopropane — Propyne — 2,2-Dibromopropane

5.35 Ketones such as 2-heptanone may be readily prepared by hydration of terminal acetylenes. Thus, if we had 1-heptyne, it could be converted to 2-heptanone.

$$HC{\equiv}C(CH_2)_4CH_3 \xrightarrow[HgSO_4]{H_2O,\ H_2SO_4} CH_3\overset{O}{\overset{\|}{C}}(CH_2)_4CH_3$$

1-Heptyne — 2-Heptanone

Acetylene, as we have seen in earlier problems, can be converted to 1-heptyne by alkylation.

$$HC{\equiv}CH \xrightarrow{NaNH_2,\ NH_3} HC{\equiv}CNa$$

Acetylene — Sodium acetylide

$$HC{\equiv}CNa + CH_3CH_2CH_2CH_2CH_2Br \longrightarrow HC{\equiv}C(CH_2)_4CH_3$$

Sodium acetylide — 1-Bromopentane — 1-Heptyne

5.36 Reasoning backwards, the final step in the synthesis is formation of the cis double bond by hydrogenation of an alkyne using Lindlar palladium as a catalyst.

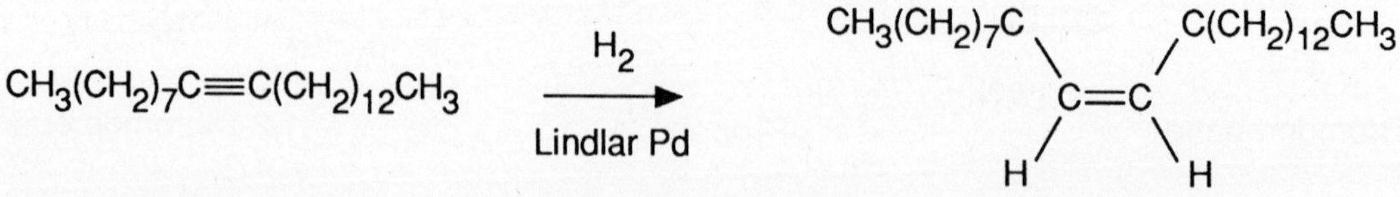

9-Tricosyne — *cis*-9-Tricosene

The necessary alkyne 9-tricosyne can be prepared by a double alkylation of acetylene.

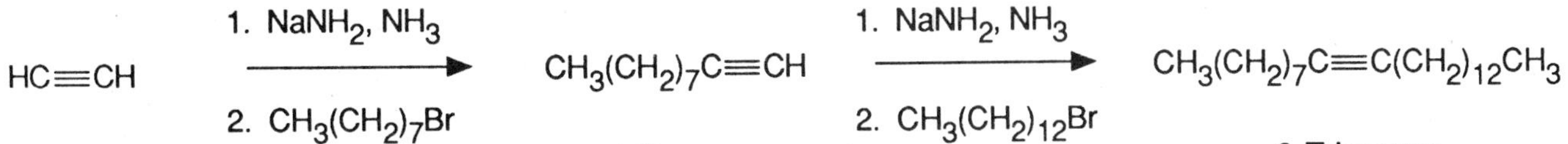

$$HC{\equiv}CH \xrightarrow[\text{2. } CH_3(CH_2)_7Br]{\text{1. } NaNH_2,\ NH_3} CH_3(CH_2)_7C{\equiv}CH \xrightarrow[\text{2. } CH_3(CH_2)_{12}Br]{\text{1. } NaNH_2,\ NH_3} CH_3(CH_2)_7C{\equiv}C(CH_2)_{12}CH_3$$

Acetylene   1-Decyne   9-Tricosyne

It does not matter which alkyl group is introduced first.

The alkyl halides are prepared from the corresponding alcohols.

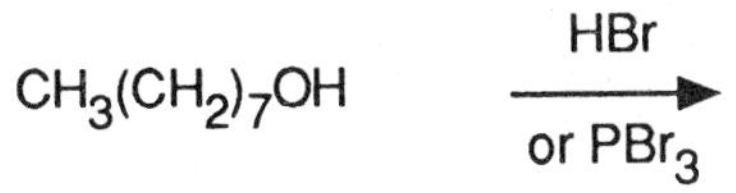

$$CH_3(CH_2)_7OH \xrightarrow[\text{or } PBr_3]{HBr} CH_3(CH_2)_7Br$$

1-Octanol   1-Bromooctane

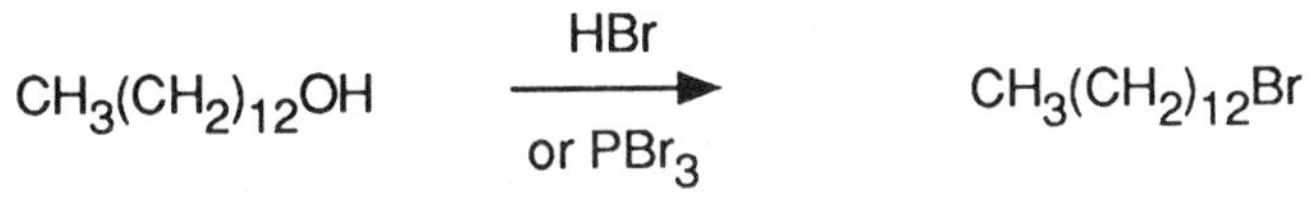

$$CH_3(CH_2)_{12}OH \xrightarrow[\text{or } PBr_3]{HBr} CH_3(CH_2)_{12}Br$$

1-Tridecanol   1-Bromotridecane

5.37 The ozonolysis data are useful in quickly identifying alkenes A and B. The carbon in the C=O group of the aldehydes and ketones formed in ozonolysis are derived from the doubly bonded carbons of the alkene. The substituents on the C=O group are the substituents on the double bond.

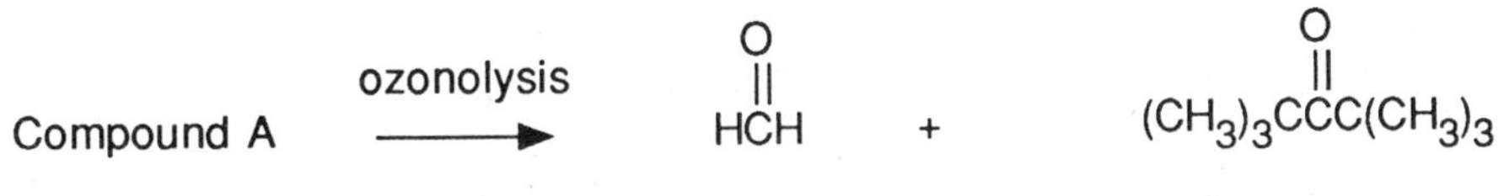

$$\text{Compound A} \xrightarrow{\text{ozonolysis}} \underset{\text{Formaldehyde}}{\overset{O}{\underset{\|}{HCH}}} + \underset{\text{2,2,4,4-Tetramethyl-3-pentanone}}{\overset{O}{\underset{\|}{(CH_3)_3CCC(CH_3)_3}}}$$

Therefore, one of the doubly bonded carbons of compound A bears two hydrogen substituents and the other bears two *tert*-butyl groups. **Compound A** is 2-*tert*-butyl-3,3-dimethyl-1-butene.

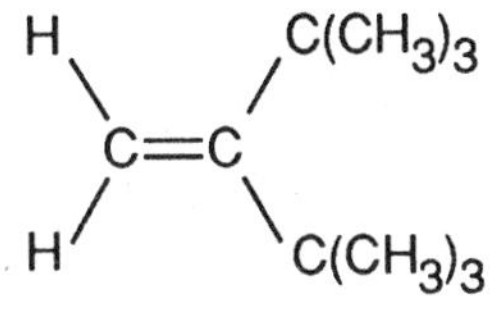

2-*tert*-Butyl-3,3-dimethyl-1-butene

Compound B gives formaldehyde and the ketone shown below on ozonolysis.

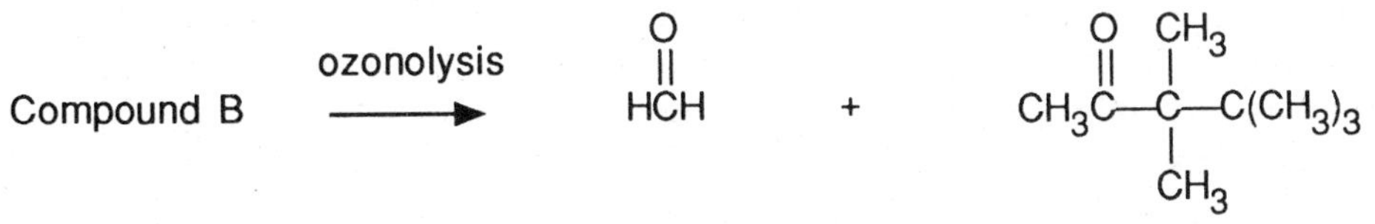

$$\text{Compound B} \xrightarrow{\text{ozonolysis}} \underset{\text{Formaldehyde}}{\overset{O}{\underset{\|}{HCH}}} + \underset{\text{3,3,4,4-Tetramethyl-2-pentanone}}{CH_3C-C-C(CH_3)_3}$$

Therefore, **compound B is** 2,3,3,4,4-pentamethyl-1-pentene.

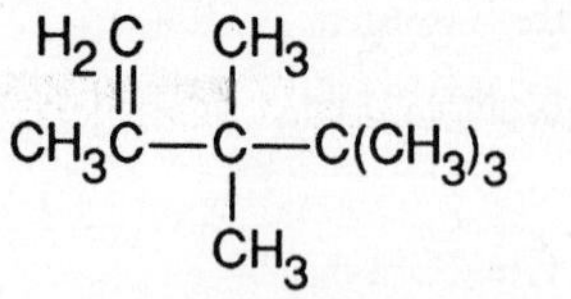
H2C CH3
CH3C—C—C(CH3)3
CH3
2,3,3,4,4-Pentamethyl-1-pentene

5.38 The important clue to deducing the structures of C and D is the ozonolysis product E. Remembering that the two carbonyl carbons of E must have been joined by a double bond in the precursor D, we write

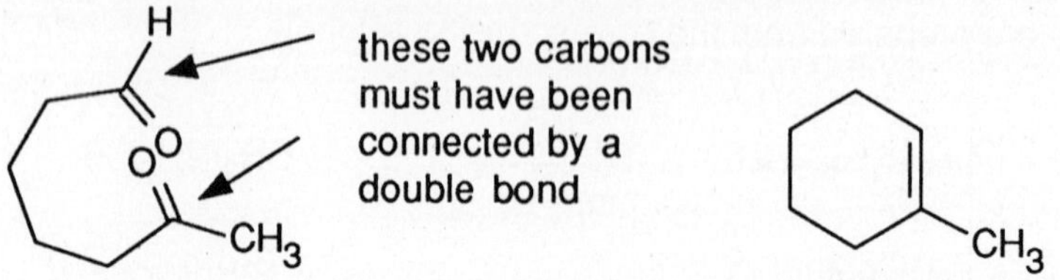
H
O
O
CH3
these two carbons
must have been
connected by a
double bond
CH3

Compound E　　　　Compound D

The tertiary bromide which gives compound D on dehydrobromination is 1-methylcyclohexyl bromide.

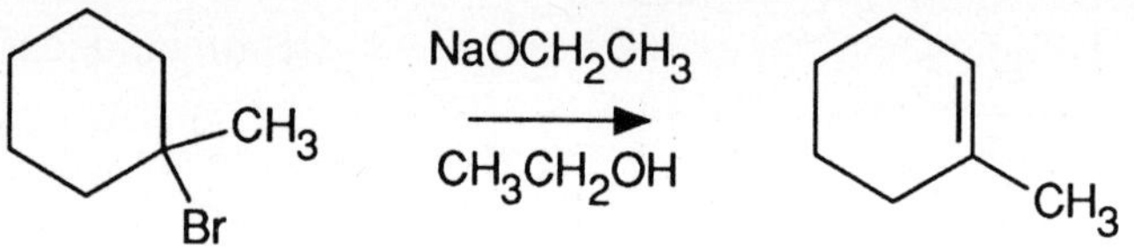
CH3
Br
NaOCH2CH3
CH3CH2OH
CH3

Compound C　　　　Compound D

When tertiary halides are treated with base, they undergo E2 elimination. The regioselectivity of elimination of tertiary halides follows the Zaitsev rule.

5.39 Since santene and 1,3-diacetylcyclopentane (compound F) contain the same number of carbon atoms, the two carbonyl carbons of the diketone must have been connected by a double bond in santene. Therefore, the structure of santene must be

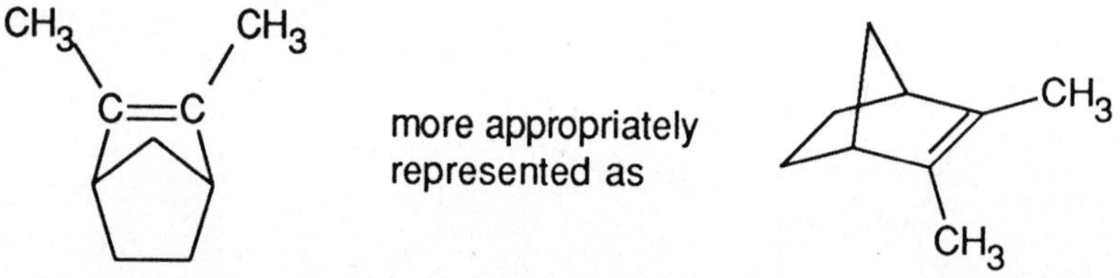
CH3
CH3
C=C
more appropriately
represented as
CH3
CH3

5.40 (a) Compound G contains 9 of the 10 carbons and 14 of the 16 hydrogens of sabinene. Ozonolysis has led to the separation of one carbon and two hydrogens from the rest of the molecule. The carbon and the two hydrogens must have been lost as formaldehyde $H_2C{=}O$. This H_2C unit was originally doubly bonded to the carbonyl carbon of compound G. Therefore, sabinene must have the structure shown in the equation representing its ozonolysis.

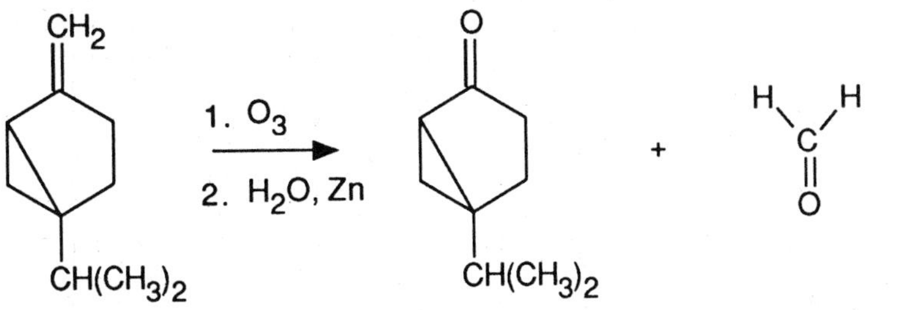

Sabinene → Compound G + Formaldehyde

(b) Compound H contains all 10 of the carbons and all 16 of the hydrogens of Δ³-carene. The two carbonyl carbons of compound H must have been linked by a double bond in Δ³-carene.

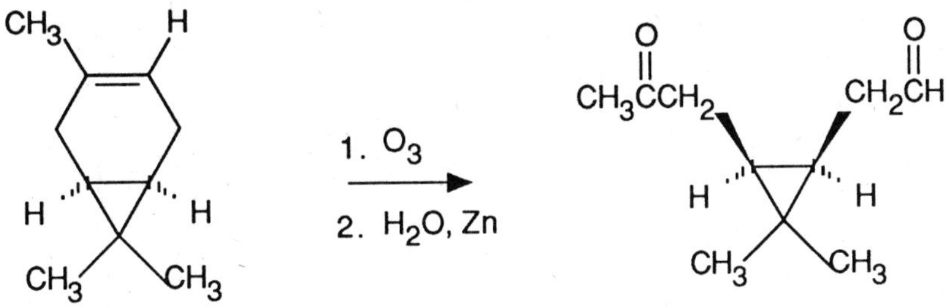

Δ³-Carene

Compound H

(c) Compound I contains all the carbon atoms of α-pinene. The two carbonyl carbons of compound I must have been doubly bonded in α-pinene, and α-pinene must be bicyclic.

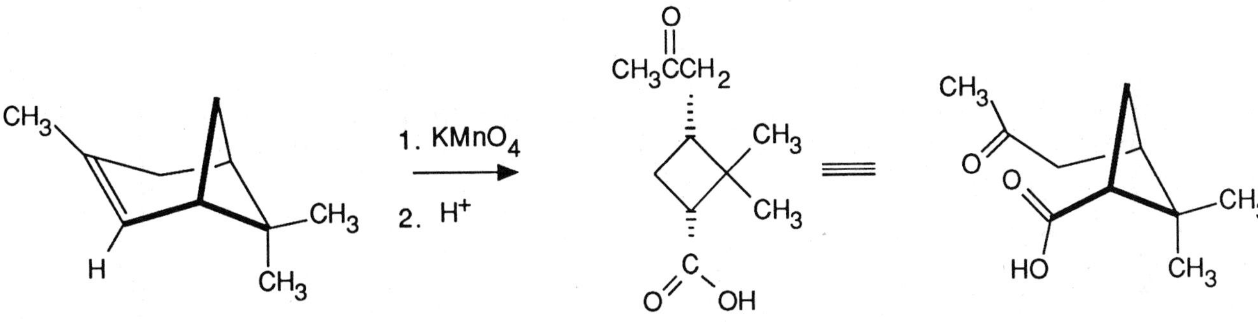

α-Pinene

Compound I

5.41 Since the sex attractant of the female housefly is $C_{23}H_{46}$, and it takes up one mole of hydrogen on catalytic hydrogenation, it must have one double bond and no rings. The position of the double bond is revealed by the ozonolysis data.

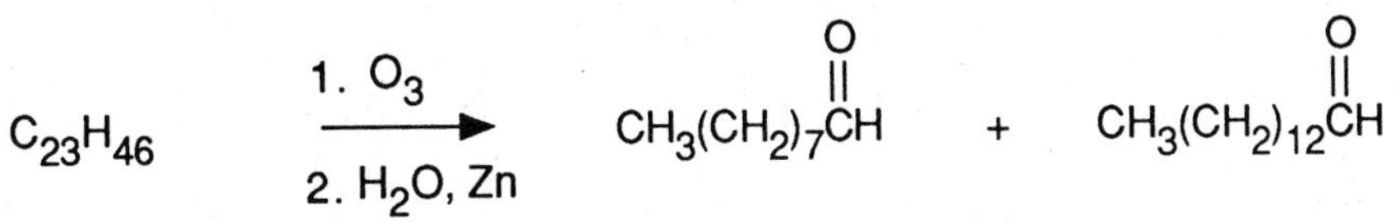

$$C_{23}H_{46} \xrightarrow[\text{2. }H_2O,\ Zn]{\text{1. }O_3} CH_3(CH_2)_7\overset{\overset{O}{\|}}{C}H \ + \ CH_3(CH_2)_{12}\overset{\overset{O}{\|}}{C}H$$

An unbranched 9-carbon unit and an unbranched 14-carbon unit comprise the carbon skeleton, and these two units must be connected by a double bond. Therefore, the housefly sex attractant has the constitution:

$$CH_3(CH_2)_7CH=CH(CH_2)_{12}CH_3$$

9-Tricosene

The data cited in the problem do not permit the stereochemistry of this natural product to be determined.

5.42 The hydrogenation data tell us that $C_{19}H_{38}$ contains one double bond and has the same carbon skeleton as 2,6,10,14-tetramethylpentadecane. We locate the double bond at C-2 on the basis of the fact that acetone, $(CH_3)_2C=O$ is obtained on ozonolysis. The structures of the natural product and the aldehyde produced on its ozonolysis are shown below.

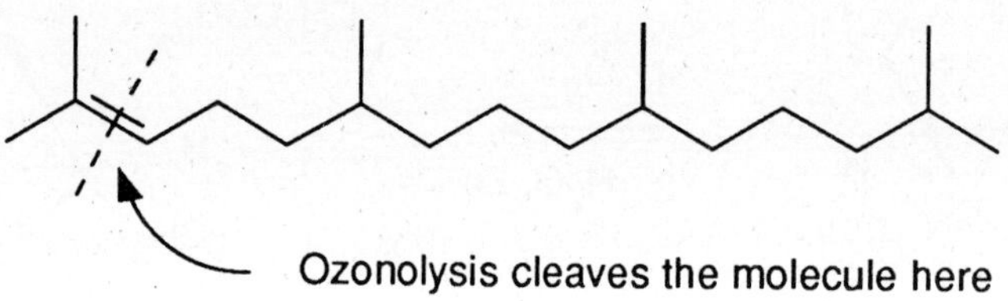

Ozonolysis cleaves the molecule here

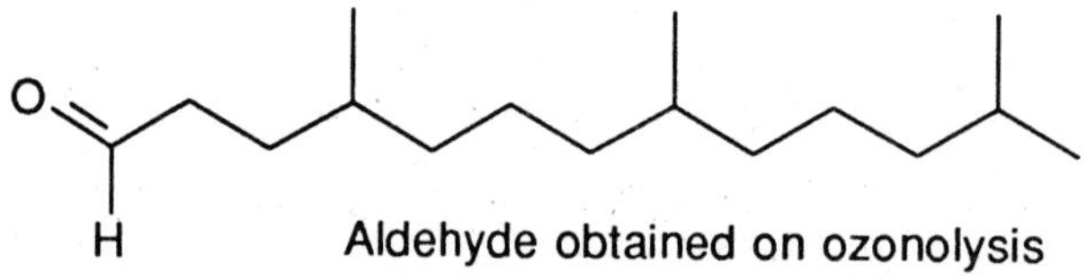

Aldehyde obtained on ozonolysis

5.43 Since $O=CHCH_2CH=O$ is one of the products of its ozonolysis, the sex attractant of the arctiid moth must contain the $=CHCH_2CH=$ unit. This unit must be bonded to an unbranched 12-carbon unit at one end and an unbranched six-carbon unit at the other in order to give $CH_3(CH_2)_{10}CH=O$ and $CH_3(CH_2)_4CH=O$ on ozonolysis.

$$CH_3(CH_2)_{10}CH\overset{|}{=}CHCH_2CH\overset{|}{=}CH(CH_2)_4CH_3$$

Sex attractant of arctiid moth

(dotted lines show positions of cleavage on ozonolysis)

1. O_3
2. H_2O, Zn

$$CH_3(CH_2)_{10}\overset{O}{\overset{||}{C}}H \quad + \quad H\overset{O}{\overset{||}{C}}CH_2\overset{O}{\overset{||}{C}}H \quad + \quad H\overset{O}{\overset{||}{C}}(CH_2)_4CH_3$$

The stereochemistry of the double bonds cannot be determined on the basis of the available information.

5.44 Compound J contains all 15 carbon atoms of cedrene. To deduce the structure of cedrene connect the carbonyl carbons by a double bond and replace the -OH group by -H. Permanganate oxidation of cedrene proceeds as shown in the equation.

Cedrene → (1. KMnO$_4$ 2. H$^+$) → Compound J

5.45 The reaction that produces compound K is reasonably straightforward. Compound K is 14-bromo-1-tetradecyne.

$$NaC\equiv CH \quad + \quad Br(CH_2)_{12}Br \quad \longrightarrow \quad Br(CH_2)_{12}C\equiv CH$$

Sodium acetylide   1,12-Dibromododecane   Compound K (C$_{14}$H$_{25}$Br)

Treatment of compound K with sodium amide converts it to compound L. Compound L on ozonolysis gives a diacid that retains all the carbon atoms of L. Therefore, compound L must be a cyclic alkyne, formed by an intramolecular alkylation.

$$Br(CH_2)_{12}C\equiv CH \xrightarrow{NaNH_2} \text{Compound L} \xrightarrow[\text{2. H}_2O]{\text{1. O}_3} HOC(CH_2)_{12}COH$$

Compound K   Compound L

Compound L is cyclotetradecyne.

Hydrogenation of compound L over Lindlar palladium yields *cis*-cyclotetradecene (compound M).

Compound L → (H$_2$, Lindlar Pd) → Compound M (C$_{14}$H$_{26}$)

Hydrogenation over platinum gives cyclotetradecane (compound N).

Compound L   Compound N ($C_{14}H_{28}$)

Sodium-ammonia reduction of compound L yields *trans*-cyclotetradecene.

Compound L   Compound O ($C_{14}H_{26}$)

The cis and trans isomers of cyclotetradecene are both converted to $HO_2(CH_2)_{12}CO_2H$ on oxidation with potassium permanganate, whereas cyclotetradecane does not react with potassium permanganate.

CHAPTER 6

ARENES AND AROMATICITY

GLOSSARY OF TERMS

Activating substituent A group which when present in place of a hydrogen substituent causes a particular reaction to occur faster. The term is most often applied to the effect of substituents on the rate of electrophilic aromatic substitution.

Acylium ion The cation

$$R-C\overset{+}{\equiv}O:$$

Arene An aromatic hydrocarbon. Often abbreviated ArH.

Aromatic compound A electron-delocalized species which is much more stable than any structure written for it in which all of the electrons are localized either in covalent bonds or as unshared electron pairs.

Aromaticity The special stability associated with aromatic compounds.

Benzene The most typical aromatic hydrocarbon.

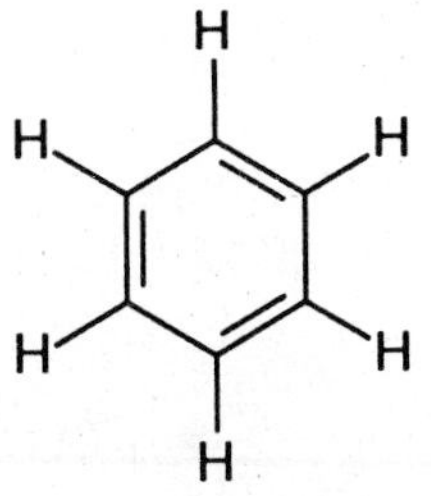

Benzyl group The group

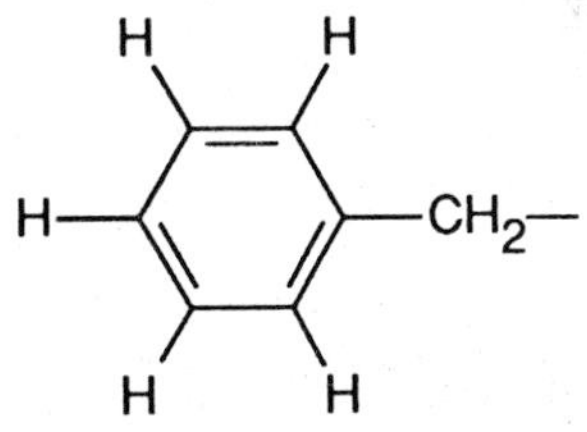

The benzyl group is often abbreviated $C_6H_5CH_2$—.

Benzylic carbocation A carbocation of the type

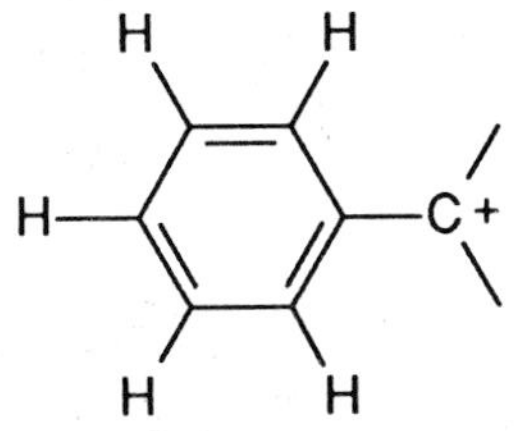

Benzylic carbocations are more stable than simple alkyl cations.

Cyclohexadienyl cation The key intermediate in electrophilic aromatic substitution reactions. Represented by the general structure:

In the general structure, "E" is derived from the electrophile which attacks the ring. Other resonance forms are possible.

Deactivating substituent   A group which when present in place of a hydrogen substituent causes a particular reaction to occur more slowly. The term is most often applied to the effect of substituents on the rate of electrophilic aromatic substitution.

Electrophilic aromatic substitution   The fundamental reaction type exhibited by aromatic compounds. An electrophilic species (E^+) attacks an aromatic ring and replaces one of the hydrogen substituents.

$$Ar\!-\!H + E\!-\!Y \longrightarrow Ar\!-\!E + H\!-\!Y$$

Friedel-Crafts acylation   An electrophilic aromatic substitution in which an aromatic compound reacts with an acyl chloride or carboxylic acid anhydride in the presence of aluminum chloride. An acyl group becomes bonded to the ring.

Friedel-Crafts alkylation   An electrophilic aromatic substitution in which an aromatic compound reacts with an alkyl halide in the presence of aluminum chloride. An alkyl group becomes bonded to the ring.

$$Ar\!-\!H + R\!-\!X \xrightarrow{AlCl_3} Ar\!-\!R$$

Halogenation   Replacement of a hydrogen substituent by a halogen. Bromination and chlorination are the most common halogenation reactions in electrophilic aromatic substitution.

$$Ar\!-\!H + X_2 \xrightarrow{FeX_3} Ar\!-\!X + HX$$

Heterocyclic   A cyclic compound in which one or more of the atoms in the ring is an element other than carbon.

Hückel's rule   Completely conjugated, planar, monocyclic hydrocarbons will possess special stability when the number of their π electrons $= 4n + 2$ where n is an integer.

Kekulé structure   A structural formula for an aromatic compound which satisfies the customary rules of bonding and is usually characterized by a pattern of alternating single and double bonds. A single Kekulé structure does not completely describe the actual bonding in the molecule.

meta   A 1,3-relationship between substituents on a benzene ring.

meta Director   A group which when present on a benzene ring directs an incoming electrophile to a position meta to itself.

Nitration   Replacement of a hydrogen substituent by an $-NO_2$ group.

$$Ar\!-\!H \xrightarrow[H_2SO_4]{HNO_3} Ar\!-\!NO_2$$

ortho   A 1,2-relationship between substituents on a benzene ring.

ortho-para Director   A group which when present on a benzene ring directs an incoming electrophile to the positions ortho and para to itself.

para   A 1,4-relationship between substituents on a benzene ring.

Phenyl group   The group

It is often abbreviated C_6H_5-.

Sulfonation   Replacement of a hydrogen substituent by an $-SO_3H$ group.

$$Ar\!-\!H \xrightarrow[H_2SO_4]{SO_3} Ar\!-\!SO_3H$$

6.1 Toluene has a methyl group attached to a benzene ring. The Kekulé structures of toluene are the two resonance forms which show alternating single and double bonds within the six-membered ring.

Kekule structures   Circle-in-a-ring formulation

6.2 (b) The parent compound is styrene, $C_6H_5CH=CH_2$. The desired compound has a chlorine in the meta position.

CH=CH₂

m-Chlorostyrene

(c) The parent compound is aniline, $C_6H_5NH_2$. *p*-Nitroaniline is therefore:

O_2N———NH_2

p-Nitroaniline

6.3 Nitration replaces one of the hydrogens directly attached to the ring with a nitro group (-NO₂). All four ring hydrogens of *p*-xylene are equivalent, so it does not matter which one is replaced.

$$\xrightarrow[\text{H}_2\text{SO}_4]{\text{HNO}_3}$$

p-Xylene   1,4-Dimethyl-2-nitrobenzene

6.4 Friedel-Crafts alkylations are effective in the preparation of alkylbenzenes. Ethylbenzene can be prepared by the following route:

$$+ \quad CH_3CH_2Cl \quad \xrightarrow{AlCl_3}$$

Benzene   Ethyl chloride   Ethylbenzene

6.5   An acyl chloride is used that has the same carbon skeleton as the acyl group in the desired product.

Benzene   Acetyl chloride   Acetophenone

6.6   (b)   The hydroxyl group (–OH) is strongly activating, therefore phenol (C_6H_5OH) reacts faster than benzene in an electrophilic aromatic substitution reaction.

(c)   Nitrobenzene is less reactive than benzene, as the nitro group (–NO_2) is deactivating. Aniline is more reactive than benzene because the amino group (–NH_2) is activating. Aniline reacts faster than nitrobenzene in a nitration reaction.

6.7   (b)   The carboxy group (–CO_2H) of benzoic acid directs an incoming electrophile meta to itself. The major product is *m*-nitrobenzoic acid.

Benzoic acid   *m*-Nitrobenzoic acid

(c)   Alkyl groups are ortho, para-directing (Table 6.1). Bromination of ethylbenzene will yield a mixture of *o*- and *p*-bromoethylbenzene.

Ethylbenzene   *o*-Bromoethylbenzene   *p*-Bromoethylbenzene

6.8   (b)   The resonance structures corresponding to attack at the position para to an alkoxy group are similar to those for attack at the ortho position. The unshared electron pair of the alkoxy group is able to stabilize the intermediate.

114

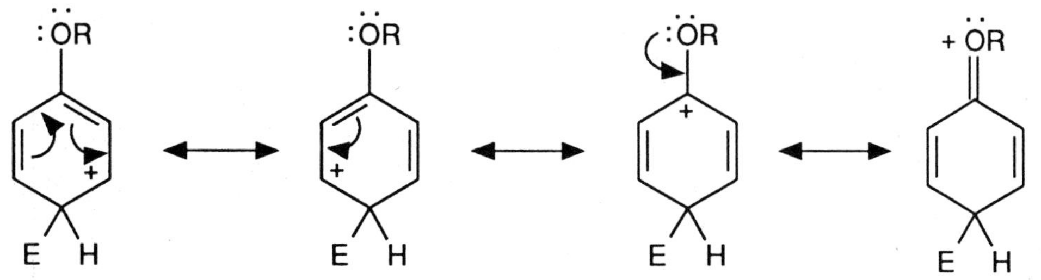

Most stable resonance form;
oxygen and all carbons have
octets of electrons

6.9 (b) The carbonyl group attached directly to the ring is a signal that the substituent is a meta-directing group. Nitration of methyl benzoate yields methyl *m*-nitrobenzoate. As actually carried out in the laboratory, this product was isolated in 81-85% yield.

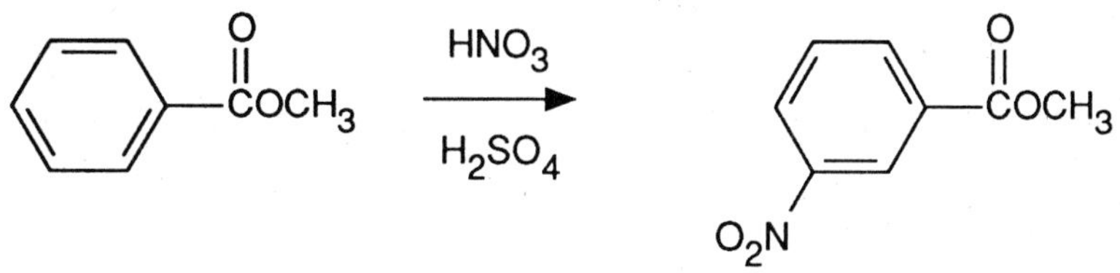

Methyl benzoate   Methyl *m*-nitrobenzoate

(c) The acyl group in propiophenone is meta-directing; as in (b) a C=O group is directly attached to the ring. The product is *m*-nitropropiophenone (isolated in 60% yield).

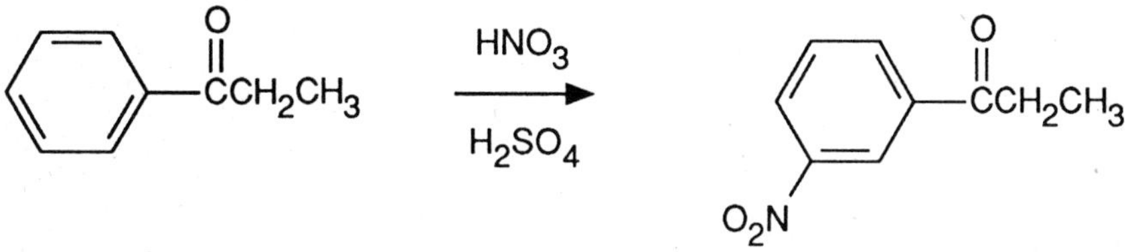

Propiophenone   *m*-Nitropropiophenone

6.10 (b) The substituents which need to be placed on the ring (Br and NO$_2$) are the same as in (a), but they are para to each other. The order in which they are introduced must be reversed; the para directing bromine is introduced first.

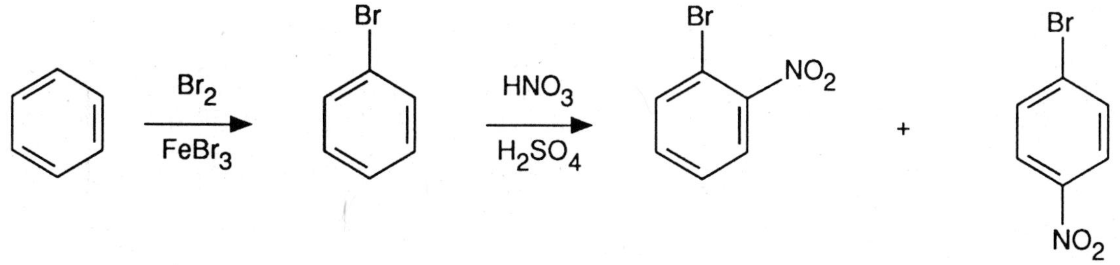

Benzene   Bromobenzene   *o*-Bromonitrobenzene   *p*-Bromonitrobenzene

115

(c)   The ortho, para-directing alkyl group (*tert*-butyl) must introduced before the meta-directing nitro group. Thus Friedel-Crafts alkylation must precede nitration.

Benzene   *tert*-Butylbenzene   *o-tert*-Butylnitrobenzene   *p-tert*-Butylnitrobenzene

6.11   (b)   While 1,3,5-cycloheptatriene contains 6 π electrons and is cyclic, it is **not aromatic**. The requirement of a *continuous* array of *p* orbitals is not met. The sp^3-hybridized carbon acts as an "insulator" and interrupts the sequence of *p* orbitals. (The sp^3-hybridized carbon is indicated below by •, all other carbons are sp^2-hybridized.)

(c)   This cation is **aromatic**. It has 6 π electrons and all carbons are sp^2-hybridized so that a cyclic system of *p* orbitals is present.

6.12   (a)   The six carbon-carbon bonds of benzene are all equivalent. They are neither pure single bonds nor pure double bonds.

(b)   The benzene ring is planar. The chair is the most stable conformation of cyclohexane.

(c)   The carbon atoms of benzene are sp^2-hybridized.

6.13   By Kekulé structure we mean alternative resonance forms for an arene that are generated by reorganization of the double bonds within the six-membered ring.

(a)   There are four Kekulé structures for biphenyl. Each of the two rings can be written in two different Kekulé formulations.

116

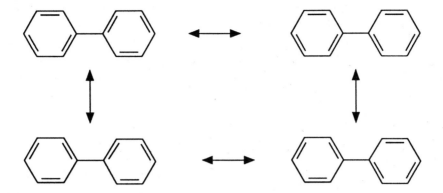

Three resonance forms are possible for naphthalene. Notice that both rings correspond to Kekulé formulations of benzene only in the first one.

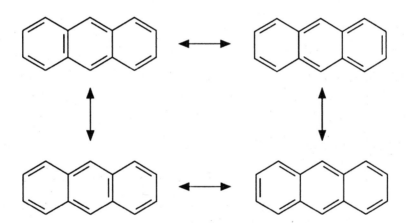

Four resonance structures are possible for anthracene. In none of the structures do all three rings correspond to Kekulé formulations of benzene.

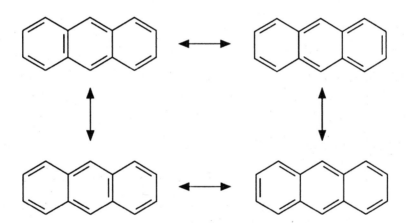

6.14 A continuous cyclic array of p orbitals is necessary for a compound to be aromatic. An sp^3-hybridized carbon does not have a p orbital available, and acts to interrupt the conjugated system.

6.15 Hückel's rule says the number of π electrons in an aromatic system must satisfy the equation $4n + 2$. The possible number of electrons are:

n	electrons	n	electrons
0	2	3	10
1	6	4	18
2	10	5	22

6.16 (a) Each of the two double bonds contributes two electrons and the structural formula indicates an additional pair of electrons on the negatively charged carbon for a total of 6 electrons distributed among the 5 carbons of a continuous π electron system. This species is **aromatic**.

(c) There are 8 π electrons in the conjugated system; six contributed by the three double bonds and two from the negatively charged carbon. This anion is **not aromatic**.

(b) This compound is called *thiophene*. It is **aromatic**. Any molecule will be aromatic if it contains 6 π electrons in a completely conjugated cyclic system of *p* orbitals. The two double bonds contribute a total of 4 π electrons. The sulfur atom has two unshared electron pairs. By involving one pair in the cyclic π system the full complement of 6 π electrons is achieved to give a very stable molecule.

(d) The positively charged carbon has an empty *p* orbital, so a continuous cyclic array of *p* orbitals is present. The only electrons in this π system are the two contributed by the double bond. Two electrons corresponds to $4n+2$ where $n = 0$, so the carbocation is **aromatic**.

6.17 Pyridoxine and nicotinamide each contain a pyridine ring. Furfural has a furan ring. Serotonin has a pyrrole ring fused to a benzene ring.

Pyridoxine   Nicotinamide   Furfural   Serotonin

6.18 (a) There are three isomeric nitrotoluenes, as the nitro group can be ortho, meta, or para to the methyl group.

o-Nitrotoluene	m-Nitrotoluene	p-Nitrotoluene
(2-Nitrotoluene)	(3-Nitrotoluene)	(4-Nitrotoluene)

(b) Benzoic acid is $C_6H_5CO_2H$. In the isomeric dichlorobenzoic acids, two of the ring hydrogens of benzoic acid have been replaced by chlorines.

2,3	2,4	2,5	2,6	3,4	3,5

The name of the first-shown isomer is **2,3-dichlorobenzoic acid**. All of the others are named in an analogous manner using the numerical locants indicated. Since the parent compound is benzoic acid, this automatically fixes the -CO_2H as being at position-1.

(c) There are six isomeric tribromophenols. They are named as derivatives of phenol (C_6H_5OH) with the hydroxyl group located at position-1. The name of the compound on the left is **2,3,4-tribromophenol**; the others are named analogously.

2,3,4	2,3,5	2,3,6	2,4,5	2,4,6	3,4,5

119

6.19 (a) There are only three monobromo derivatives of anthracene. Any other structure which might be written would be equivalent to one of the three shown. (A different resonance form is not an isomer!)

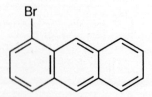 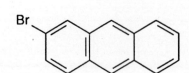 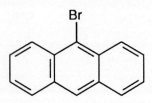

(b) There are eight hydrogen atoms in naphthalene, divided into two sets of four equivalent hydrogens. Those at C-1, C-4, C-5, and C-8 are equivalent, as are those at C-2, C-3, C-6, and C-7. Thus, only two monobromo derivatives of naphthalene are possible.

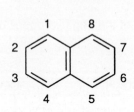

(d) Phenanthrene can form a total of five monobromo derivatives.

6.20 (a) The phenyl group is attached to C-2 of a five-carbon chain

$$CH_3CHCH_2CH_2CH_3 \quad \text{2-Phenylpentane}$$
$$| $$
$$C_6H_5$$

(b) The phenyl substituent is attached to C-2 of 1-butanol.

$$CH_3CH_2CHCH_2OH \quad \text{2-Phenyl-1-butanol}$$
$$|$$
$$C_6H_5$$

(c) A phenyl group is a substituent on C-1 of propene. (The stereochemistry of the double bond is not given in the problem.)

$$C_6H_5CH=CHCH_3 \qquad \text{1-Phenylpropene}$$

(d) This compound is named as a phenyl-substituted cyclohexanol, so the hydroxyl-bearing carbon is C-1. This compound is 4-phenylcyclohexanol. As in (c), the stereochemistry is not indicated.

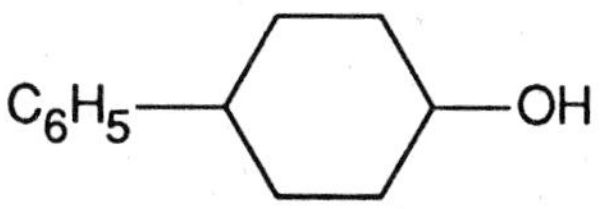

C_6H_5—⬡—OH   4-Phenylcyclohexanol

6.21 (a) In *p*-chlorophenol the benzene ring bears a chlorine substituent para to the hydroxyl group of C_6H_5OH.

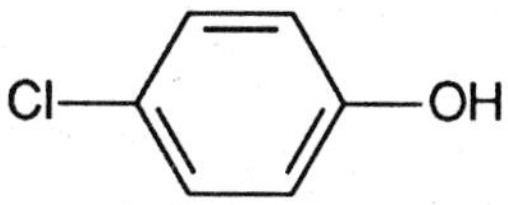

Cl—⬡—OH   *p*-Chlorophenol

(b) The structure of acetophenone is given in Table 6.1. It has a $CH_3C=O$ group attached to a benzene ring. Therefore:

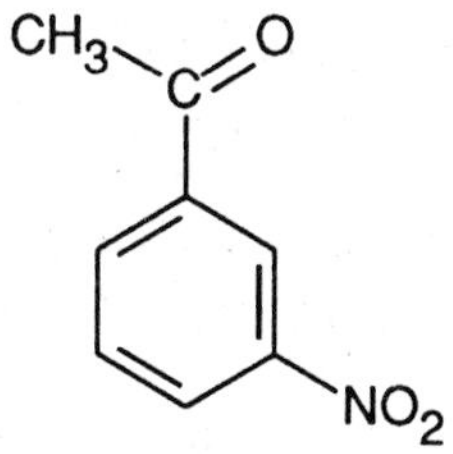

m-Nitroacetophenone

(c) 1-Butene bears a phenyl group (C_6H_5-) at C-1 and an ethyl group at C-2.

$$C_6H_5CH=C(CH_2CH_3)_2 \qquad \text{2-Ethyl-1-phenyl-1-butene}$$

(d) Two phenyl groups are attached directly to the double bond of cyclohexene.

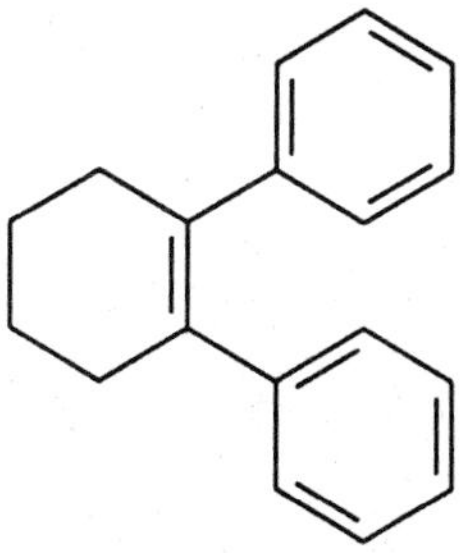

1,2-Diphenylcyclohexene

(e) Two isopropyl groups are in a 1,4-relationship on a benzene ring.

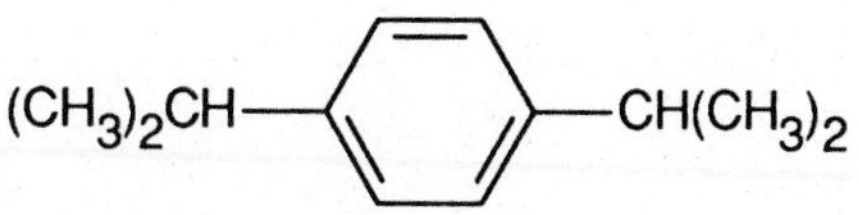

$(CH_3)_2CH$—⬡—$CH(CH_3)_2$ *p*-Diisopropylbenzene

(f)   Aniline ($C_6H_5NH_2$) is substituted with bromines at positions-2,4, and 6. Since the compound is named as a derivative of aniline, the -NH_2 group is at C-1.

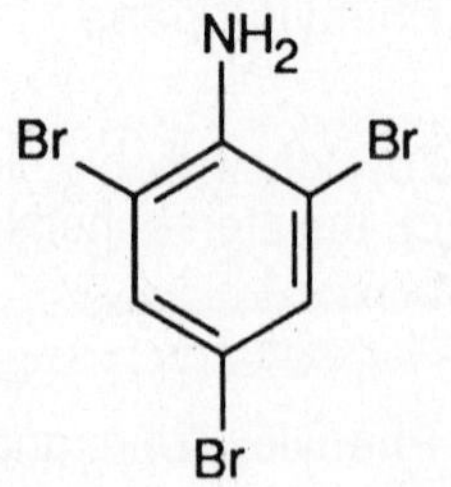

2,4,6-Tribromoaniline

(g)   Benzoic acid ($C_6H_5CO_2H$) is substituted with bromines at C-3 and C-5.

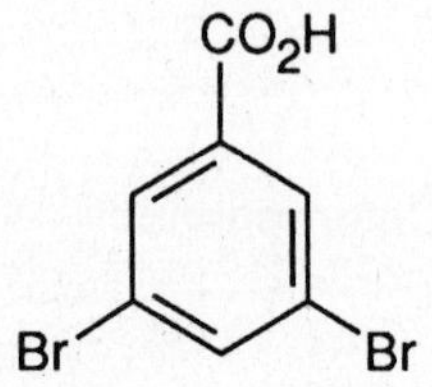

3,5-Dibromobenzoic acid

6.22 (a)   Because there is a methyl substituent directly bonded to a benzene ring, this compound can be named as a derivative of toluene ($C_6H_5CH_3$). Both **m-bromotoluene** and **3-bromotoluene** are correct names.

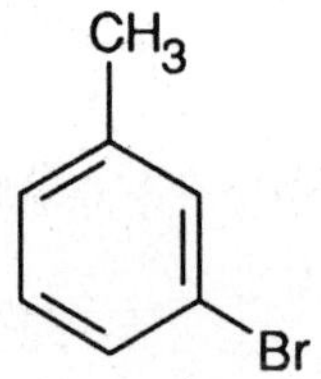

(b)   As noted in Table 6.1 the compound $C_6H_5OCH_3$ is named *anisole*. Thus the structure given is a dimethyl derivative of anisole. The carbons must be numbered; ortho, meta, and para can only be used for disubstituted aromatics. The compound is **3,4-dimethylanisole**.

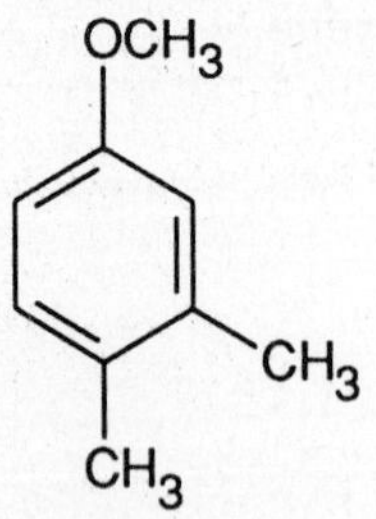

(c)   The parent compound from Table 6.1 is acetophenone (a benzene ring with a $CH_3C=O$ substituent). This compound is **o-chloroacetophenone** or **2-chloroacetophenone**.

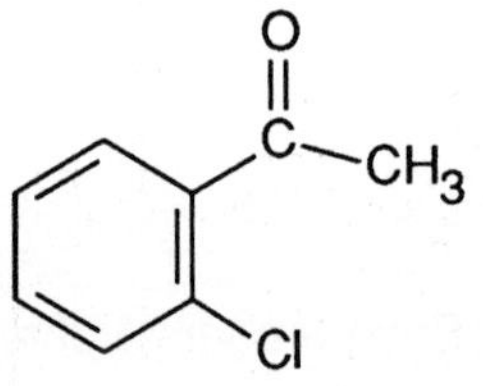

(d)   This compound is named as a derivative of phenol (C_6H_5OH). It is **2,4-dinitrophenol**.

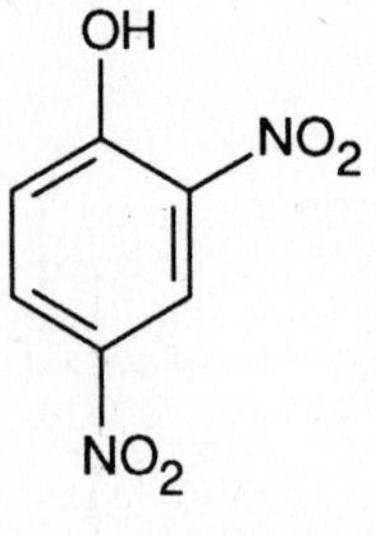

122

6.23 (a) Nitration of benzene is carried out with a mixture of nitric acid and sulfuric acid.

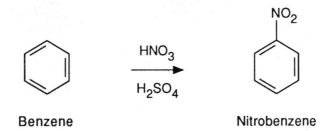

Benzene   Nitrobenzene

(b) Nitrobenzene is much less reactive than benzene toward electrophilic aromatic substitution. The nitro group on the ring is a meta director.

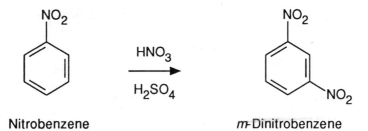

Nitrobenzene   *m*-Dinitrobenzene

(c) Toluene is more reactive than benzene in electrophilic aromatic substitution. A methyl group is an ortho-para director.

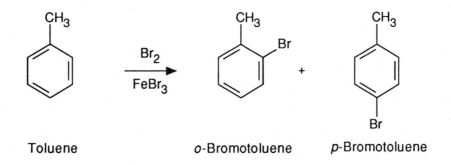

Toluene   *o*-Bromotoluene   *p*-Bromotoluene

(d) Trifluoromethyl is deactivating and meta directing.

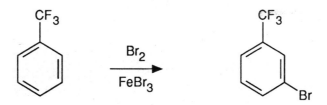

(Trifluoromethyl)benzene   *m*-Bromo(trifluoromethyl)benzene

(e) Anisole is ortho, para-directing, strongly activated toward electrophilic aromatic substitution, and readily sulfonated in sulfuric acid.

Anisole   o-Methoxybenzenesulfonic acid   p-Methoxybenzenesulfonic acid

(f) Acetanilide is quite similar to anisole in its behavior toward electrophilic aromatic substitution.

Acetanilide   o-Acetamidobenzenesulfonic acid   p-Acetamidobenzenesulfonic acid

(g) Bromobenzene is less reactive than benzene. A bromine substituent is ortho, para-directing.

Bromobenzene   o-Bromochlorobenzene   p-Bromochlorobenzene

6.24 (a) The product of this Friedel-Crafts alkylation is diphenylmethane.

Benzene   Benzyl chloride   Diphenylmethane

(b) Anisole is a reactive substrate toward Friedel-Crafts alkylation and yields a mixture of ortho and para benzylated products when treated with benzyl chloride and aluminum chloride.

CH_3O—⟨ ⟩ + ⟨ ⟩—CH_2Cl $\xrightarrow{AlCl_3}$ o-Benzylanisole (OCH$_3$, CH$_2$C$_6$H$_5$) + p-Benzylanisole (OCH$_3$, CH$_2$C$_6$H$_5$)

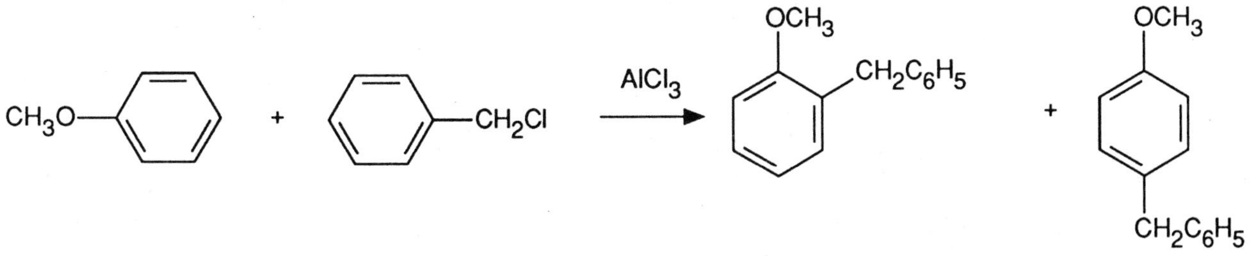

Anisole   Benzyl chloride   o-Benzylanisole   p-Benzylanisole

(c) Benzene will undergo acylation with benzoyl chloride and aluminum chloride.

⟨ ⟩ + ⟨ ⟩—$\overset{O}{\overset{\|}{C}}Cl$ $\xrightarrow{AlCl_3}$ ⟨ ⟩—$\overset{O}{\overset{\|}{C}}$—⟨ ⟩

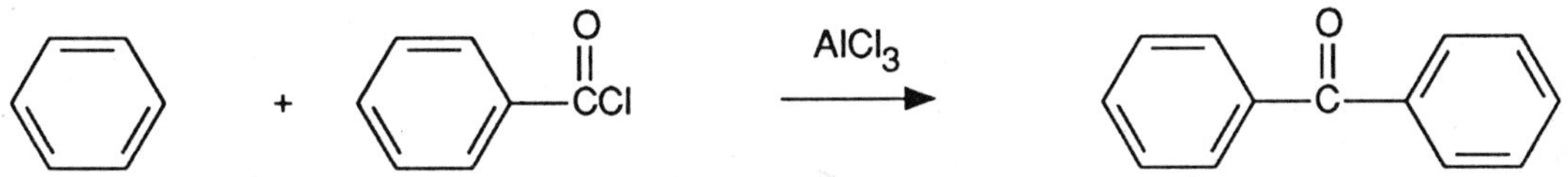

Benzene   Benzoyl chloride   Benzophenone

(d) The product of (a) can be thought of as a benzyl-substituted benzene. The benzyl group acts as an alkyl group would, and directs an incoming electrophile to the ortho and para positions.

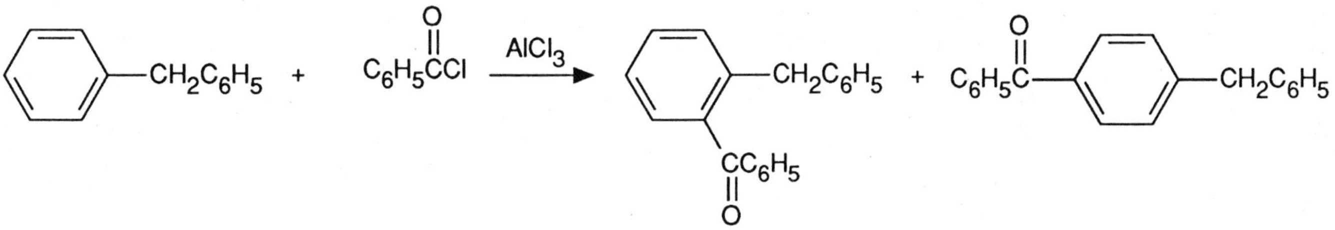

⟨ ⟩—$CH_2C_6H_5$ + $C_6H_5\overset{O}{\overset{\|}{C}}Cl$ $\xrightarrow{AlCl_3}$ o-Benzylbenzophenone (—CH$_2$C$_6$H$_5$, $\overset{}{\underset{O}{\overset{\|}{C}C_6H_5}}$) + $C_6H_5\overset{O}{\overset{\|}{C}}$—⟨ ⟩—$CH_2C_6H_5$

Diphenylmethane   Benzoyl chloride   o-Benzylbenzophenone   p-Benzylbenzophenone

6.25 Each ring is influenced by the substituent directly attached to it. Ring A in the structure below is activated by its –NH substituent (nitrogen bears an unshared electron pair), while ring B is deactivated by the C=O group directly bonded to it. Bromination will occur in the more reactive ring ortho and para to the nitrogen.

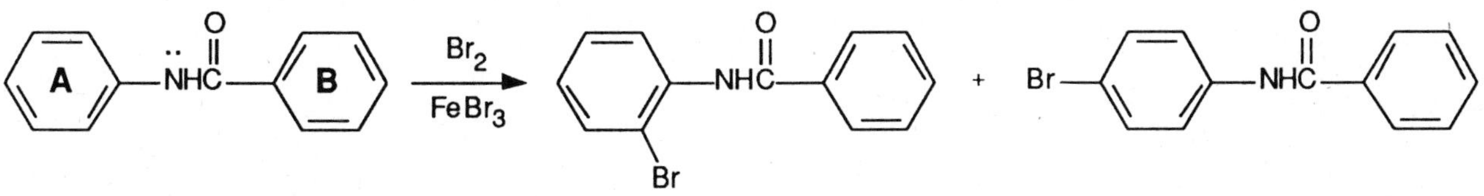

⟨A⟩—$N\overset{..}{H}\overset{O}{\overset{\|}{C}}$—⟨B⟩ $\xrightarrow[FeBr_3]{Br_2}$ ⟨ ⟩(Br)—$NH\overset{O}{\overset{\|}{C}}$—⟨ ⟩ + Br—⟨ ⟩—$NH\overset{O}{\overset{\|}{C}}$—⟨ ⟩

125

6.26 In a Friedel-Crafts acylation reaction an acyl chloride or acid anhydride reacts with an arene to yield an aryl ketone.

$$ArH \ + \ \overset{\displaystyle O}{\underset{\displaystyle}{\overset{\|}{R\text{–}C}}}Cl \ \xrightarrow{AlCl_3} \ \overset{\displaystyle O}{\overset{\|}{Ar\text{–}C\text{–}R}}$$

The ketone carbonyl is bonded directly to the ring. Therefore, identify the bond between the aromatic ring and the carbonyl group and realize that it arises as shown in the above general equation.

arises from ... and

6.27 The electrophile in **halogenation** is a Lewis acid-Lewis base complex between the halogen and the ferric chloride catalyst.

$$:\!\overset{+}{\overset{..}{\underset{..}{X}}}\!-\!\overset{..}{\underset{..}{X}}\!-\!\overset{-}{FeX_3} \qquad (X = Br, Cl)$$

The electrophile in **nitration** is the nitronium ion.

$$:\overset{+}{\underset{..}{O}}\!=\!N\!=\!\overset{..}{\underset{..}{O}}:$$

Sulfonation occurs by attack of sulfur trioxide, SO_3, on the aromatic ring.

The intermediate in a **Friedel-Crafts alkylation** is a carbocation formed by abstraction of a halide ion from the corresponding alkyl halide.

$$R^+ \qquad AlCl_4^-$$

An acylium ion is the intermediate in a **Friedel-Crafts acylation**.

$$R\!-\!\overset{+}{C}\!\equiv\!\overset{..}{O}: \qquad AlCl_4^-$$

6.28 (a) **Toluene** is more reactive than chlorobenzene in electrophilic aromatic substitution because a methyl substituent is activating while a halogen is deactivating. Both are ortho, para-directing, however. Nitration of toluene is faster than nitration of benzene.

Faster:

| Toluene | o-Nitrotoluene | p-Nitrotoluene |

(b)   A fluorine substituent is not nearly as strongly deactivating as is a trifluoromethyl group. The reaction that takes place is Friedel-Crafts alkylation of **fluorobenzene**.

Fluorobenzene   Benzyl chloride   o-Fluorodiphenylmethane   p-Fluorodiphenylmethane

(c)   A carbonyl group (C=O) directly bonded to a benzene ring strongly *deactivates* it toward electrophilic aromatic substitution. An oxygen substituent directly attached to the ring strongly *activates* it toward electrophilic aromatic substitution. **Phenyl acetate** is much more reactive than methyl benzoate.

Phenyl acetate   o-Bromophenyl acetate   p-Bromophenyl acetate

(d)   **Acetanilide** is strongly activated toward electrophilic aromatic substitution and reacts faster than nitrobenzene which is strongly deactivated. The nitrogen atom attached to the ring in acetanilide bears an unshared electron pair which can be donated into the ring to stabilize the positive charge that develops during electrophilic aromatic substitution. The nitrogen attached to the ring in nitrobenzene is positively charged and attracts electrons away from the ring.

Acetanilide   Nitrobenzene

Acetanilide   o-Acetamidobenzenesulfonic acid   p-Acetamidobenzenesulfonic acid

6.29 The electrophilic species is formed by reaction of chlorine with ferric chloride in a Lewis acid-Lewis base reaction.

Benzene and the chlorine-ferric chloride complex react to form the cyclohexadienyl cation intermediate.

The cyclohexadienyl cation intermediate is a resonance-stabilized carbocation.

Deprotonation of the carbocation intermediate gives the product, chlorobenzene.

6.30 (a)   There are three resonance structures for the cyclohexadienyl cation in which the positive charge is on one of the ring carbons. The positively charged carbon has six electrons. The fourth resonance structure—the most stable one—has its positive charge on nitrogen. In this resonance form, all of the atoms have octets of electrons.

Most stable
resonance structure

(b)   Attack by bromine at the position meta to the amino group gives a cyclohexadienyl cation intermediate in which delocalization of the nitrogen lone pair cannot participate in dispersal of the positive charge. The nitrogen lone pair can only be used to stabilize a positive charge when the NH_2 group is directly attached to a positively charged carbon.

(c)   Attack at the position para to the amino group yields a cyclohexadienyl cation intermediate that is stabilized by delocalization of the electron pair of the amino group.

6.31 (a)   An isopropyl group is an activating substituent and is ortho, para-directing. The most stable intermediates are the ones corresponding to ortho and para attack. Because it is less crowded, the intermediate corresponding to para attack is slightly more stable.

(b)   A nitro substituent is deactivating and meta-directing. One resonance form of the most stable cyclohexadienyl cation formed in the bromination of nitrobenzene is:

This ion is less stable than the cyclohexadienyl cation formed during bromination of benzene.

6.32   (a)   The unshared electron pairs on sulfur can stabilize the cyclohexadienyl cation intermediate in electrophilic aromatic substitution. The $-SH$ substituent is an **activating ortho-para director**.

(b)   The positively charged nitrogen of the trimethylammonium group destabilizes the cyclohexadienyl cation intermediate. Trimethylammonium is a **deactivating meta director**.

(c)   Dimethylsulfonium resembles trimethylammonium in that the atom attached to the ring is positively charged. It is a **deactivating meta director**.

6.33   (a)   Alkyl-substituted aromatic compounds are oxidized to benzoic acids with potassium permanganate.

1,2,3,4-Tetrahydronaphthalene   Benzene-1,2-dicarboxylic acid

(b)   Hydrogen bromide adds to the alkene double bond in the direction that proceeds through the more stable benzylic carbocation intermediate.

1,2-Dihydronaphthalene   1-Bromo-1,2,3,4-Tetrahydronaphthalene

The carbocation intermediate is:

positive charge at benzylic carbon

130

(c)   As in (b), the alkene double bond is hydrated in the direction that corresponds to formation of the more stable carbocation intermediate.

1,2-Dihydronaphthalene   1,2,3,4-Tetrahydro-1-naphthol

(d)   One mole of hydrogen will reduce the alkene double double bond. The aromatic π system is reduced much more slowly than an alkene double bond.

1,2-Dihydronaphthalene   1,2,3,4-Tetrahydronaphthalene

6.34   o-Phthalic acid is obtained from o-xylene as follows:

o-Xylene   o-Phthalic acid

6.35   When carrying out each of the following syntheses, evaluate how the structure of the product differs from that of benzene. That is, determine which groups have to be introduced onto the benzene ring or altered in some way. The sequence of reaction steps when multiple substitution is desired is important; recall that some groups direct ortho, para and others meta.

(a)   Isopropylbenzene may be prepared by a Friedel-Crafts alkylation of benzene with isopropyl chloride (or bromide, or iodide).

Benzene   Isopropyl chloride   p-Isopropylbenzene

(b)   As the isopropyl and sulfonic acid groups are para to each other, the first group introduced on the ring must be the ortho, para director, i.e., the isopropyl group. Therefore we may use the product of (a), isopropylbenzene, in this synthesis. An isopropyl group is a fairly bulky ortho, para director, so sulfonation of isopropylbenzene gives mainly p-isopropylbenzenesulfonic acid.

131

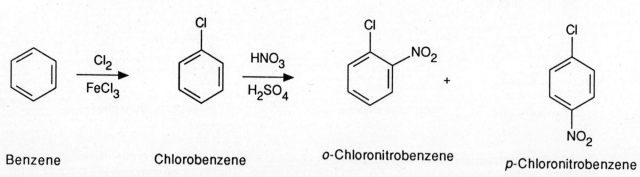

p-Isopropylbenzene → p-Isopropylbenzenesulfonic acid (SO$_3$, H$_2$SO$_4$)

A sulfonic acid group is meta-directing so the order of steps must be alkylation followed by sulfonation rather than the reverse.

(c)   Two electrophilic aromatic substitution reactions need to be performed, chlorination and Friedel-Crafts acylation. The order in which the reactions are carried out is important; chlorine is an ortho, para director, and the acetyl group is a meta director. Since the groups are meta in the desired compound, introduce the acetyl group first.

Benzene   Acetyl chloride   Acetophenone   m-Chloroacetophenone

(d)   Reverse the order of steps in (e) above to prepare p-chloroacetophenone.

Benzene   Chlorobenzene   p-Chloroacetophenone

6.36   The order of substitution is important! The directing influence of the *first* group on the ring determines the orientation of the *second* group to go on. Doing the chlorination step first leads to a mixture of o- and p-chloronitrobenzenes.

Benzene   Chlorobenzene   o-Chloronitrobenzene   p-Chloronitrobenzene

Nitration followed by chlorination leads to the desired *m*-chloronitrobenzene as the major product. Remember, nitro is a meta-directing substituent.

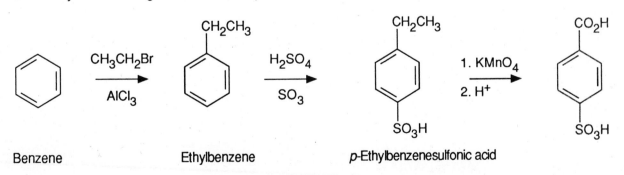

6.37 (a) A meta director must be present on the ring to achieve the desired orientation in the product. Oxidation of toluene with potassium permanganate converts the ortho, para-directing alkyl group to a meta-directing carboxyl group. Bromination completes the synthesis.

Toluene → Benzoic acid → *m*-Bromobenzoic acid

(b) To prepare the para isomer, brominate toluene *before* oxidizing the side chain.

Toluene → *p*-Bromotoluene → *p*-Bromobenzoic acid

(separate from ortho isomer)

6.38 Only **Scheme II** gives the desired product.

Benzene → Ethylbenzene → *p*-Ethylbenzenesulfonic acid

Scheme I fails. If the sulfonic acid group is introduced first, subsequent electrophilic aromatic substitution reactions will not give the desired para-substitution pattern. The group -SO$_3$H is a meta-director. (Indeed, while it has not been discussed in the text, one cannot carry out Friedel-Crafts reactions on benzenesulfonic acid.)

Scheme III fails because sulfonation of benzoic acid gives the product of meta substitution.

CHAPTER 7

STEREOCHEMISTRY

GLOSSARY OF TERMS

Absolute configuration The spatial arrangement of atoms or groups at a chiral center.

Achiral The opposite of chiral. An achiral object is superimposable on its mirror image.

Cahn-Ingold-Prelog notation A system for specifying absolute configuration as *R* or *S* based on the order in which atoms or groups are attached to a chiral center. Groups are ranked in order of precedence according to rules based on atomic number. The molecule is oriented so that the lowest-ranked group is held away from the viewer. If the order of decreasing precedence of the three remaining groups traces a clockwise path, the configuration is *R*. If the path is counterclockwise, the configuration is *S*.

Chiral An object is chiral if it is not superimposable on its mirror image.

Chiral carbon atom A carbon that is bonded to four groups, all of which are different from one another. Also called an *asymmetric carbon atom*. A more modern term is *stereogenic center*.

Diastereomers Stereoisomers that are not enantiomers. Stereoisomers that are not mirror images of one another.

E, Z notation for alkenes A system for specifying double bond configuration that is an alternative to cis, trans notation. When higher-ranked substituents are on the same side of the double bond, the configuration is *Z*.

When higher-ranked substituents are on opposite sides, the configuration is *E*. Rank is determined by the Cahn-Ingold-Prelog system.

Enantiomers Stereoisomers which are related as an object and its nonsuperimposable mirror image.

Fischer projection A method for representing stereochemical relationships. The four bonds to a chiral carbon are represented by a cross. The horizontal bonds are understood to project toward the viewer and the vertical bonds away from the viewer.

Meso stereoisomer An achiral molecule which has chiral centers is called a *meso* form. The most common kind of *meso* compound is a molecule with two chiral centers and a plane of symmetry.

Optical activity The ability of a substance to rotate the plane of polarized light. In order to be optically active, a substance must be chiral and one enantiomer must be present in excess of the other.

Plane of symmetry A plane which bisects an object, such as a molecule, into two mirror image halves. Also called a *mirror plane*. When a line is drawn from any element in the object perpendicular to such a plane and extended an equal distance in the opposite direction, a duplicate of the element is encountered.

Polarimeter   An instrument used to measure optical activity.

Racemic mixture   A mixture containing equal quantities of enantiomers.

Resolution   The separation of a racemic mixture into its enantiomeric constituents.

Specific rotation   The optical activity of a substance per unit concentration per unit path length:

$$[\alpha] = \frac{100\alpha}{cl}$$

α is the observed rotation in degrees, c is the concentration in g/100 mL, and l is the path length in decimeters.

Stereochemistry   Chemistry in three dimensions. The relationship of physical and chemical properties to the spatial arrangement of the atoms in a molecule.

Stereoisomers   Isomers that have the same constitution but which differ in respect to the arrangement of their atoms in space.

SOLUTIONS TO TEXT PROBLEMS

7.1   (a)   An object is chiral if it is not superimposable on its mirror image. The mirror image of a glove for a right hand is a left-handed glove. A glove is **chiral**.

(b)   The mirror image of a shoe for a right foot is a shoe for the left foot. A shoe is **chiral**.

(c)   The mirror image of an undecorated cup and saucer is superimposable on the original objects. They are **achiral**.

7.2   (b)   None of the carbon atoms in 3-bromopentane has four different substituents, so none of its atoms are chiral centers.

$$CH_3CH_2CHCH_2CH_3$$
$$|$$
$$Br$$

(c)   The carbon bearing the ethyl and methyl substituents is a chiral center.

chiral center

$$CH_3$$
$$|$$
$$CH_3CH_2CH_2-C-CH_2CH_2CH_2CH_3$$
$$|$$
$$CH_2CH_3$$

4-Ethyl-4-methyloctane

(d)   There is one chiral center, C-2, in 1,2-dibromobutane.

chiral center

$$H$$
$$|$$
$$CH_3CH_2-C-CH_2Br$$
$$|$$
$$Br$$

1,2-Dibromobutane

136

7.3 (b) As shown in the solution to (a), the perspective views may be drawn with two bonds pointing "back," represented by dashed lines, and two bonds pointing "forward," represented by wedges.

4-Ethyl-4-methyloctane:

$$CH_3 \blacktriangleright \overset{\displaystyle CH_2CH_2CH_3}{\underset{\displaystyle CH_2CH_2CH_2CH_3}{C}} \blacktriangleleft CH_2CH_3 \qquad CH_3CH_2 \blacktriangleright \overset{\displaystyle CH_2CH_2CH_3}{\underset{\displaystyle CH_2CH_2CH_2CH_3}{C}} \blacktriangleleft CH_3$$

1,2-Dibromobutane:

$$H \blacktriangleright \overset{\displaystyle CH_2Br}{\underset{\displaystyle CH_2CH_3}{C}} \blacktriangleleft Br \qquad Br \blacktriangleright \overset{\displaystyle CH_2Br}{\underset{\displaystyle CH_2CH_3}{C}} \blacktriangleleft H$$

7.4 (b) The molecules drawn in Problem 7.3 are in the correct orientation to draw Fischer projections. The Fischer projections are drawn by representing the wedges are horizontal lines and the dashes as vertical lines. The chiral carbon is not shown.

4-Ethyl-4-methyloctane:

$$CH_3 - \overset{\displaystyle CH_2CH_2CH_3}{\underset{\displaystyle CH_2CH_2CH_2CH_3}{|}} - CH_2CH_3 \qquad CH_3CH_2 - \overset{\displaystyle CH_2CH_2CH_3}{\underset{\displaystyle CH_2CH_2CH_2CH_3}{|}} - CH_3$$

1,2-Dibromobutane:

$$H - \overset{\displaystyle CH_2Br}{\underset{\displaystyle CH_2CH_3}{|}} - Br \qquad Br - \overset{\displaystyle CH_2Br}{\underset{\displaystyle CH_2CH_3}{|}} - H$$

137

7.5 The equation relating specific rotation $[\alpha]$ to the observed rotation α is

$$[\alpha] = \frac{100\alpha}{c\,l}$$

The concentration c is expressed in grams per 100 mL and the length l of the polarimeter tube in decimeters. Since the problem specifies the concentration as 0.3 g per 15 mL and the path length as 10 cm, the specific rotation $[\alpha]$ is:

$$[\alpha] = \frac{100\,(-0.78°)}{100\left(\dfrac{0.3\ g}{15\ mL}\right)\left(\dfrac{10\ cm}{10\ cm/dm}\right)}$$

$$[\alpha] = -39°$$

7.6 (b) The higher priority group is CH_2OH $[-C(O,H,H)]$. Although $-C(CH_3)_3$ has three carbon substituents on its attached atom $[-C(C,C,C)$, the highest ranked atom of (O,H,H) is oxygen, which has a higher atomic number than any of the atoms of (C,C,C).

(c) The higher priority group is $-CH(CH_3)_2$. The group of lower priority is $-CH_2CH_2OH$. The substituent $[-C(C,C,H)]$ ranks higher than $[-C(C,H,H)]$. Since precedence is determined by the **first** point of difference, the OH group does not affect the priority ranking.

(d) The group $-CH_2F$ $[-C(F,H,H)]$ has a higher priority ranking than $-CH(CH_3)_2$ $[-C(C,C,H)]$. Again, ties are broken at the first point of difference and F is higher atomic number than C. Atoms are considered one-by-one, not added together.

7.7 (b) The solution to this problem is exactly analogous to the sample solution to part (a) given in the text.

(+)-1-Fluoro-2-methylbutane

Order of precedence: CH_2F > CH_2CH_3 > CH_3 > H

The lowest ranked substituent (H) at the chiral center points away from the reader so the molecule is oriented properly as drawn. The three higher-ranked substituents trace a clockwise path from CH_2F to CH_2CH_3 to CH_3.

The absolute configuration is R; the compound is (R)-(+)-1-fluoro-2-methylbutane.

(c) The highest-ranked substituent at the chiral center of 1-bromo-2-methylbutane is CH_2Br, and the lowest-ranked substituent is H. Of the remaining two, ethyl outranks methyl.

Order of precedence: CH_2Br > CH_2CH_3 > CH_3 > H

The lowest-ranking substituent (H) at the chiral center is directed toward the reader and therefore the molecule needs to be reoriented so that H points in the opposite direction.

turn 180°

(+)–1-Bromo-2-methylbutane

The three highest-ranking substituents trace a counterclockwise path when the lowest-ranked substituent is held away from the reader.

The absolute configuration is S, and thus the compound is (S)–(+)-1-bromo-2-methylbutane.

(d)   The highest-ranked substituent at the chiral center of 3-buten-2-ol is the hydroxyl (–OH) group, and the lowest-ranked substituent is H. Of the remaining two, vinyl (–CH=CH$_2$) outranks methyl (–CH$_3$).

Order of precedence: OH >   CH=CH$_2$ >   CH$_3$ >   H

The lowest-ranking substituent (H) at the chiral center is directed away from the reader. We see that the order of decreasing precedence appears in a counterclockwise manner.

(+)-3-Buten-2-ol

The absolute configuration is S, and the compound is (S)–(+)-3-buten-2-ol.

7.8   (b)   One of the carbons bears a methyl group and a hydrogen. Methyl is of higher rank than hydrogen. The other carbon bears the groups –CH$_2$CH$_2$F and –CH$_2$CH$_2$CH$_2$CH$_3$. At the first point of difference between these two, fluorine is of higher atomic number than carbon, so CH$_2$CH$_2$F is of higher precedence.

Higher-ranked substituents are on the same side of the double bond; the alkene has the Z configuration.

(c)   One of the carbons bears a methyl group and a hydrogen. As we have seen, methyl is of higher priority. The other carbon bears –CH$_2$CH$_2$OH and –C(CH$_3$)$_3$. Let us analyze these two groups so that we may determine their order of precedence.

–CH$_2$CH$_2$OH [–**C(C,H,H)**] is of lower rank than –C(CH$_3$)$_3$ [–**C(C,C,C)**]

139

We examine the atoms one by one at the point of attachment before proceeding down the chain. Therefore, $-C(CH_3)_3$ outranks $-CH_2CH_2OH$.

Higher-ranked substituents are on opposite sides; the configuration of the alkene is *E*.

(d) The cyclopropyl ring is attached to the double bond by a carbon that bears the substituents (C,C,H) and is therefore of higher precedence than an ethyl group $-C(C,H,H)$.

Higher-ranked substituents are on opposite sides; the configuration of the alkene is *E*.

7.9 The sample solution describes the conversion of the ball-and-stick model of stereoisomer I in Figure 7.9 (page 209) to a perspective drawing, then to a Fischer projection. The same procedure is used to generate the Fischer projections of stereoisomers II, III, and IV.

7.10 In the text we are told that the (2*R*, 3*S*) stereoisomer is a crystalline solid. Its enantiomer (2*S*, 3*R*) has the same physical properties, and so must also be a solid.

7.11 As in problem, 7.9, use the ball-and-stick drawings of Figure 7.10 as a guide. Represent the wedges by horizontal lines and the dashed bonds as vertical lines.

CH₃
HO—C—H   HO———H
H—C—OH   H———OH
CH₃   CH₃
2R, 3R

CH₃   CH₃
H—C—OH   H———OH
HO—C—H   HO———H
CH₃   CH₃
2S, 3S

CH₃   CH₃
H—C—OH   H———OH
H—C—OH   H———OH
CH₃   CH₃
2R, 3S

7.12 2-Ketohexoses have three chiral centers. They are marked with asterisks in the structural formula.

$$\underset{\text{OH} \quad \text{OH} \quad \text{OH}}{\text{HOCH}_2\overset{\text{O}}{\overset{\|}{\text{C}}}\overset{*}{\text{CH}}-\overset{*}{\text{CH}}-\overset{*}{\text{CH}}\text{CH}_2\text{OH}}$$

No meso forms are possible, so there are a total of 2^3, or 8 stereoisomeric 2-ketohexoses.

7.13 The more soluble salt must have the opposite configuration at the chiral center of 1-phenylethylamine, *i.e.*, the *S* configuration. The malic acid used in the resolution is a single enantiomer, *S*. Therefore the more soluble salt in this particular case is (*S*)-1-phenylethylammonium (*S*)-malate.

7.14 There are eight isomeric halides having the formula $C_5H_{11}Br$. Three are chiral (the chiral centers are marked with an asterisk). The remaining five are achiral.

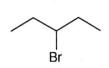
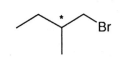

1-Bromopentane   2-Bromopentane   3-Bromopentane   1-Bromo-2-methylbutane
(achiral)   (chiral)   (achiral)   (chiral)

141

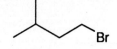

1-Bromo-3-methylbutane
(achiral)

2-Bromo-3-methylbutane
(chiral)

2-Bromo-2-methylbutane
(achiral)

1-Bromo-2,2-
dimethylbutane
(achiral)

7.15 Use wedges to show bonds coming up from the paper, and dashed lines to show bonds going back behind the paper.

7.16 The wedges in Problem 7.15 are represented by horizontal lines in Fischer projections. The dashed bonds are represented as vertical lines.

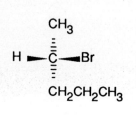

2-Bromopentane

2-Bromopentane

1-Bromo-2-methylbutane

1-Bromo-2-methylbutane

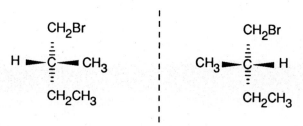

2-Bromo-3-methylbutane

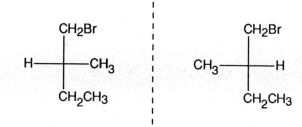

2-Bromo-3-methylbutane

7.17 (a) The two compounds are **constitutional isomers** because the orders in which their atoms are connected are different. Their IUPAC names clearly reflect this difference.

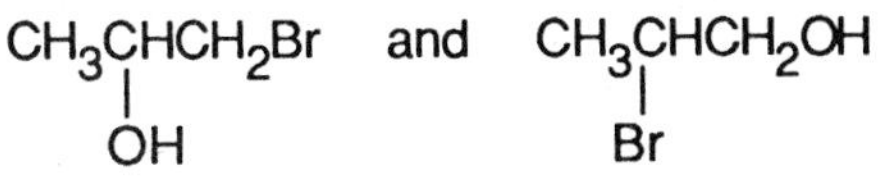

$$CH_3CHCH_2Br \quad and \quad CH_3CHCH_2OH$$
$$\quad\,|\qquad\qquad\qquad\,|$$
$$\quad OH \qquad\qquad\qquad Br$$

1-Bromo-2-propanol   2-Bromo-1-propanol

(b) The two structures have the same constitution. Test them for superimposability. To do this we need to place them in comparable orientations. For example, draw them both in perspective views that have the carbon chain vertical and with the vertical bonds pointing away from you.

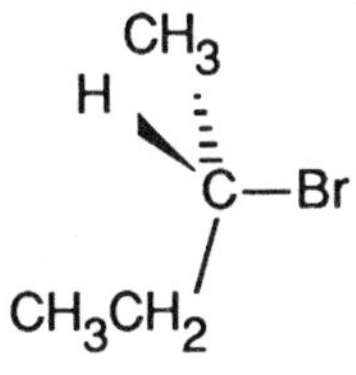

is equivalent to

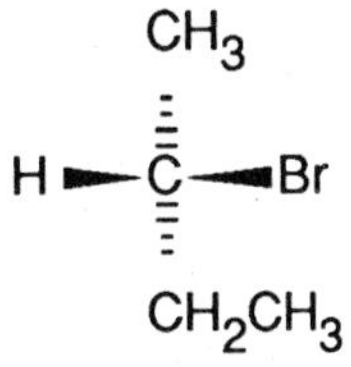

and

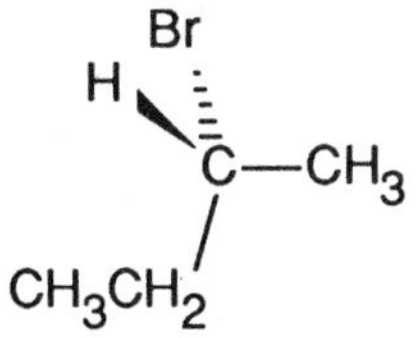

is equivalent to

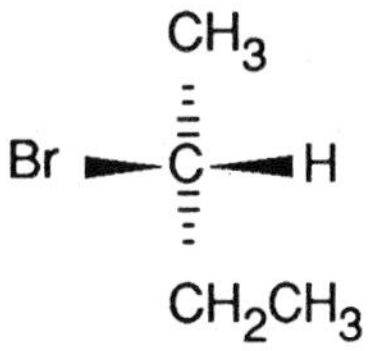

The two structures are nonsuperimposable mirror images of each other. They are **enantiomers**.

(c) Again, place the structures in comparable orientations and examine them for superimposability.

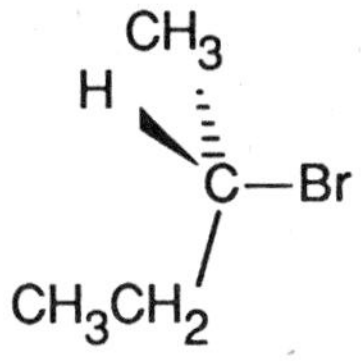

is equivalent to

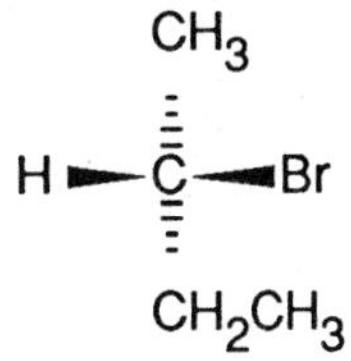

and

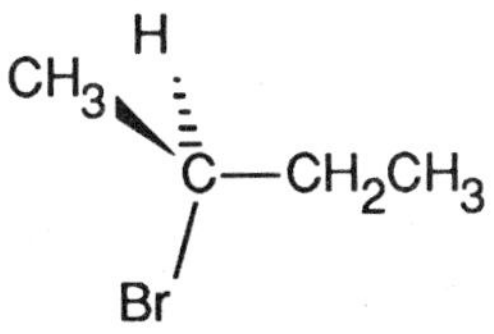

is equivalent to

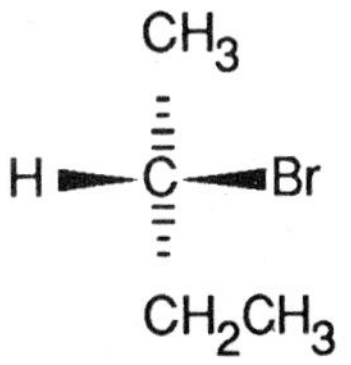

The two structures are **identical**; they are superimposable upon one another.

7.18 When we drew the perspective views of the compounds in parts (b) and (c) of Problem 7.17, we arranged the atoms in a manner suitable for easy translation to a Fischer projection. The carbon chains are vertical and the vertical bonds are directed away from us.

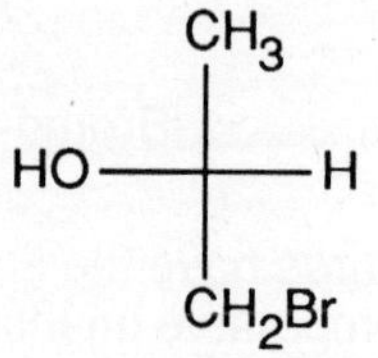

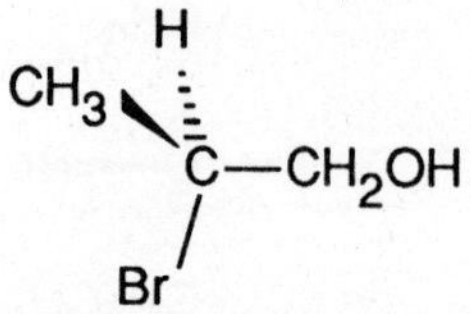

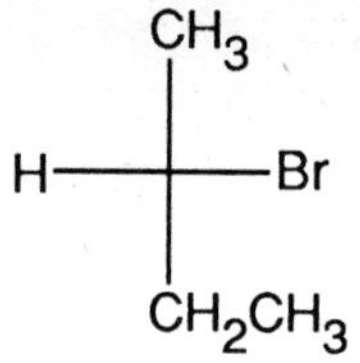

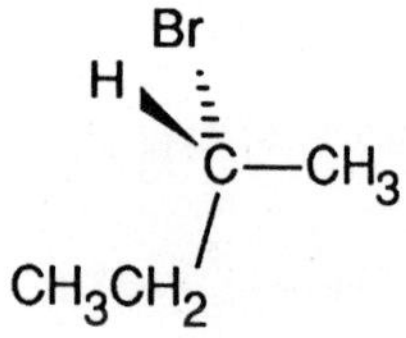

(a)

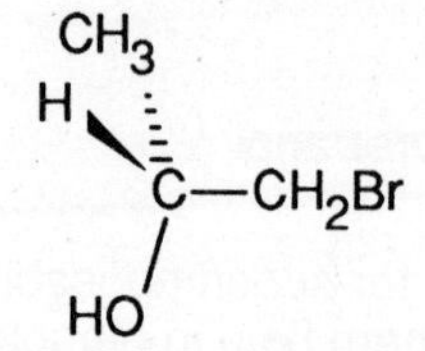

$$CH_3 \quad \text{C—CH}_2Br \quad \text{is equivalent to} \quad HO—\!\!\!\!\!\!\!—H$$

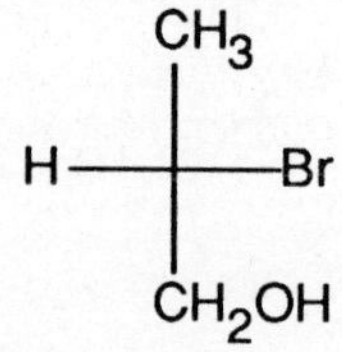

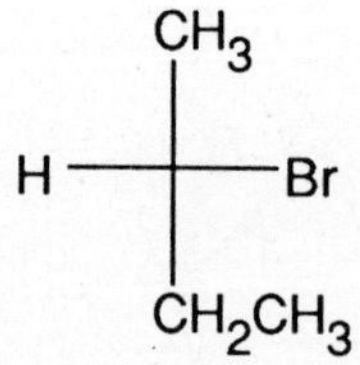

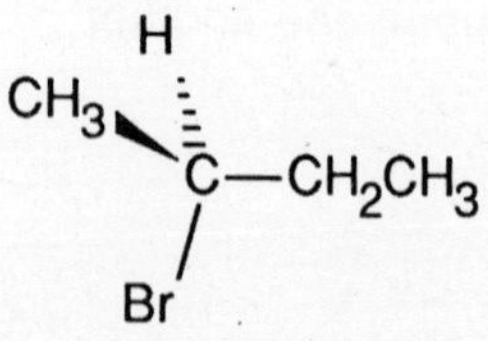

(b)

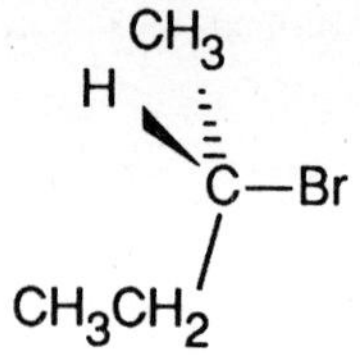

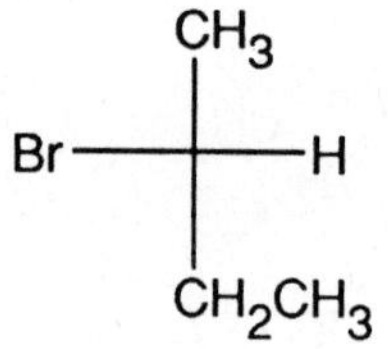

(c)

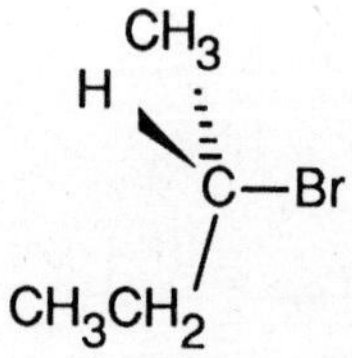

144

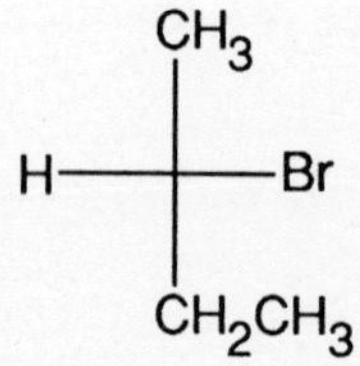

7.19 (a) There is a plane of symmetry in *cis*-1,2-dichlorocyclopropane which bisects the C(1)-C(2) bond and passes through C-3. The molecule is **achiral**.

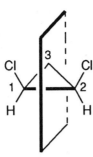

(b) *trans*-1,2-Dichlorocyclopropane lacks a plane of symmetry. It is not superimposable on its mirror image and is, therefore, **chiral**.

(c) and (d) The cis and trans isomers of 2-chlorocyclopropanol are both **chiral**. Neither one has a plane of symmetry.

7.20 There are four isomers of trichlorocyclopropane. They are:

Enantiomeric forms of 1,1,2-trichlorocyclopropane

cis–1,2,3-Trichlorocyclopropane   *trans*-1,2,3-Trichlorocyclopropane

(achiral, vertical plane of symmetry bisects molecule)   (achiral, vertical plane of symmetry bisects molecule)

7.21 Stereoisomers I and II are enantiomers of each other. Isomers III and IV are enantiomers. Isomers I and II are diastereomers of III and IV.

7.22 Both 2,3-dibromopentane and 2,4-dibromopentane have two chiral centers. The 2^n formula gives the maximum number of stereoisomers if no *meso* forms are possible. Such is the case for 2,3-dibromopentane and it has four stereoisomers. An achiral meso isomer is possible, however, for 2,4-dibromopentane.

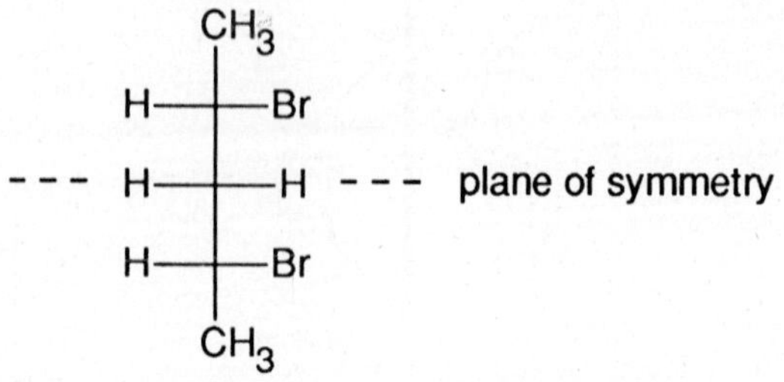

Therefore, 2,4-dibromopentane has a total of three stereoisomers. Two are chiral and enantiomeric to one another; the third one is the achiral *meso* form shown above.

7.23 There is a plane of symmetry in the cis stereoisomer of 1,3-dimethylcyclohexane. It is an achiral meso isomer.

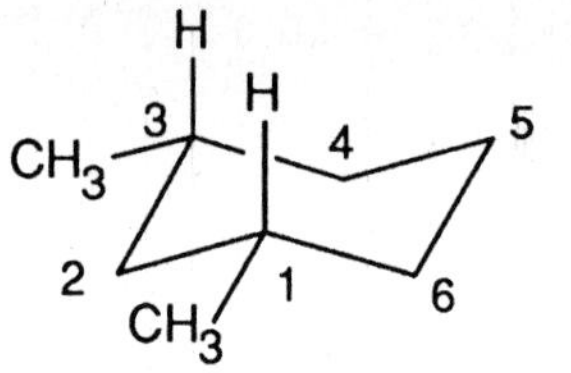

Plane of symmetry passes through C-2 and C-5 and bisects the ring

The trans stereoisomer is chiral.

7.24 A molecule with eight chiral centers has 2^8 or 256 stereoisomers. One of these is cholesterol, and the other 255 are stereoisomers of cholesterol.

7.25 (a) There is one chiral center in limonene.

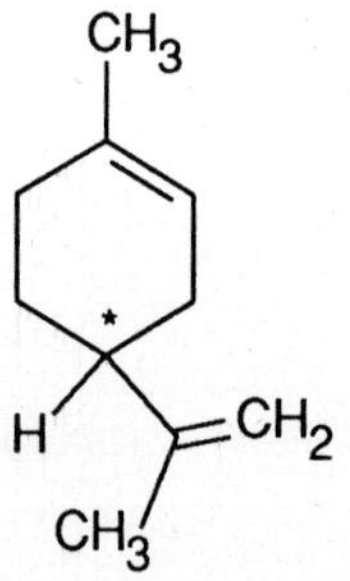

(b) There are three chiral centers in biotin. Naturally occurring (+)-biotin is the stereoisomer shown.

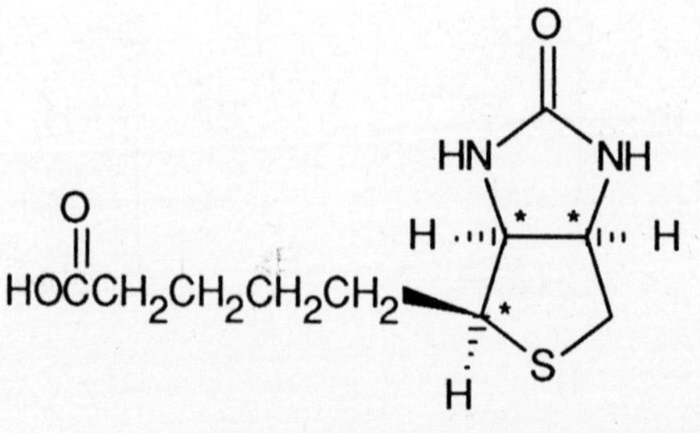

146

(c) There are four chiral centers in periplanone B.

(d) Calciferol has six chiral centers.

7.26 There are two sites of stereochemical variation in 3-penten-2-ol. The double bond may be either cis or trans, and the chiral center may be either *R* or *S*.

trans, R   *trans, S*   *cis, R*   *cis, S*

7.27 The two chiral centers of an aldotetrose give rise to four stereoisomers (two pairs of enantiomers). There are no meso isomers.

147

7.28 A molecule with three chiral centers has 2^3, or 8, stereoisomers. The eight combinations of R and S chiral centers are:

	Chiral center					Chiral center		
	1	2	3			1	2	3
Isomer 1	R	R	R		Isomer 5	S	S	S
Isomer 2	R	R	S		Isomer 6	S	S	R
Isomer 3	R	S	R		Isomer 7	S	R	S
Isomer 4	S	R	R		Isomer 8	R	S	S

7.29 From Problem 7.14 recall that the chiral isomers are:

(a) 2-Bromopentane; (b) 1-Bromo-2-methylbutane; and (c) 2-Bromo-3-methylbutane

In Problem 7.15, perspective (wedge-and-dash) representations were generated for both enantiomers of each chiral molecule. In the present problem, we need to determine the absolute configuration of each one according to the R,S notational system. Rather than showing both enantiomers of the three compounds, we will describe only one of each with the understanding that the enantiomer which is omitted must have a configuration opposite to the one shown. In all three cases, we will examine the structure in which C-1 is at top and the hydrogen substituent at C-2 is on the left in the perspective view.

(a) 2-Bromopentane priorities: $Br > CH_2CH_2CH_3 > CH_3 > H$

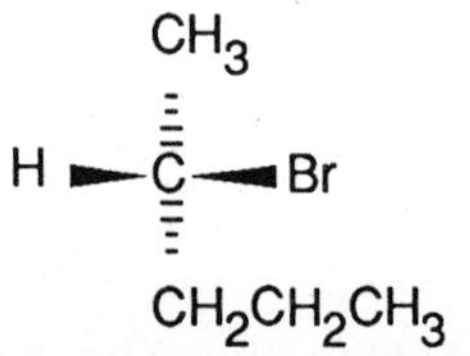

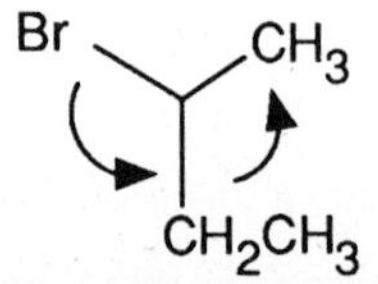

The lowest-ranked substituent (H) is directed toward us. We need to examine the molecule (or a molecular model of it) from a position in which the hydrogen is away from us. When we do that we see that the order of decreasing substituent precedence traces a counterclockwise path. The absolute configuration of the enantiomer shown above is S.

(b) 1-Bromo-2-methylbutane priorities: $CH_2Br > CH_2CH_3 > CH_3 > H$

Here again, the lowest-ranked substituent (H) is toward us, so we need to look at the molecule from the side opposite the C–H bond. When viewed from this perspective, the order of decreasing precedence appears in a clockwise fashion. The configuration is R.

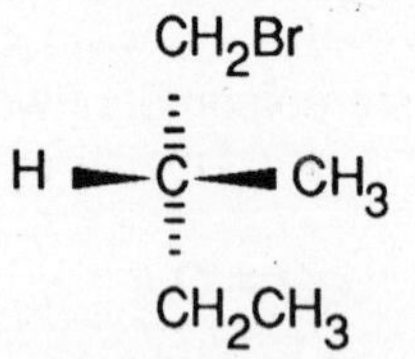

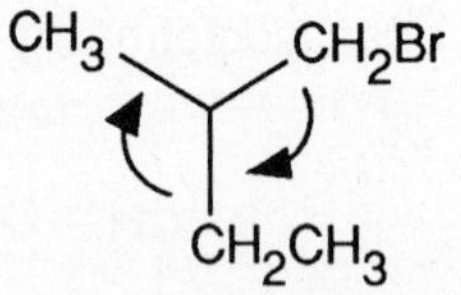

148

(c) 2-Bromo-3-methylbutane priorities: Br > CH(CH₃)₂ > CH₃ > H

The order of decreasing precedence is counterclockwise. The absolute configuration is *S*.

7.30 As in the previous problem, you must orient the molecule so that the lowest-ranked substituent (usually H) is away from you.

(a) The stereoisomer of 1-bromo-2-propanol given is *R*. Hydroxyl is the highest-ranked group, followed by CH₂Br, followed by CH₃. These substituents appear in a clockwise fashion when viewed from the side opposite the C–H bond.

The stereoisomer of 2-bromo-1-propanol given in Problem 7.17 is *S*.

(b) In Problem 7.17 we determined that the two compounds shown were enantiomers. In keeping with that analysis, the two enantiomers have opposite configurations. The first structure has the *S* configuration at its chiral center.

The second structure has the *R* configuration.

c) This compound is the same as the first structure in (b), and has the *S* configuration.

7.31 Structures I-IV have two chiral centers. Redraw the Fischer projections of Problem 7.21 as perspective views and determine the configuration of each chiral center independently.

$$CH_3$$

I
$$H \blacktriangleright \overset{S}{C} \blacktriangleleft Br$$
$$H \blacktriangleright \underset{R}{C} \blacktriangleleft Br$$
$$CH_2CH_3$$

II
$$CH_3$$
$$Br \blacktriangleright \overset{R}{C} \blacktriangleleft H$$
$$Br \blacktriangleright \underset{S}{C} \blacktriangleleft H$$
$$CH_2CH_3$$

III
$$CH_3$$
$$Br \blacktriangleright \overset{R}{C} \blacktriangleleft H$$
$$H \blacktriangleright \underset{R}{C} \blacktriangleleft Br$$
$$CH_2CH_3$$

IV
$$CH_3$$
$$H \blacktriangleright \overset{S}{C} \blacktriangleleft Br$$
$$Br \blacktriangleright \underset{S}{C} \blacktriangleleft H$$
$$CH_2CH_3$$

7.32 First write the constitution of the molecule in order to locate the chiral center and identify its substituents. Then determine the priority rankings for the substituents and arrange the groups in the correct sequence in a perspective view or a Fischer projection.

(a) (S)–2-Pentanol priorities: $OH > CH_2CH_2CH_3 > CH_3 > H$

$$CH_3$$
$$H \blacktriangleright \overset{\vdots}{C} \blacktriangleleft OH$$
$$CH_2CH_2CH_3$$

$$CH_3$$
$$H \rule{1cm}{0.4pt} OH$$
$$CH_2CH_2CH_3$$

(R)-3-Chloro-2-methylpentane priorities: $Cl > CH(CH_3)_2 > CH_2CH_3 > H$

$$CH(CH_3)_2$$
$$H \blacktriangleright \overset{\vdots}{C} \blacktriangleleft Cl$$
$$CH_2CH_3$$

$$CH(CH_3)_2$$
$$H \rule{1cm}{0.4pt} Cl$$
$$CH_2CH_3$$

(2S, 3S)-2-Bromo-3-pentanol priorities:

C-2: $Br > (C\text{-}3) > CH_3 > H$
C-3: $OH > (C\text{-}2) > CH_2CH_3 > H$

$$CH_3$$
$$H \blacktriangleright \overset{\vdots}{C} \blacktriangleleft Br$$
$$HO \blacktriangleright C \blacktriangleleft H$$
$$CH_2CH_3$$

$$CH_3$$
$$H \rule{1cm}{0.4pt} Br$$
$$HO \rule{1cm}{0.4pt} H$$
$$CH_2CH_3$$

(R)-3-Bromo-1-pentene priorities: $Br > CH=CH_2 > CH_2CH_3 > H$

$$CH=CH_2$$
$$H \blacktriangleright \overset{\vdots}{C} \blacktriangleleft Br$$
$$CH_2CH_3$$

$$CH=CH_2$$
$$H \rule{1cm}{0.4pt} Br$$
$$CH_2CH_3$$

150

7.33 Since the molecule is a six-carbon alkene and is chiral, the substituents on the chiral center are, in order of decreasing precedence: $CH=CH_2$ > CH_2CH_3 > CH_3 > H

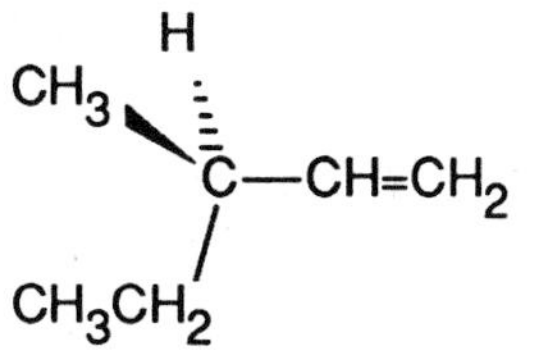

$$\underset{R}{} \qquad\qquad \underset{S}{}$$

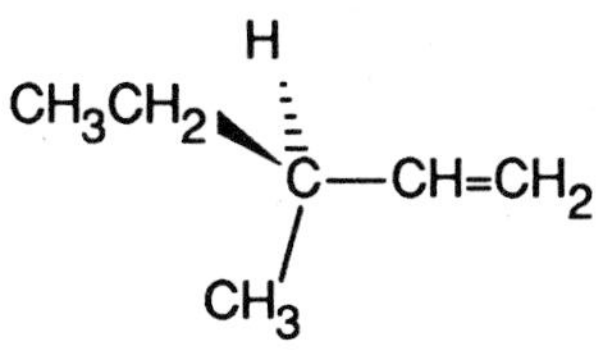

7.34 **(a)** The *Z* configuration means the higher priority substituents are on the same side of the double bond. At the left side of the double bond, CH_3 outranks H. At the right side, $CH_2CH_2CH_2CH_3$ outranks CH_3.

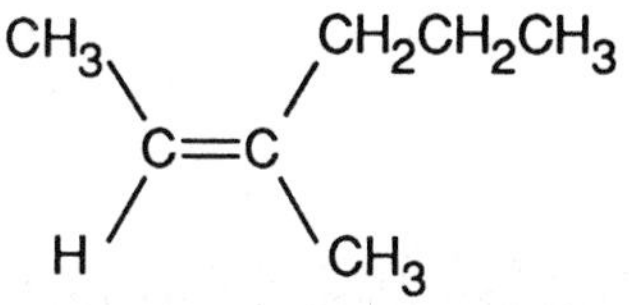

(*Z*)-3-Methyl-2-hexene

(b) The *E* configuration means the higher priority substituents are on opposite sides of the double bond. The higher-ranked substituent on the right side of the double bond is Cl.

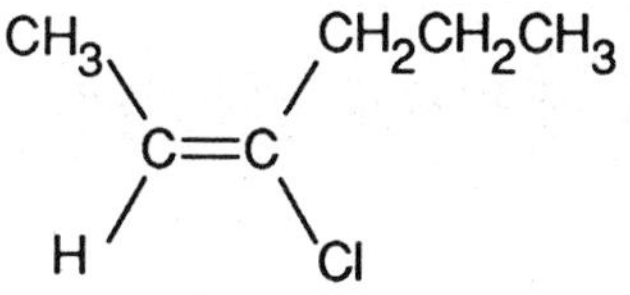

(*E*)-3-Chloro-2-hexene

(c) The higher-ranked substituents are Br and $CH_2CH_2CH_2CH_3$.

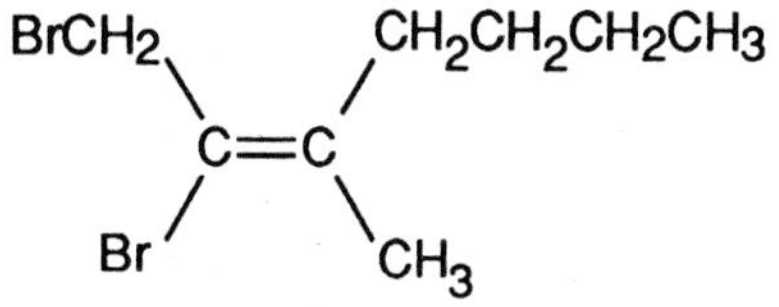

(*E*)-1,2-Dibromo-3-methyl-2-heptene

(d) (*Z*)-4-Ethyl-3-methyl-3-heptene is

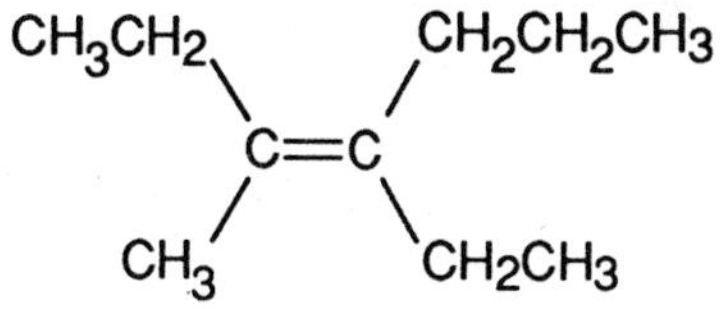

Ethyl is the higher-ranked substituent on the left side of the double bond, but the lower-ranked substituent on the right side.

7.35 **(a)** The *E* configuration means that the higher-priority substituents are on opposite sides of the double bond.

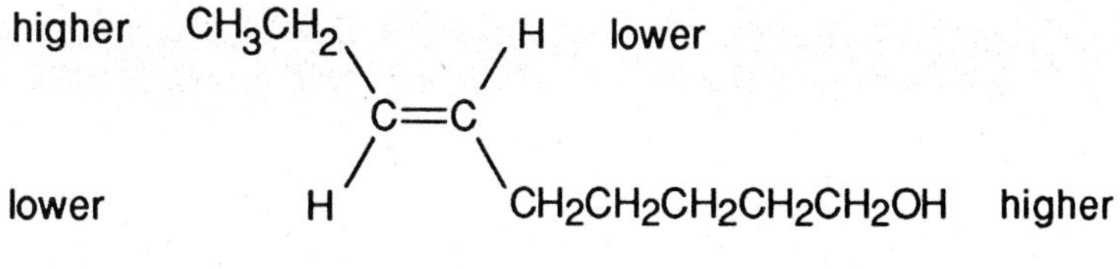

(b) Geraniol has two double bonds, but only one of them, the one between C-2 and C-3, is capable of stereochemical variation. Of the substituents at C-2, CH_2OH is of higher priority than H. At C-3 CH_2CH_2 outranks CH_3. Higher-priority substituents are on opposite sides of the double bond in the *E* isomer. Therefore geraniol has the structure shown:

151

$$(CH_3)_2C=CHCH_2CH_2 \diagdown \qquad \diagup H$$
$$C=C$$
$$CH_3 \diagup \qquad \diagdown CH_2OH$$

(c) Beginning at the 6,7 double bond, we see that the propyl group is of higher priority than the methyl group at C-7. Since the 6,7 double bond is *E*, the propyl group must be on the opposite side of the the higher-ranked substituent at C-6 where the CH_2 fragment outranks hydrogen. Therefore, we write for the stereochemistry of the 6,7 double bond:

$$\begin{array}{ccc} & 10 \quad 9 \quad 8 & \\ \text{higher} & CH_3CH_2CH_2 \diagdown & \diagup H \quad \text{lower} \\ & 7 \; C=C \; 6 & \\ & \diagup \qquad \diagdown 5 & \\ \text{lower} \quad CH_3 & CH_2- \quad \text{higher} \end{array}$$

E

At C-2 CH_2OH is of higher priority than H, and at C-3 CH_2CH_2C- is of higher priority than CH_2CH_3. The double-bond configuration at C-2 is *Z*. Therefore

$$\begin{array}{ccc} & 6\;5 \quad 4 & 1 \\ \text{higher} & -CCH_2CH_2 \diagdown & \diagup CH_2OH \quad \text{higher} \\ & 3 \quad 2 & \\ & C=C & \\ & \diagup \qquad \diagdown & \\ \text{lower} \quad CH_3CH_2 & H \quad \text{lower} \end{array}$$

Combining the two partial structures, we obtain for the full structure of the codling moth's sex pheromone

$$CH_3CH_2CH_2 \diagdown \qquad \diagup H$$
$$C=C$$
$$CH_3 \diagup \qquad \diagdown CH_2CH_2 \diagdown \qquad \diagup CH_2OH$$
$$C=C$$
$$CH_3CH_2 \diagup \qquad \diagdown H$$

The compound is (2*Z*, 6*E*)-3-ethyl-7-methyl-2,6-decadien-1-ol.

(d) The sex pheromone of the honeybee is (*E*)-9-oxo-2-decenoic acid, with the structure

$$\begin{array}{ccc} & O & \\ & \| & \\ \text{higher} & CH_3C(CH_2)_4CH_2 \diagdown & \diagup H \quad \text{lower} \\ & C=C & \\ & \diagup \qquad \diagdown & \\ \text{lower} \quad H & CO_2H \quad \text{higher} \end{array}$$

7.36 The equation which relates the specific rotation $[\alpha]$ to the observed rotation α is

$$[\alpha] = \frac{100\alpha}{cl}$$

where c is the concentration in grams per 100 mL and l is the path length in decimeters.

$$[\alpha] = \frac{100\,(-5.20°)}{\left(\dfrac{2.0\text{ g}}{100\text{ mL}}\right)(2\text{dm})}$$

$$[\alpha] = -130°$$

7.37 (a) The two other stereoisomeric tartaric acids, in addition to the (2R, 3R) isomer, are (2S, 3S) and (2R, 3S).

2R, 3R 2S, 3S 2R, 3S

(b) No. Pasteur separated an optically inactive racemic mixture into two optically active enantiomers. A meso isomer is achiral, and is incapable of being separated into optically active forms.

7.38 The first step in acid-catalyzed hydration of an alkene is formation of a carbocation. The carbocation is planar, and can be attacked from either the top or bottom face by a nucleophile, in this case water. The 2-butanol formed by hydration of 1-butene is produced as a racemic mixture of enantiomers.

153

7.39 The product from the reaction of (S)-3-chloro-1-butene with hydrogen chloride is a mixture of two stereoisomers, (2R, 3S) -dichlorobutane and (2S, 3S) -dichlorobutane. These stereoisomers are **diastereomers** of each other. The stereochemistry at C-3 remains unchanged when HCl adds to the double bond. Both the R and S configurations are possible at the new chiral center.

(S)–3-Chloro-1-butene   (2S,3S)–2,3-Dichlorobutane   (2R,3S)–2,3-Dichlorobutane

154

CHAPTER 8

NUCLEOPHILIC SUBSTITUTION REACTIONS

GLOSSARY OF TERMS

Concerted reaction A reaction which occurs in a single step. Bond-making and bond-breaking both contribute to the transition state.

Inversion of configuration The reversal of the three-dimensional arrangement of the four bonds to sp^3-hybridized carbon. The representation shown below illustrates inversion of configuration in a nucleophilic substitution reaction where LG becomes the leaving group and Nu is the nucleophile.

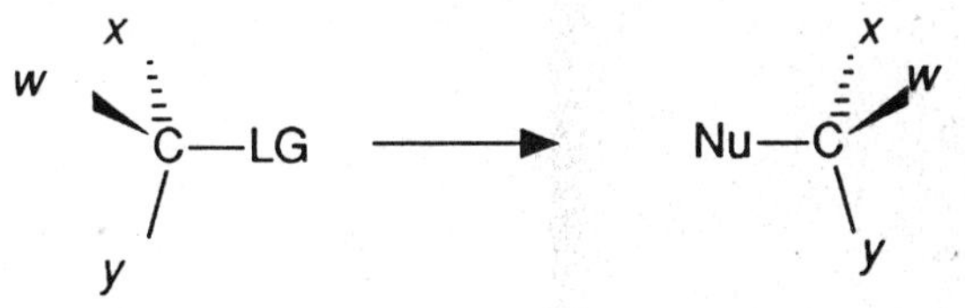

Leaving group The group which is lost from carbon in a nucleophilic substitution (or elimination). It is most normally a halogen. The pair of electrons in the bond between carbon and the leaving group becomes an unshared electron pair of the leaving group when the bond to carbon is broken.

Nucleophile An atom that has an unshared electron pair which can be used to form a bond to carbon. Nucleophiles are Lewis bases.

Nucleophilic substitution A reaction in which an attacking atom uses an unshared electron pair to bond to carbon and a leaving group is lost from that carbon.

Nucleophilicity A measure of the reactivity of a Lewis base in a nucleophilic substitution reaction.

Potential energy diagram A plot of potential energy versus some arbitrary measure of the degree to which a reaction has proceeded (the reaction coordinate). The point of maximum potential energy is the transition state.

S$_N$1 (*substitution-nucleophilic-unimolecular*) A mechanism for nucleophilic substitution characterized by a two-step process. The first step is rate determining and is the ionization of an alkyl halide to a carbocation and a halide ion.

S$_N$2 (*substitution-nucleophilic-bimolecular*) A concerted mechanism for nucleophilic substitution in which the nucleophile attacks carbon from the side opposite the bond to the leaving group and assists the departure of the leaving group.

Solvolysis reaction Nucleophilic substitution in a medium in which the only nucleophile present is the solvent.

Steric hindrance In nucleophilic substitution, the resistance to nucleophilic attack caused by the presence of large (or numerous) atoms or groups near the carbon that bears the leaving group.

SOLUTIONS TO TEXT PROBLEMS

8.1 The halogen of an alkyl halide is directly bonded to an sp^3 hybridized carbon. In an aryl halide the halogen is directly bonded to an aromatic (benzene) ring. The halogen of a vinyl halide is directly attached to a double bond.

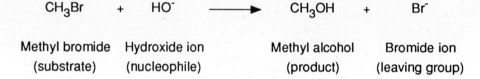

| Alkyl halide | Aryl halide | Vinyl halide |

8.2 The Lewis base (hydroxide ion) acts as a nucleophile and reacts with the alkyl halide substrate (methyl bromide) to form the product (methyl alcohol) and releasing the leaving group (bromide ion)

$$CH_3Br \quad + \quad HO^- \quad \longrightarrow \quad CH_3OH \quad + \quad Br^-$$

| Methyl bromide | Hydroxide ion | Methyl alcohol | Bromide ion |
| (substrate) | (nucleophile) | (product) | (leaving group) |

8.3 (b) Potassium *tert*-butoxide serves as a source of the nucleophilic anion $(CH_3)_3CO^-$.

$$(CH_3)_3CO^- \quad + \quad CH_3CH_2Br \quad \longrightarrow \quad (CH_3)_3COCH_2CH_3 \quad + \quad Br^-$$

| *tert*-Butoxide ion | Bromoethane | *tert*-Butyl ethyl ether | Bromide ion |
| (nucleophile) | (substrate) | (product) | (leaving group) |

(c) The nucleophilic anion in KCN is cyanide ion ^-CN. The carbon atom is negatively charged and is normally the site of nucleophilic reactivity.

$$:N{\equiv}C:^- \quad + \quad CH_3CH_2Br \quad \longrightarrow \quad :N{\equiv}CCH_2CH_3 \quad + \quad Br^-$$

| Cyanide ion | Bromoethane | Ethyl cyanide | Bromide ion |
| (nucleophile) | (substrate) | (product) | (leaving group) |

(d) The anion in sodium hydrogen sulfide (NaSH) is ^-SH.

$$HS^- \quad + \quad CH_3CH_2Br \quad \longrightarrow \quad CH_3CH_2SH \quad + \quad Br^-$$

| Hydrogen sulfide ion | Bromoethane | Ethanethiol | Bromide ion |
| (nucleophile) | (substrate) | (product) | (leaving group) |

8.4 Write out the structure of the starting material. Notice that it contains a primary bromide and a primary chloride. Bromide is a better leaving group than chloride and is the one that is displaced faster by the nucleophilic cyanide ion.

$$\text{ClCH}_2\text{CH}_2\text{CH}_2\text{Br} \xrightarrow{\text{NaCN}} \text{ClCH}_2\text{CH}_2\text{CH}_2\text{C}\equiv\text{N}$$

1-Bromo-3-chloropropane   4-Chlorobutanenitrile

8.5 The example given in the text illustrates inversion of configuration in the S_N2 hydrolysis of (*S*)-(+)-2-bromooctane, which yields (*R*)-(-)-2-octanol. The hydrolysis of (*R*)-(-)-2-bromooctane exactly mirrors that of its enantiomer and yields (*S*)-(+)-2-octanol.

Hydrolysis of racemic 2-bromooctane gives racemic 2-octanol. Remember, optically inactive reactants must yield optically inactive products.

8.6 Sodium iodide in acetone is a reagent that converts alkyl chlorides and bromides into alkyl iodides by an S_N2 mechanism. Pick the alkyl halide that is most reactive toward S_N2 displacement.

(b) **1-Bromopentane** is a primary alkyl halide and so is more reactive than 3-bromopentane, which is secondary.

$$\text{BrCH}_2\text{CH}_2\text{CH}_2\text{CH}_2\text{CH}_3 \qquad\qquad \underset{\underset{\text{Br}}{|}}{\text{CH}_3\text{CH}_2\text{CHCH}_2\text{CH}_3}$$

1-Bromopentane   3-Bromopentane

(primary; more reactive in S_N2)   (secondary; less reactive in S_N2)

Crowding increases at the transition state for S_N2 reactions. The less crowded alkyl halide reacts faster.

(c) Both halides are secondary, but fluoride is quite a poor leaving group in nucleophilic substitution reactions. Alkyl chlorides are more reactive than alkyl fluorides. **2-Chloropentane** is more reactive.

$$\underset{\underset{\text{Cl}}{|}}{\text{CH}_3\text{CHCH}_2\text{CH}_2\text{CH}_3} \qquad\qquad \underset{\underset{\text{F}}{|}}{\text{CH}_3\text{CHCH}_2\text{CH}_2\text{CH}_3}$$

2-Chloropentane   2-Fluoropentane

(more reactive)   (less reactive)

(d) A secondary alkyl bromide reacts faster under S_N2 conditions than a tertiary one. **2-Bromo-5-methylhexane** is more reactive.

$$\underset{\underset{\text{CH}_3}{|}\underset{\text{Br}}{|}}{\text{CH}_3\text{CHCH}_2\text{CH}_2\text{CHCH}_3} \qquad\qquad \overset{\overset{\text{CH}_3}{|}}{\underset{\underset{\text{Br}}{|}}{\text{CH}_3\text{CCH}_2\text{CH}_2\text{CH}_2\text{CH}_3}}$$

2-Bromo-5-methylhexane   2-Bromo-2-methylhexane

(secondary; more reactive in S_N2)   (tertiary; less reactive in S_N2)

8.7 The reactivity of an alkyl halide in an S_N1 reaction is dictated by the ease with which it ionizes to form a carbocation. Tertiary alkyl halides are the most reactive, methyl halides the least reactive.

(b) Cyclopentyl iodide ionizes to form a secondary carbocation while the carbocation from 1-methylcyclopentyl iodide is tertiary. The **tertiary halide** is more reactive.

1-Methylcyclopentyl iodide
(tertiary; more reactive in S_N1)

Cyclopentyl iodide
(secondary; less reactive in S_N1)

(c) **Cyclopentyl bromide** ionizes to a secondary carbocation. 1-Bromo-2,2-dimethylpropane is a primary alkyl halide and is therefore less reactive.

Cyclopentyl bromide
(secondary; more reactive in S_N1)

$$CH_3-\underset{\underset{CH_3}{|}}{\overset{\overset{CH_3}{|}}{C}}-CH_2Br$$

1-Bromo-2,2-dimethylpropane
(primary; less reactive in S_N1)

(d) Iodide is a better leaving group than chloride in both S_N1 and S_N2 reactions. **tert-Butyl iodide** is more reactive.

$(CH_3)_3C-I$

tert-Butyl iodide
(more reactive)

$(CH_3)_3C-Cl$

tert-Butyl chloride
(less reactive)

8.8 Since the hydrolysis of a second compound gives the same product mixture, the most reasonable conclusion is that it and 3-chloro-3-methyl-1-butene must ionize to form the same carbocation. The second compound is therefore 1-chloro-3-methyl-2-butene.

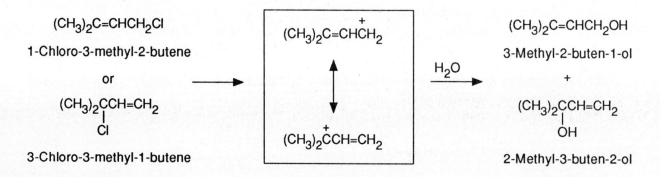

$(CH_3)_2C=CHCH_2Cl$

1-Chloro-3-methyl-2-butene

or

$(CH_3)_2\underset{\underset{Cl}{|}}{C}CH=CH_2$

3-Chloro-3-methyl-1-butene

$(CH_3)_2C=CHCH_2$ +

$(CH_3)_2\overset{+}{C}CH=CH_2$

$\xrightarrow{H_2O}$

$(CH_3)_2C=CHCH_2OH$

3-Methyl-2-buten-1-ol

+

$(CH_3)_2\underset{\underset{OH}{|}}{C}CH=CH_2$

2-Methyl-3-buten-2-ol

8.9 (b) Ethyl bromide is a primary alkyl halide and reacts with the potassium salt of cyclohexanol by substitution.

CH_3CH_2Br + ⬡—OK ⟶ ⬡—OCH$_2$CH$_3$

Ethyl bromide   Potassium cyclohexanolate   Cyclohexyl ethyl ether

(c)   No strong base is present in this reaction; the nucleophile is methanol itself, not methoxide. It reacts with *sec*-butyl bromide by substitution, not elimination.

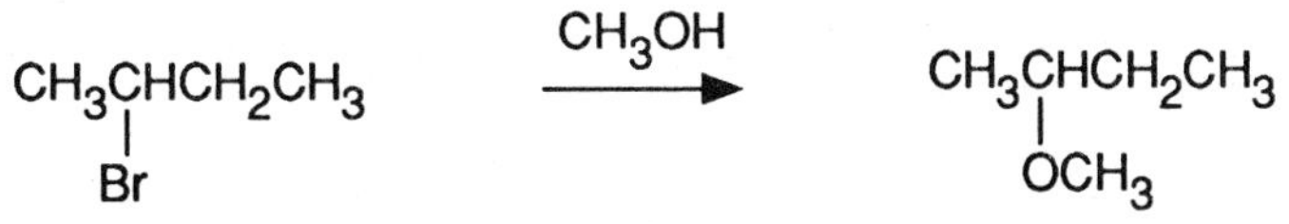

$$CH_3CHCH_2CH_3 \xrightarrow{\ CH_3OH\ } CH_3CHCH_2CH_3$$

$$\underset{\text{Br}}{|} \qquad\qquad\qquad \underset{\text{OCH}_3}{|}$$

sec-Butyl bromide   *sec*-Butyl methyl ether

(d)   Secondary alkyl halides react with alkoxide bases by E2 elimination.

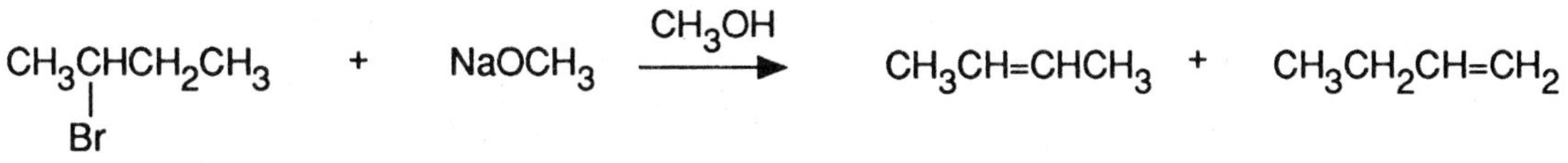

$$CH_3CHCH_2CH_3 \ + \ NaOCH_3 \xrightarrow{\ CH_3OH\ } CH_3CH=CHCH_3 \ + \ CH_3CH_2CH=CH_2$$

$$\underset{\text{Br}}{|}$$

sec-Butyl bromide   Sodium methoxide   2-Butene   1-Butene

(major product; cis + trans)

8.10   All the reactions are S_N2 displacements on a primary alkyl halide, so they should proceed cleanly.

(a)

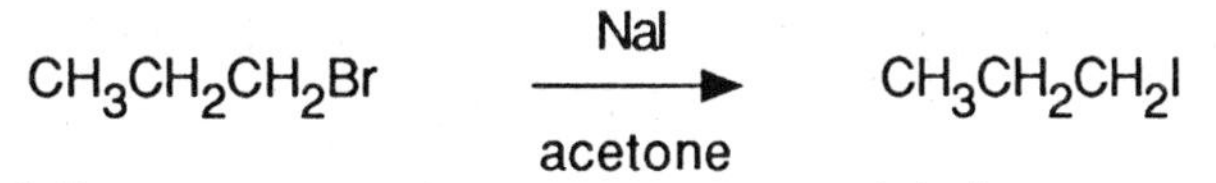

$$CH_3CH_2CH_2Br \xrightarrow[\text{acetone}]{\text{NaI}} CH_3CH_2CH_2I$$

1-Bromopropane   1-Iodopropane

(b)

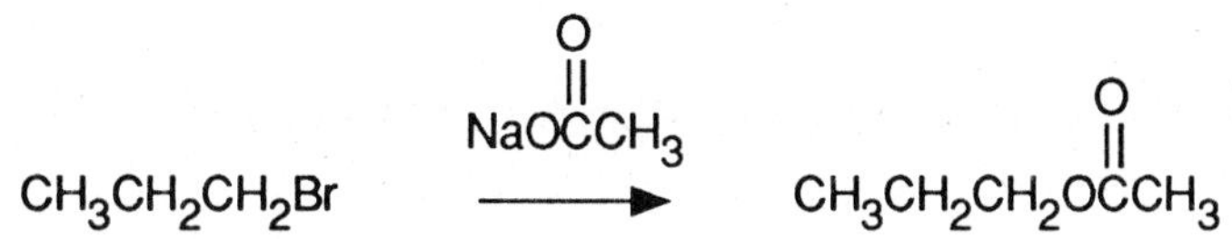

$$CH_3CH_2CH_2Br \xrightarrow{\overset{\overset{\text{O}}{\|}}{\text{NaOCCH}_3}} CH_3CH_2CH_2O\overset{\overset{\text{O}}{\|}}{C}CH_3$$

1-Bromopropane   Propyl acetate

(c)

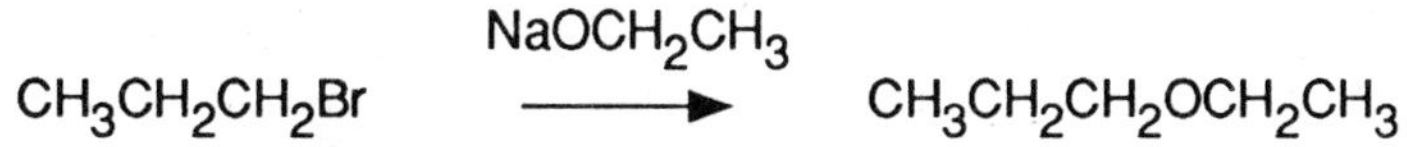

$$CH_3CH_2CH_2Br \xrightarrow{\text{NaOCH}_2CH_3} CH_3CH_2CH_2OCH_2CH_3$$

1-Bromopropane   Ethyl propyl ether

(d)

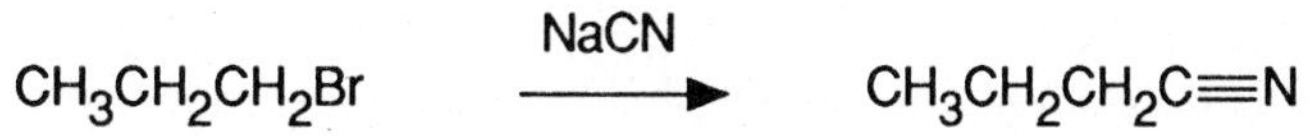

$$CH_3CH_2CH_2Br \xrightarrow{\text{NaCN}} CH_3CH_2CH_2C\equiv N$$

1-Bromopropane   Propyl cyanide

(e)

$$CH_3CH_2CH_2Br \xrightarrow{\text{NaN}_3} CH_3CH_2CH_2N_3$$

1-Bromopropane   Propyl azide

159

(f)

$$CH_3CH_2CH_2Br \xrightarrow{\text{NaSH}} CH_3CH_2CH_2SH$$

1-Bromopropane   1-Propanethiol

(g)

$$CH_3CH_2CH_2Br \xrightarrow{\text{NaSCH}_3} CH_3CH_2CH_2SCH_3$$

1-Bromopropane   Methyl propyl thioether

8.11 In the corresponding reactions of 2-bromopropane, the possibility of elimination is more pronounced.

(a) Iodide is weakly basic and very nucleophilic, so substitution is the principal reaction.

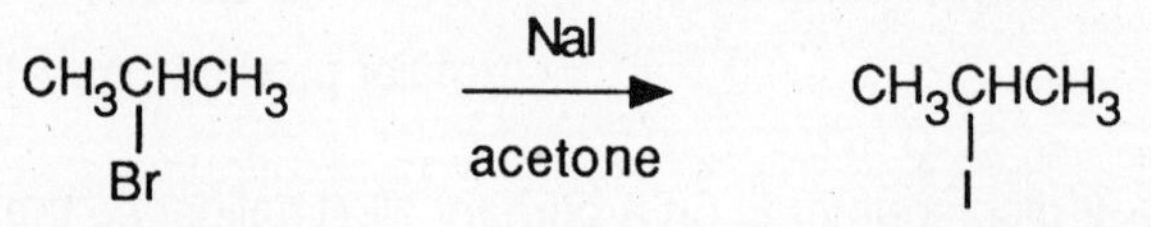

$$\underset{\overset{|}{Br}}{CH_3CHCH_3} \xrightarrow[\text{acetone}]{\text{NaI}} \underset{\overset{|}{I}}{CH_3CHCH_3}$$

2-Bromopropane   2-Iodopropane

(b) Sodium acetate is weakly basic so the product should be isopropyl acetate.

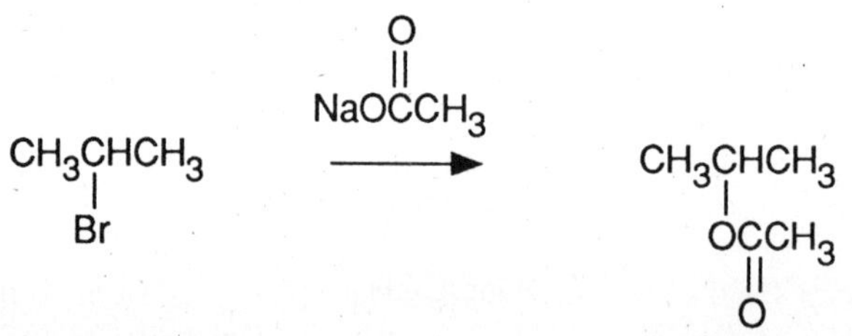

$$\underset{\overset{|}{Br}}{CH_3CHCH_3} \xrightarrow{\overset{\overset{O}{\|}}{NaOCCH_3}} CH_3CHCH_3$$

with OCCH₃ group

2-Bromopropane   Isopropyl acetate

(c) Bases as strong as or stronger than hydroxide react with secondary alkyl halides by E2 elimination. Ethoxide is comparable with hydroxide in basicity.

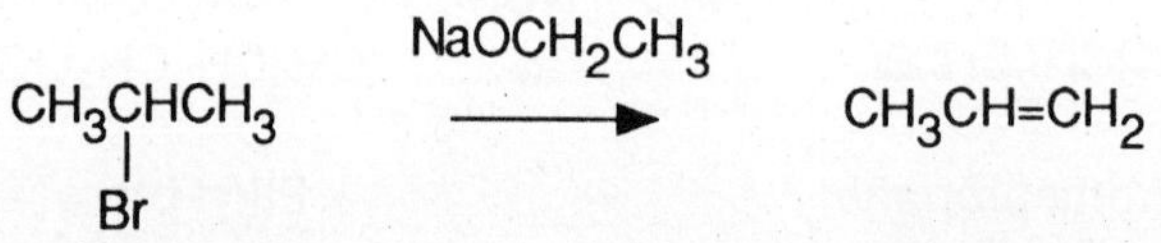

$$\underset{\overset{|}{Br}}{CH_3CHCH_3} \xrightarrow{\text{NaOCH}_2CH_3} CH_3CH=CH_2$$

2-Bromopropane   Propene

(By way of confirmation, it is known that the elimination substitution ratio is 87:13 for the reaction of isopropyl bromide with KOH in ethanol-water.)

(d) Sodium cyanide is a weak enough base and a good enough nucleophile that substitution is favored with most secondary alkyl halides.

$$\underset{\overset{|}{Br}}{CH_3CHCH_3} \xrightarrow{\text{NaCN}} \underset{\overset{|}{CN}}{CH_3CHCH_3}$$

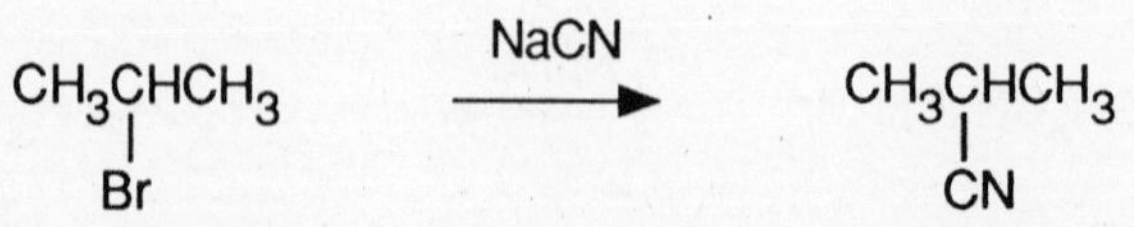

2-Bromopropane   2-Cyanopropane

(e) Sodium azide is an even weaker base than cyanide and a better nucleophile. Substitution occurs.

$$CH_3CHCH_3 \ (Br) \xrightarrow{NaN_3} CH_3CHCH_3 \ (N_3)$$

2-Bromopropane   Isopropyl azide

(f) Sodium hydrogen sulfide reacts with secondary alkyl halides by substitution.

$$CH_3CHCH_3 \ (Br) \xrightarrow{NaSH} CH_3CHCH_3 \ (SH)$$

2-Bromopropane   2-Propanethiol

(g) Substitution occurs with sodium methanethiolate.

$$CH_3CHCH_3 \ (Br) \xrightarrow{NaSCH_3} CH_3CHCH_3 \ (SCH_3)$$

2-Bromopropane   Isopropyl methyl thioether

8.12 The alkyl halide is tertiary and reacts with all anionic nucleophiles by elimination.

$$CH_3CCH_3 \ (CH_3)(Br) \xrightarrow{NaX} CH_2=C(CH_3)_2$$

2-Bromo-2-methylpropane   2-Methylpropene

8.13 (a) Reactions of primary and secondary alkyl halides with sodium iodide in acetone proceed by an S_N2 mechanism, and the product is formed with inversion of configuration.

$$\xrightarrow{NaI \ acetone}$$

(R)-2-Bromopentane   (S)-2-Iodopentane

(b) Sodium cyanide also reacts by an S_N2 mechanism, giving the product with complete inversion of configuration.

$$\xrightarrow{NaCN}$$

(R)-2-Bromopentane   (S)-2-Cyanopentane

161

(c) Solvolysis in alcohol proceeds by an S_N1 mechanism. The product will be a mixture of enantiomers.

(R)-2-Bromopentane   (S)-2-Ethoxypentane   (R)-2-Ethoxypentane

(d) The major reaction of a strong base such as sodium ethoxide and a secondary halide is elimination.

(R)-2-Bromopentane   1-Pentene   2-Pentene
(major product)

8.14 (a) The starting material incorporates both a primary chloride and a secondary chloride. The nucleophile (iodide) attacks the less hindered primary position.

ClCH$_2$CH$_2$CHCH$_2$CH$_3$ —NaI/acetone→ ICH$_2$CH$_2$CHCH$_2$CH$_3$

1,3-Dichloropentane   3-Chloro-1-iodopentane

(b) Nucleophilic substitution of the first bromide by sulfur occurs in the usual way.

$^-SCH_2CH_2S^-$ + CH$_2$—Br → $^-SCH_2CH_2SCH_2CH_2Br$

The product of this step cyclizes by way of an intramolecular nucleophilic substitution to give a six-membered ring containing two sulfur atoms.

(c)   The nucleophile is a dianion (S^{2-}). Two nucleophilic substitution reactions take place; the second of the two leads to intramolecular cyclization.

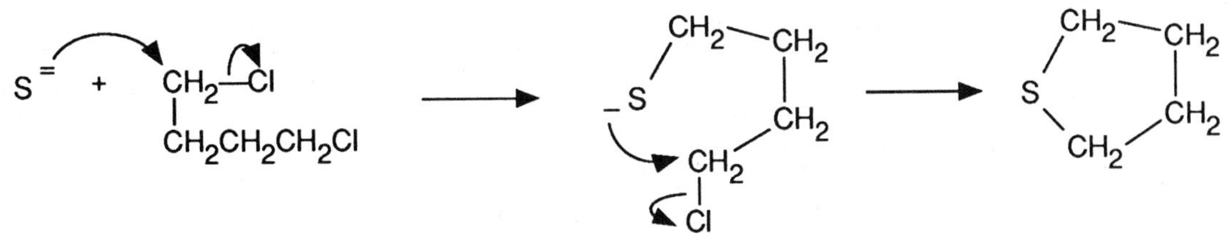

8.15   The alkyl halide is tertiary and so undergoes hydrolysis by an S_N1 mechanism. The carbocation can be captured by water at either face. A mixture of the axial and the equatorial alcohols is formed.

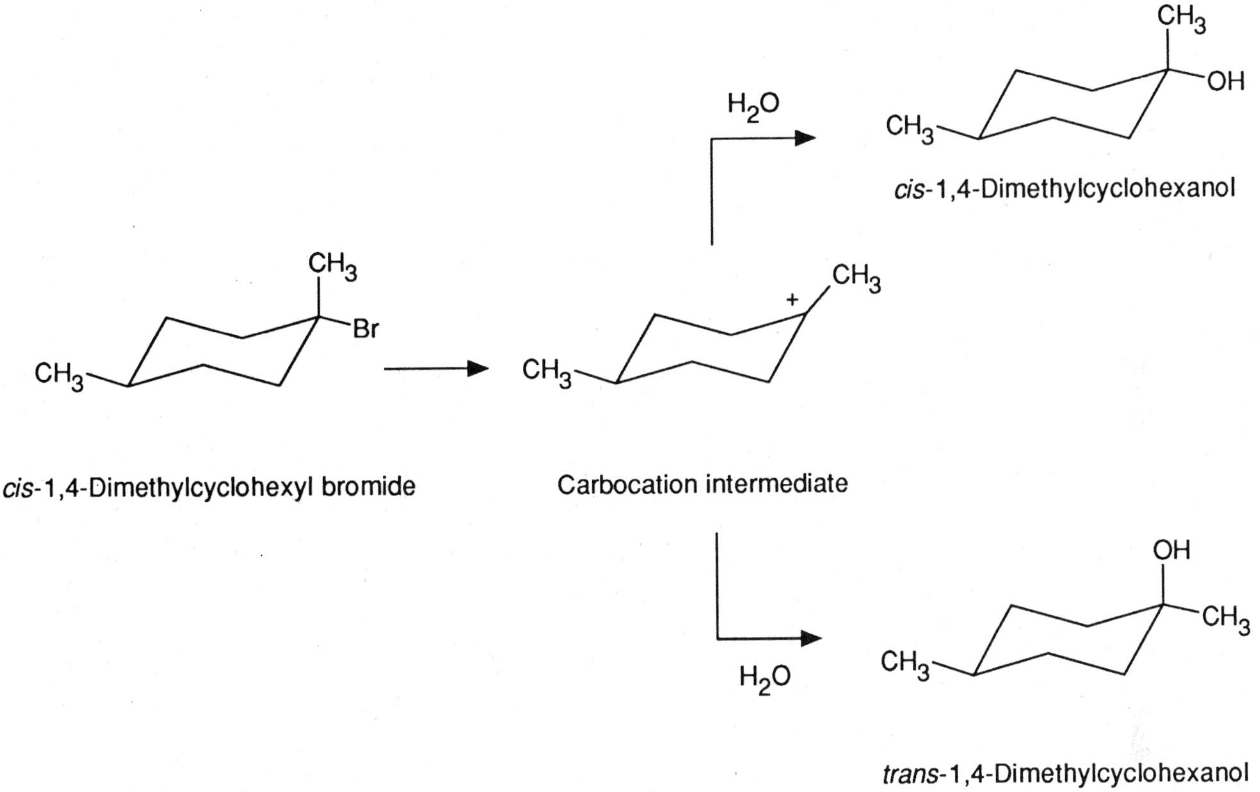

cis-1,4-Dimethylcyclohexanol

cis-1,4-Dimethylcyclohexyl bromide   Carbocation intermediate

trans-1,4-Dimethylcyclohexanol

The same carbocation is formed from *trans*-1,4-dimethylcyclohexyl bromide. Therefore, a mixture of the same two alcohols is formed as well.

8.16   (a)   The substrate is a primary alkyl bromide and reacts with sodium iodide in acetone to give the corresponding iodide.

Ethyl bromoacetate   Ethyl iodoacetate

163

(b) Primary alkyl chlorides react readily with sodium iodide in acetone to yield the corresponding iodides.

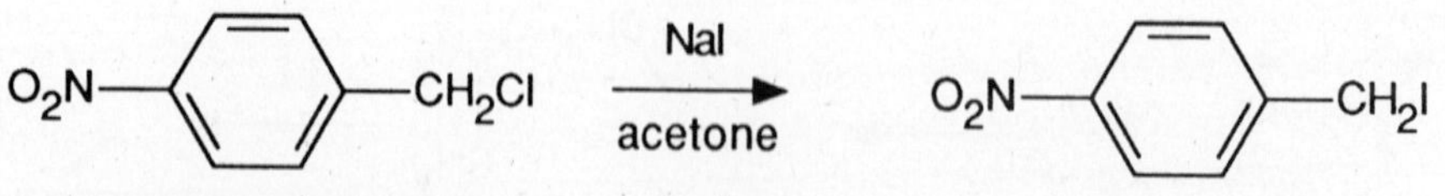

$$O_2N-\text{C}_6H_4-CH_2Cl \xrightarrow[\text{acetone}]{\text{NaI}} O_2N-\text{C}_6H_4-CH_2I$$

p-Nitrobenzyl chloride   p-Nitrobenzyl iodide

(c) An analogous reaction occurs with sodium acetate to yield an acetate ester.

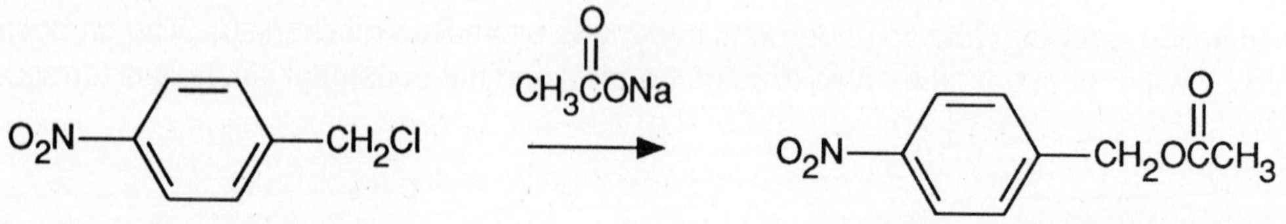

$$O_2N-\text{C}_6H_4-CH_2Cl \xrightarrow{CH_3\overset{O}{\overset{\|}{C}}ONa} O_2N-\text{C}_6H_4-CH_2O\overset{O}{\overset{\|}{C}}CH_3$$

p-Nitrobenzyl chloride   p-Nitrobenzyl acetate

(d) The only leaving group in the substrate is bromide. Neither of the carbon-oxygen bonds is susceptible to cleavage by nucleophilic attack.

$$CH_3CH_2OCH_2CH_2Br \xrightarrow{\text{NaCN}} CH_3CH_2OCH_2CH_2CN$$

2-Bromoethyl ethyl ether   2-Ethoxypropanenitrile

8.17 The isomers of C_4H_9Cl are:

$CH_3CH_2CH_2CH_2Cl$

$CH_3\underset{\overset{|}{CH_3}}{CH}CH_2Cl$

$CH_3\underset{\overset{|}{Cl}}{CH}CH_2CH_3$

$CH_3\underset{\overset{|}{CH_3}}{\overset{\overset{CH_3}{|}}{C}}Cl$

1-Chlorobutane
(n-butyl chloride)

1-Chloro-2-methylpropane
(isobutyl chloride)

2-Chlorobutane
(sec-butyl chloride)

2-Chloro-2-methylpropane
(tert-butyl chloride)

The reaction conditions (sodium iodide in acetone) are typical for an S_N2 process. The order of S_N2 reactivity is primary > secondary > tertiary, and branching of the chain close to the site of substitution hinders reaction. The unbranched primary halide n-butyl chloride will be the most reactive and the tertiary halide tert-butyl chloride the least.

Therefore the order of reactivity is:

1-chlorobutane > 1-chloro-2-methylpropane > 2-chlorobutane > 2-chloro-2-methylpropane
(primary)   (crowded primary)   (secondary)   (tertiary)

8.18 In each of these problems the reaction is specified to follow an S_N2 mechanism.

(a) Iodide is a better leaving group than bromide irrespective of whether the mechanism is S_N1 or S_N2, so **$CH_3CH_2CH_2CH_2I$** reacts faster than $CH_3CH_2CH_2CH_2Br$.

(b)   1-Bromobutane is a primary alkyl halide and 2-bromobutane is secondary. Primary alkyl halides are less hindered and thus more reactive than secondary halides toward substitution by the S_N2 mechanism. 1-Bromobutane **CH₃CH₂CH₂CH₂Br** reacts faster.

(c)   Although both alkyl chlorides are primary, 1-chloro-2,2-dimethylbutane has methyl groups on the carbon adjacent to the leaving group that will hinder attack by a nucleophile at C-1. **1-Chlorohexane** is an unbranched primary alkyl halide and reacts faster.

$$CH_3CH_2\overset{\overset{\displaystyle CH_3}{|}}{\underset{\underset{\displaystyle CH_3}{|}}{C}}CH_2Cl$$

1-Chloro-2,2,-dimethylbutane
(crowded, less reactive)

$$CH_3CH_2CH_2CH_2CH_2CH_2Cl$$

1-Chlorohexene
(unhindered, reacts faster)

8.19   In each of these problems the reaction is specified to follow an S_N1 mechanism. The rate of S_N1 solvolysis is governed by the stability of the intermediate carbocation - the more stable the cation, the faster the rate of reaction.

(a)   2-Bromobutane is a secondary alkyl halide while 1-bromo-2-methylpropane is primary. An S_N1 reaction of **2-bromobutane** will proceed via a secondary carbocation, so will be faster than an S_N1 reaction of 1-bromo-2-methylpropane which would involve an unstable primary carbocation.

$$CH_3\underset{\underset{\displaystyle CH_3}{|}}{CH}CH_2Br$$

1-Bromo-2-methylbutane
(primary; less reactive)

$$CH_3\underset{\underset{\displaystyle Br}{|}}{CH}CH_2CH_3$$

2-Bromobutane
(secondary; more reactive)

(b)   Reaction of 1-chlorocyclohexene by the S_N1 mechanism gives an unstable vinyl carbocation. Reaction of **3-chlorocyclohexene** by the S_N1 mechanism gives a more stable allylic carbocation (Section 5.7), and will proceed at the faster rate.

1-Chlorocyclohexene
(slower S_N1 reaction)

3-Chlorocyclohexene
(faster S_N1 reaction)

(c)   Reaction of chlorocyclohexane by the S_N1 mechanism gives a secondary carbocation. The secondary allylic carbocation formed by S_N1 reaction of 3-chlorocyclohexene is more stable, however, and is formed faster. **3-Chlorocyclohexene** is more reactive.

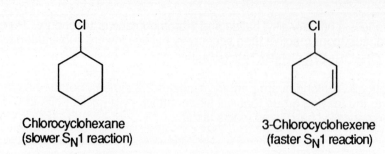

Chlorocyclohexane
(slower S_N1 reaction)

3-Chlorocyclohexene
(faster S_N1 reaction)

(d)   An aryl halide is far less reactive toward nucleophilic substitution than an alkyl halide irrespective of the mechanism. **Benzyl bromide** reacts faster than *p*-bromotoluene.

p-Bromotoluene

(an aryl halide; less reactive)

Benzyl bromide

(a benzylic halide; more reactive)

Even though the carbocation from benzyl bromide is primary, it is relatively stable because it is benzylic and stabilized by electron delocalization.

(e)   Both compounds form secondary carbocations on ionization. Because benzylic carbocations are more stable than simple alkyl cations, however, the benzylic iodide **1-iodo-1-phenylpropane** reacts faster.

2-Iodo-1-phenylpropane
(not benzylic; less reactive)

1-Iodo-1-phenylpropane
(benzylic; more reactive)

8.20   (a)   Methyl halides undergo substitution only by the S_N2 mechanism. Methyl cation $^+CH_3$ is too unstable to be an intermediate in any nucleophilic substitution reaction.

(b)   Unhindered primary halides react with good nucleophiles by S_N2 mechanism.

(c)   Secondary halides undergo elimination upon reaction with strong bases by the **E2** mechanism.

(d)   Tertiary halides form stable carbocations and undergo solvolysis by the S_N1 mechanism.

(e)   **S_N2 and E2** reactions proceed in a single step. An intermediate is not involved in either one. Both are concerted bimolecular processes.

(f)   Carbocations are intermediates in reactions proceeding by the S_N1 mechanism.

8.21   (a)   Tertiary alkyl halides undergo nucleophilic substitution only by way of carbocations: S_N1 is the most likely mechanism for solvolysis of the 2-halo-2-methylbutanes.

$$\underset{\underset{\displaystyle CH_3}{|}}{\overset{\overset{\displaystyle X}{|}}{CH_3CCH_2CH_3}}$$ 2-Halo-2-methylbutanes are tertiary alkyl halides

(b)   Tertiary alkyl halides can undergo elimination by either the E1 or E2 mechanism. Since no alkoxide base is present, elimination most likely occurs in this case by an **E1** mechanism.

(c), (d) Iodides react faster than bromides in substitution and elimination reactions irrespective of whether the mechanism is E1, E2, S_N1, or S_N2. **2-Iodo-2-methylbutane** reacts faster.

(e)   Solvolysis in aqueous ethanol can give rise to an alcohol or an ether as product, depending on whether the carbocation is captured by water or ethanol.

$$\underset{\underset{\displaystyle CH_3}{|}}{\overset{\overset{\displaystyle X}{|}}{CH_3CCH_2CH_3}} \quad \xrightarrow[H_2O]{CH_3CH_2OH} \quad \underset{\underset{\displaystyle CH_3}{|}}{\overset{\overset{\displaystyle OH}{|}}{CH_3CCH_2CH_3}} \quad + \quad \underset{\underset{\displaystyle CH_3}{|}}{\overset{\overset{\displaystyle OCH_2CH_3}{|}}{CH_3CCH_2CH_3}}$$

2-Methyl-2-pentanol   Ethyl 1,1-dimethylpropyl ether

(f)   Elimination can yield either of the two isomeric alkenes.

$$\underset{\underset{\displaystyle CH_3}{|}}{\overset{\overset{\displaystyle X}{|}}{CH_3CCH_2CH_3}} \quad \longrightarrow \quad (CH_3)_2C=CHCH_3 \quad + \quad \underset{\underset{\displaystyle CH_3}{|}}{CH_2=CCH_2CH_3}$$

2-Methyl-2-butene   2-Methyl-1-butene

Zaitsev's rule predicts that 2-methyl-2-butene should be the major alkene.

(g)   The product distribution is determined by what happens to the carbocation intermediate. If the carbocation is free of its leaving group, its fate will be the same no matter whether the leaving group is bromide or iodide.

8.22   The reaction to be considered is the formation of an ether by S_N2 displacement by a potassium alkoxide ion on an alkyl bromide. Therefore, choose the less crowded alkyl bromide so as to maximize substitution over elimination. A crowded alkyl bromide is more prone to undergo elimination rather than substitution.

(a)   *tert*- Butyl methyl ether is best prepared by reaction of methyl bromide with potassium *tert*-butoxide.

$$CH_3Br \quad + \quad (CH_3)_3CO^- \, K^+ \quad \longrightarrow \quad CH_3OC(CH_3)_3$$

Methyl bromide   Potassium *tert*-butoxide   *tert*-Butyl methyl ether

The reaction of potassium methoxide with *tert*-butyl bromide would have given an alkene by E2 elimination of the the alkyl halide rather than substitution.

(b)   Again, the best alternative is to choose the less hindered alkyl halide to promote substitution as the major reaction.

CH_3Br +   [cyclopentyl]$-O^-K^+$ $\longrightarrow$ [cyclopentyl]$-OCH_3$

Methyl bromide   Potassium cyclopentoxide   Cyclopentyl methyl ether

(c)   The bond to oxygen does not involve a chiral center, so the configuration at the chiral center remains unchanged during the conversion of the alkyl bromide to the ether.

$(R)-CH_3CH_2\overset{*}{C}HCH_2Br$ + $(CH_3)_2CHO^-K^+$ $\longrightarrow$ $(R)-CH_3CH_2\overset{*}{C}HCH_2OCH(CH_3)_2$
$\quad\quad\quad\;\; |$ $\quad\quad\quad\quad\;\; |$
$\quad\quad\quad\;\; CH_3$ $\quad\quad\quad\quad\;\; CH_3$

(R)-1-Bromo-2-methylbutane   Potassium isopropoxide   (R)-1-Isopropoxy-2-methylbutane

(d)   The least hindered candidate for the alkyl bromide is bromoethane.

CH_3CH_2Br + $(CH_3)_3CCH_2O^-K^+$ $\longrightarrow$ $(CH_3)_3CCH_2OCH_2CH_3$

Bromoethane   Potassium 2,2-dimethyl-1-propoxide   1-Ethoxy-2,2-dimethylpropane

8.23   The wrong choice of alkyl bromide and potassium alkoxide in parts (a)-(c) of the preceding problem would ave produced an alkene by an elimination reaction instead of the desired ether.

(a)   Reaction of an alkoxide ion with a secondary halide leads for E2 elimination as the favored reaction pathway.

$CH_3O^-K^+$ + $(CH_3)_3CBr$ $\longrightarrow$ CH_3OH + $CH_2=C(CH_3)_2$

Potassium methoxide   tert-Butyl bromide   Methanol   2-Methylpropene

(b)

$CH_3O^-K^+$ +   [cyclopentyl]$-Br$ $\longrightarrow$ CH_3OH +   [cyclopentene]

Potassium methoxide   Cyclopentyl bromide   Methanol   Cyclopentene

(c)

$(R)-CH_3CH_2\overset{*}{C}HCHO^-K^+$ + $(CH_3)_2CHBr$ $\longrightarrow$ $(R)-CH_3CH_2\overset{*}{C}HCH_2OH$ + $CH_3CH=CH_2$
$\quad\quad\quad\quad\;\; |$ $\quad\quad\quad\quad\;\; |$
$\quad\quad\quad\quad\;\; CH_3$ $\quad\quad\quad\quad\;\; CH_3$

Potassium (R)-2-methyl-1-butoxide   Isopropyl bromide   (R)-2-Methyl-1-butanol   Propene

168

8.24 (a)   Thiols cannot be prepared directly from alcohols, but alcohols can be converted to alkyl halides which can, in turn, be converted to thiols by nucleophilic substitution with the hydrogen sulfide ion, HS⁻.

$$CH_3CH_2OH \xrightarrow[\text{or } PBr_3]{\text{HBr, heat}} CH_3CH_2Br \xrightarrow{\text{NaSH}} CH_3CH_2SH$$

Ethanol   Ethyl bromide   Ethanethiol

(b)   Cyclopentyl cyanide can be prepared from a cyclopentyl halide by a nucleophilic substitution reaction. The first task, therefore, is to convert cyclopentane to a cyclopentyl halide.

Cyclopentane   Chlorine   Cyclopentyl chloride   Hydrogen chloride

Once prepared, cyclopentyl chloride is then treated with a source of cyanide ion.

Cyclopentyl chloride   Sodium cyanide   Cyclopentyl cyanide   Sodium chloride

(c)   Cyclopentene can serve as a precursor to a cyclopentyl halide. The equation shows the addition of hydrogen bromide. Alternatively, the chloride or iodide could be prepared by an analogous reaction of cyclopentene with HCl or with HI.

Cyclopentene   Hydrogen Bromide   Cyclopentyl bromide

Once cyclopentyl bromide has been prepared, it is converted to cyclopentyl cyanide by nucleophilic substitution, as shown in (b).

(d)   Dehydration of cyclopentanol yields cyclopentene, which can then be converted to cyclopentyl cyanide, as shown in (c).

Cyclopentanol   Cyclopentene

An equally correct approach would be to convert cyclopentanol to cyclopentyl bromide (HBr, heat), then treat cyclopentyl bromide with sodium cyanide (NaCN).

(e) Two cyano groups are required here, both of which must be introduced in nucleophilic substitution reactions. The substrate in the key reaction is $BrCH_2CH_2Br$ which is prepared from ethylene. The overall synthesis from ethyl alcohol is formulated as shown:

$$CH_3CH_2OH \xrightarrow[\text{heat}]{H_2SO_4} CH_2=CH_2 \xrightarrow{Br_2} BrCH_2CH_2Br \xrightarrow{NaCN} NCCH_2CH_2CN$$

Ethanol   Ethylene   1,2-Dibromoethane   1,2-Dicyanoethane

(f) In this synthesis a primary alkyl chloride must be converted to a primary alkyl iodide. This is precisely the kind of transformation for which sodium iodide in acetone is used.

$$(CH_3)_2CHCH_2Cl \xrightarrow[\text{acetone}]{NaI} (CH_3)_2CHCH_2I$$

Isobutyl chloride   Isobutyl iodide

8.25 The conjugate bases of acetylene and terminal alkynes are good nucleophiles, and are formed by reaction of the appropriate alkyne with a strong base such as sodium amide ($NaNH_2$). The alkynide ions react with methyl and primary alkyl halides in alkylation reactions.

(a) 1-Bromobutane is the desired alkyl halide to prepare 1-hexyne by reaction with sodium acetylide.

$$HC{\equiv}CH \xrightarrow{NaNH_2} HC{\equiv}CNa \xrightarrow{CH_3CH_2CH_2CH_2Br} HC{\equiv}CCH_2CH_2CH_2CH_3$$

Acetylene   Sodium acetylide   1-Hexyne

(b) Two successive alkylations must be carried out to prepare 2-hexyne from acetylene. The order in which the alkylations are carried out is not important. In the solution shown below, the methyl group is attached first followed by the propyl group. The reverse order is also acceptable (add the propyl group first, then the methyl group).

$$HC{\equiv}CH \xrightarrow{NaNH_2} HC{\equiv}CNa \xrightarrow{CH_3Br} HC{\equiv}CCH_3$$

Acetylene   Sodium acetylide   Propyne

$$HC{\equiv}CCH_3 \xrightarrow{NaNH_2} NaC{\equiv}CCH_3 \xrightarrow{CH_3CH_2CH_2Br} CH_3CH_2CH_2C{\equiv}CCH_3$$

Propyne   Sodium propynide   2-Hexyne

(c) The preparation of 3-hexyne is similar to the scheme used for the preparation of 2-hexyne. Two separate alkylations with ethyl bromide are required.

$$HC{\equiv}CH \xrightarrow{NaNH_2} HC{\equiv}CNa \xrightarrow{CH_3CH_2Br} HC{\equiv}CCH_2CH_3$$

Acetylene   Sodium acetylide   1-Butyne

$$HC{\equiv}CCH_2CH_3 \xrightarrow{\text{NaNH}_2} NaC{\equiv}CCH_2CH_3 \xrightarrow{\text{CH}_3\text{CH}_2\text{Br}} CH_3CH_2C{\equiv}CCH_2CH_3$$

1-Butyne   Sodium 1-butynide   3-Hexyne

(d)   Alkylation of the acetylide ion (as sodium acetylide) with 2-methyl-1-bromopropane gives rise to 4-methyl-1-pentyne.

$$HC{\equiv}CH \xrightarrow{\text{NaNH}_2} HC{\equiv}CNa \xrightarrow{(CH_3)_2CHCH_2Br} HC{\equiv}CCH_2CH(CH_3)_2$$

Acetylene   Sodium acetylide   4-Methyl-1-pentyne

CHAPTER 9

SPECTROSCOPY

GLOSSARY OF TERMS

Base peak   The most intense peak in a mass spectrum. The base peak is assigned a relative intensity of 100 and the intensities of all other peaks are cited as a percentage of the base peak.

Bending vibration   The regular, repetitive motion of an atom or group along an arc the radius of which is the bond connecting the atom or group to the rest of the molecule. Bending vibrations are one type of molecular motion that gives rise to a peak in the infrared spectrum.

^{13}C NMR   Nuclear magnetic resonance spectroscopy in which the environments of individual carbon atoms are examined via their mass-13 isotope.

Chemical shift   A measure of how shielded the nucleus of a particular atom is. Nuclei of different atoms have different chemical shifts, and nuclei of the same atom have chemical shifts which are sensitive to their molecular environment. In proton and carbon-13 nmr, chemical shifts are cited as δ, or parts per million (ppm) ,from the hydrogens or carbons, respectively, of tetramethylsilane.

Chromophore   The structural unit of a molecule principally responsible for absorption of radiation of a particular frequency. A term usually applied to ultraviolet-visible spectroscopy.

Coupling   In nmr, the interaction of two nuclei so as to communicate nuclear spin information to one another. The observable manifestation of coupling is spin-spin splitting.

Downfield   The low-field region of an nmr spectrum. A signal which is downfield with respect to another lies to its left on the spectrum.

Electromagnetic radiation   Various forms of radiation propagated at the speed of light. Electromagnetic radiation includes (among others) visible light, infrared, ultraviolet, and microwave radiation, radio waves, cosmic rays, and X-rays.

Electron impact   A method for producing positive ions in mass spectrometry whereby a molecule is bombarded by high-energy electrons.

Elements of unsaturation   (*see SODAR*)

Extinction coefficient   (*see Molar absorptivity*)

Fingerprint region   The region 1400-625 cm^{-1} of an infrared spectrum. This region is less characteristic of functional groups than others, but varies so much from one molecule to another that it can be used to determine whether two substances are identical or not.

Fragmentation pattern   Dissociation of the molecular ion produces a variety of smaller ions referred to as its fragmentation pattern.

Frequency   The number of waves per unit time. The symbol for frequency is s^{-1}.

Index of hydrogen deficiency   (*see SODAR*)

Infrared (IR) spectroscopy An analytical technique based on energy absorbed by a molecule as it vibrates by stretching and bending bonds. Infrared spectroscopy is useful for analyzing the functional groups in a molecule.

Integrated area The relative area of a signal in an nmr spectrum. Areas are proportional to the number of equivalent protons responsible for the peak.

Magnetic resonance imaging (MRI) A diagnostic method in medicine in which tissues are examined by nmr.

Mass spectrometry An analytical method in which a molecule is ionized and the various ions examined on the basis of their mass-to-charge ratio.

Molar absorptivity A measure of the intensity of a peak, usually in UV-VIS spectroscopy.

Molecular Ion The species produced by loss of an electron from a molecule in mass spectrometry.

Multiplicity The number of peaks into which a signal is split in nuclear magnetic resonance spectroscopy. Signals are described as *singlets, doublets, triplets, etc.* according to the number of peaks into which they are split.

Nuclear magnetic resonance (NMR) spectroscopy A method for structure determination based on the effect of molecular environment on the energy to promote a given nucleus from a lower-energy spin state to a higher one.

Planck's constant A constant of proportionality (h) which relates the energy (E) and the frequency (v) of electromagnetic radiation.

$$E = hv$$

Shielding The effect of a molecule's electrons that decreases the strength of an external magnetic field felt by a proton or other nucleus.

SODAR Acronym for "sum of double bonds and rings." Examination of the molecular formula can provide information about a substance if one re-

members that each double bond or ring in a molecule has causes its molecular formula to contain two fewer hydrogens than the corresponding alkane.

$$SODAR = \frac{1}{2}[C_nH_{2n+2} - C_nH_x]$$

Spectrometer A device designed to measure absorption of electromagnetic radiation by a sample.

Spectrum The output, usually in chart form, of a spectrometer. Analysis of a spectrum provides information about molecular structure.

Spin-spin splitting The splitting of nmr signals caused by the coupling of nuclear spins. Only non-equivalent nuclei (such as protons with different chemical shifts) can split one another.

Stretching vibration A regular, repetitive motion of two atoms or groups along the bond that connects them.

Tetramethylsilane (TMS) The molecule $(CH_3)_4Si$ used as a standard to calibrate proton and carbon-13 nmr spectra.

Ultraviolet-visible (UV-VIS) spectroscopy An analytical method based on transitions between electronic energy states in molecules. Useful in studying conjugated systems such as polyenes.

Upfield The high-field region of an nmr spectrum. A signal which is upfield with respect to another lies to its right on the spectrum.

Wave numbers Conventional units in infrared spectroscopy that are proportional to frequency. Wavenumbers are reciprocal centimeters (cm^{-1}).

Wavelength The distance between two successive maxima (peaks) or two successive minima (troughs) of a wave.

SOLUTIONS TO TEXT PROBLEMS

9.1 (b) Frequency and wavelength are inversely proportional. Red light has a longer wavelength than violet light, and has a lower frequency. The frequency may be calculated as in (a):

$$v = \frac{c}{\lambda}$$

Substituting the value for the wavelength of red light,

$$\nu = \frac{3 \times 10^8 \text{ m/s}}{4 \times 10^{-7} \text{ m}}$$

$$\nu = 3.8 \times 10^{14} \text{ s}^{-1}$$

The frequency of red light is indeed less than the frequency of violet light (7.5×10^{14} s^{-1}).

9.2   The frequency and energy of electromagnetic radiation are directly proportional. A photon of the higher frequency radiation (violet light) has the higher energy.

9.3   The infrared spectrum of Figure 9.6 is void of absorption in the 1600 to 1800 cm^{-1} region, so the unknown compound cannot contain a carbonyl (C=O) group. Therefore, it cannot be acetophenone or benzoic acid. There is a broad, intense absorption at 3300 cm^{-1} attributable to a hydroxyl (OH) group. The infrared spectrum is that of benzyl alcohol.

9.4   The same equation is employed to calculate the chemical shifts as was used in the sample solution to (a).

(b)   Iodoform (CHI$_3$), 322 Hz at 60 MHz

$$\delta = \frac{322 \text{ Hz}}{60 \times 10^6 \text{ Hz}} \times 10^6 = 5.37 \text{ ppm}$$

(c)   Methyl chloride (CH$_3$Cl), 184 Hz at 60 MHz

$$\delta = \frac{184 \text{ Hz}}{60 \times 10^6 \text{ Hz}} \times 10^6 = 3.07 \text{ ppm}$$

9.5   (b)   Diethyl ether has two sets or "types" of protons: the protons of the methyl groups and the protons of the methylene groups. The two ethyl groups are equivalent, and so there are two sets of protons, not four. The nmr spectrum of diethyl ether has two signals.

(c)   All three methyl groups of *tert*-butyl bromide are equivalent, and therefore all nine protons appear as one signal.

(d)   The two methyl groups of 2-chloropropane are equivalent. The nmr spectrum of 2-chloropropane has two signals, one due to the methyl protons and one from the methine (CH) proton.

9.6   (b)   The signal for the methyl protons would be found between δ = 0.9 and 1.8 ppm. The methylene protons, because of the adjacent oxygen atom, would be expected to absorb in the region δ = 3.3 to 3.7 ppm. The actual chemical shifts for diethyl ether are δ = 1.2 ppm (CH$_3$) and 3.5 ppm (CH$_2$).

(c)   The signal for *tert*-butyl bromide would be found in the region δ = 0.9 to 1.8 ppm. The actual chemical shift is δ = 1.8 ppm.

(d)   The methyl protons of 2-chloropropane are expected to absorb between δ = 0.9 and 1.8 ppm. The methine hydrogen, due to the presence of the adjacent chlorine atom, is expected to absorb in the region δ = 1.8 ppm (CH$_3$) and 3.7 ppm (CH).

9.7   (b)   The relative intensities of the nmr signals for diethyl ether are 3:3 for the CH$_3$ and CH$_2$ protons, respectively. The ratio of signal intensities is a lowest-whole-number ratio. In the case of diethyl ether the 3:2 ratio represents 6 and 4 protons, respectively.

(c)   There is only one signal in the nmr spectrum of *tert*-butyl bromide, so relative intensity has no meaning.

(d)   The relative intensities of the protons in 2-chloropropane are 6:1 for the CH_3 and CH protons, respectively.

9.8   (b)   The three methyl protons of 1,1,1-trichloroethane (Cl_3CCH_3) are equivalent. They have the same chemical shift and do not split each other. The ^1H nmr spectrum of Cl_3CCH_3 consists of a single sharp peak.

(c)   Separate signals will be seen for the methylene protons (2H) and for the methine proton (1H) of 1,1,2-trichloroethane.

$Cl_2C–CH_2Cl$ 1,1,2-Trichloroethane

triplet   doublet

The methine proton splits the signal for the methylene protons into a doublet. The two methylene protons split the methine proton's signal into a triplet.

(d)   Examine the structure of 1,1,2-trichloropropane.

$ClCH_2CCH_3$ 1,1,2-Trichloropropane

The ^1H nmr spectrum exhibits a signal for the two equivalent methylene protons and one for the three equivalent methyl protons. Both these signals are sharp singlets. The protons of the methyl group and the methylene group are separated by more than three bonds and do not split each other.

(e)   The methine proton of 1,1,1,2-tetrachloropropane splits the signal of the methyl protons into a doublet; it is split into a quartet by the three methyl protons.

$Cl_3CC–CH_3$ 1,1,1,2-Tetrachloropropane

quartet   doublet

9.9   (b)   The two ethyl groups of diethyl ether are equivalent to each other. The two methyl groups appear as a single triplet and the two methylene groups as a single quartet.

triplet   triplet
$CH_3CH_2OCH_2CH_3$
quartet   quartet

(c)   As all the hydrogens of *tert*-butyl bromide are equivalent, the nmr spectrum appears as a sharp singlet.

(d)   The signal for the protons of the two methyl groups of 2-chloropropane appear as a single doublet. The methine hydrogen is split by six hydrogens, and appears as a heptet (seven lines).

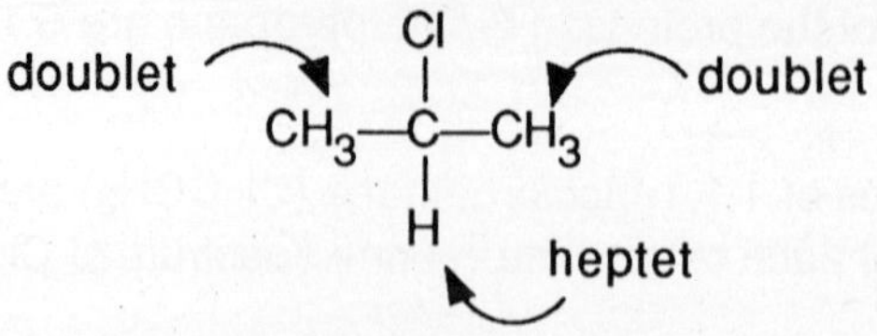
doublet
doublet
heptet
Cl
CH_3—C—CH_3
H

9.10 The ^{1}H nmr spectrum shown in the figure is that of $Cl_2CHCH(OCH_2CH_3)_2$. All three compounds contain two equivalent ethyl groups, but only $Cl_2CHCH(OCH_2CH_3)_2$ has its remaining two protons in different environments. The two methine protons split each other's signal into a doublet.

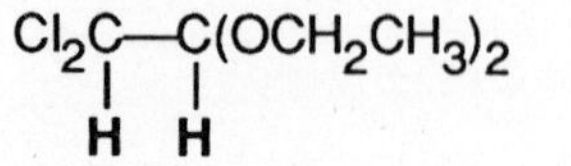
Cl_2C—$C(OCH_2CH_3)_2$
H H

nonequivalent;
each is a doublet

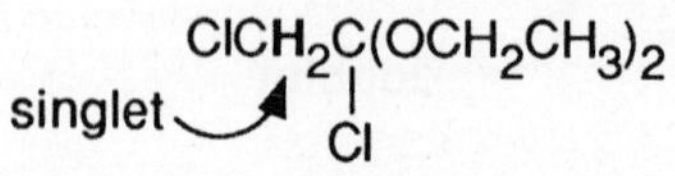
$ClCH_2C(OCH_2CH_3)_2$
singlet
Cl

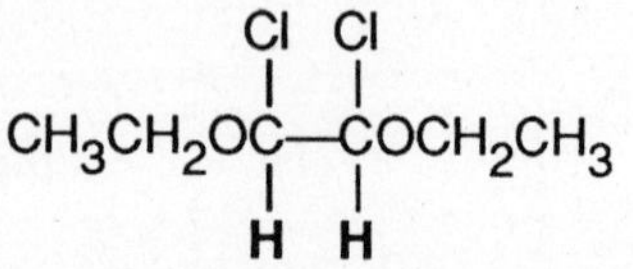
Cl Cl
CH_3CH_2OC—$COCH_2CH_3$
H H

equivalent;
therefore a singlet

9.11 (b) The two methyl carbons of the isopropyl group are equivalent. There are four different types of carbons in the aromatic ring and two different types in the isopropyl group. The ^{13}C nmr spectrum of isopropylbenzene contains six signals.

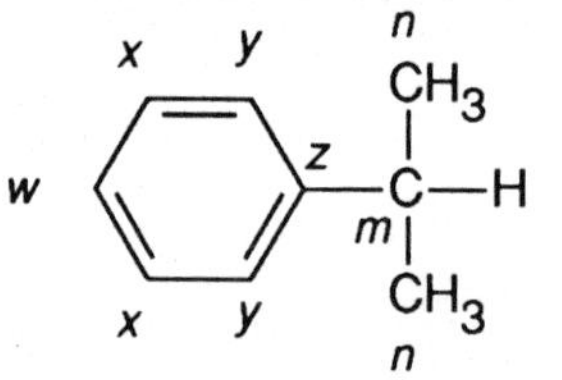
x
y
w
z
x
y
n
CH_3
C—H
m
CH_3
n

(c) All the methyl carbons of 1,3,5-trimethylbenzene are equivalent. Because of its high symmetry 1,3,5-trimethylbenzene has only three signals in its ^{13}C nmr spectrum.

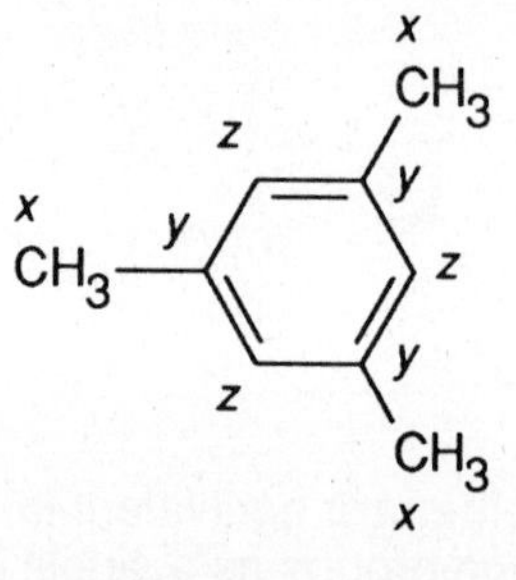
x
CH_3
z
y
x
y
CH_3
z
y
z
CH_3
x

9.12 The base peak in the mass spectrum of alkylbenzenes corresponds to carbon-carbon bond cleavage at the benzylic carbon.

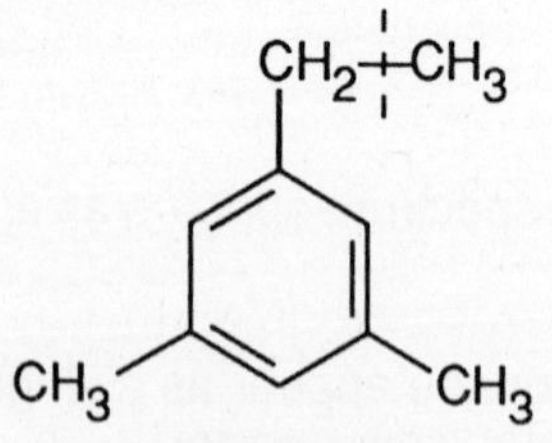
CH_2 | CH_3
CH_3
CH_3

Base peak: $C_9H_{11}^+$

m/z 119

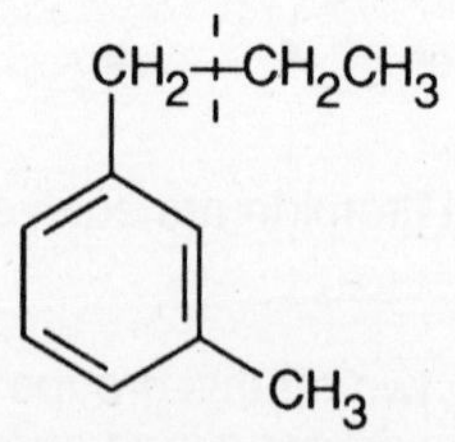
CH_2 | CH_2CH_3
CH_3

Base peak: $C_8H_9^+$

m/z 105

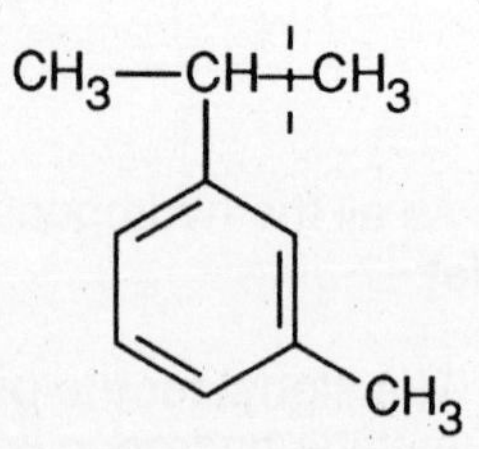
CH_3—CH | CH_3
CH_3

Base peak: $C_9H_{11}^+$

m/z 119

9.13 (b) The sum of double bonds and rings (SODAR) is given by the formula:

$$SODAR = 1/2 \, (C_nH_{2n+2} - C_nH_x)$$

The compound given contains eight carbons (C_8H_8). Therefore:

$$SODAR = 1/2 \, (C_8H_{18} - C_8H_8)$$
$$SODAR = 5$$

The problem specifies that the compound consumes two moles of hydrogen, so it must contains two double bonds (or one triple bond). Since the SODAR is equal to 5, there must be three rings.

(c) Chlorine substituents are equivalent to hydrogens when calculating the SODAR. Therefore, consider $C_8H_8Cl_2$ as equivalent to C_8H_{10}. Thus, the SODAR of this compound is four.

$$SODAR = 1/2 \, (C_8H_{18} - C_8H_{10})$$
$$SODAR = 4$$

If the compound consumes two moles of hydrogen on catalytic hydrogenation, it must therefore contain two rings.

(d) Oxygen atoms are ignored when calculating the SODAR. Thus, C_8H_8O is treated as if it were C_8H_8.

$$SODAR = 1/2 \, (C_8H_{18} - C_8H_8)$$
$$SODAR = 5$$

Since the problem specifies that two moles of hydrogen are consumed on catalytic hydrogenation, this compound contains three rings.

(e) Ignoring the oxygen atoms in $C_8H_{10}O_2$, we treat this compound as if it were C_8H_{10}.

$$SODAR = 1/2 \, (C_8H_{18} - C_8H_{10})$$
$$SODAR = 4$$

If it consumes two moles of hydrogen on catalytic hydrogenation, there must be two rings.

(f) Ignore the oxygen and treat the chlorine as if it were hydrogen. Thus, C_8H_9ClO is treated as if it were C_8H_{10}. Its SODAR is 4, and it contains two rings.

9.14 (a) The wavelength of electromagnetic radiation is given by

$$\lambda = \frac{c}{\nu}$$

Substituting the frequency of microwave radiation gives

$$\lambda = \frac{3 \times 10^8 \text{ m s}^{-1}}{1.5 \times 10^{10} \text{ s}^{-1}} = 2 \times 10^{-2} \text{ m} = 2.0 \text{ cm}$$

(b) Rearranging the equation in part (a)

$$c = \nu \times \lambda$$

Substituting the frequency and wavelength of high C

$$\text{speed of sound} = (1024 \text{ s}^{-1}) \times (0.324 \text{ m}) = 332 \text{ m/s}$$

9.15 From the data in Table 9.1 the broad band between 3000 and 3500 cm^{-1} is best attributed to a hydroxyl (OH) group. The strong peak at 1710 cm^{-1} is explained by the presence of a carbonyl (C=O) group. Of the three compounds given, the best choice to explain the spectral data is the phenylacetic acid as the carboxyl group contains both a carbonyl group and a hydroxyl group.

Phenylacetic acid

9.16 Conjugated π systems absorb at longer wavelength in the ultraviolet region than isolated double bonds. The conjugated diene 2,4-pentadiene absorbs at the longest wavelength.

9.17 (a) The chemical shift in ppm is calculated from the shift in Hz and the spectrometer frequency by

$$\delta = \frac{366\ Hz}{60 \times 10^6\ Hz} = 6.1\ ppm$$

(b) Rearranging the equation gives the chemical shift in Hz.

$$chemical\ shift = \frac{4.35\ Hz}{1 \times 10^6\ Hz}(220 \times 10^6\ Hz) = 957\ Hz$$

(c) The spectrometer frequency is found by

$$frequency = \frac{210\ Hz}{3.50\ ppm} = 60 \times 10^6\ Hz = 60\ MHz$$

9.18 (a) All the protons on a single carbon atom of $CH_3CH_2CH_2CH_2Br$ are equivalent to one another, but are not equivalent to any of the protons on a different carbon. There will be separate signals for the protons of C-1, C-2, C-3, and C-4. The ^1H nmr spectrum of 1-bromobutane has four signals.

(b) All the carbons in 2,2-dibromobutane are different from each other, so protons attached to one carbon are not equivalent to the protons attached to any of the other carbons. This compound should have three signals in its ^1H nmr spectrum.

2,2-Dibromobutane

(c) 1,4-Dibromobutane has two different types of protons. Those marked in bold face in BrCH$_2$C*H$_2$C*H$_2$CH$_2$Br comprise one set of four equivalent protons. Those marked in italics comprise a second set of protons which are equivalent to one another, but distinct from those marked in bold. The ^1H nmr spectrum exhibits two signals.

(d) Each carbon is unique in Br$_2$CHCH$_2$CH$_2$CH$_3$ so there are four nonequivalent sets of protons in 1,1,4-tribromobutane. It will exhibit four signals in its ^1H nmr spectrum.

(e) The seven protons of 1,1,1-tribromobutane (Br$_3$CCH$_2$CH$_2$CH$_3$) belong to three nonequivalent sets, and hence the ^1H nmr spectrum will consist of three signals.

9.19 (a) The ethyl group appears as a triplet-quartet pattern and the methyl group as a singlet.

$$CH_3CH_2OCH_3$$

triplet ——⟋ ↑ ⟍—— singlet; not vicinal to any
             quartet   other protons in molecule

(b) The two ethyl groups of diethyl ether are equivalent to each other. The two methyl groups appear as a single triplet and the two methylene groups as a single quartet.

$$CH_3CH_2OCH_2CH_3$$

triplet ——⟋ ↑ ↑ ⟍—— triplet
       quartet   quartet

(c) The ^1H nmr spectrum should have three signals. The two methylene groups are not chemically equivalent, so each one would be expected to appear as a triplet. However, the chemical shifts of the two CH_2 groups happen to coincide, and so the signals appear as a singlet in the actual ^1H nmr spectrum. Protons having the same chemical shift do not split each other. The methyl group is not adjacent (vicinal) to a carbon bearing any hydrogens, and so it appears as a singlet.

$$ClCH_2CH_2OCH_3$$

triplet ——⟋ ↑ ⟍—— singlet
         triplet

(d) This molecule has two signals in its ^1H nmr spectrum. Both signals are singlets as the hydrogens are not on adjacent carbons.

$$ClCH_2{-}\overset{\overset{\displaystyle Cl}{|}}{\underset{\underset{\displaystyle CH_3}{|}}{C}}{-}CH_3$$

singlet ——⟋   ⟍—— singlet

(e) The ^1H nmr spectrum of 1,3-dichloropropane has two signals. The hydrogens on carbons 1 and 3 (shown in bold in $ClCH_2CH_2CH_2Cl$) are equivalent, and give one signal which is split into a triplet by the two protons at C-2 (shown in italics). The signal for the protons at C-2 appears as a quintet (five peaks) because it is split by the four equivalent protons at C-1 and C-3.

(f) There are four sets of nonequivalent hydrogens in this molecule, and thus there are four signals in its ^1H nmr spectrum. The splitting of each set of hydrogens is as shown.

$$Cl_2CHCH_2\overset{\overset{\displaystyle O}{\|}}{C}CH_2CH_3$$

triplet ——⟋ ↑ ↑ ⟍—— triplet
       doublet quartet

(g) The ^1H nmr spectrum of this $CH_3OCH_2CH_2OCH_3$ has two signals, and both are singlets. The CH_2 groups do not split each other. These four protons are equivalent and equivalent protons do not split one another. There is not splitting of the CH_3 groups by the CH_2 groups and vice versa because the protons involved are not connected to adjacent carbons.

9.20 The first column to the right of each structure lists the approximate chemical shift predicted from Table 9.3 for each set of hydrogens. The second column lists the actual chemical shift.

(a)	$CH_3CH_2OCH_3$	(a)	0.9-1.8	1.2
		(b)	3.3-3.7	3.5
		(c)	3.3-3.7	3.2
(b)	$CH_3CH_2OCH_2CH_3$	(a)	0.9-1.8	1.2
		(b)	3.3-3.7	3.5
(c)	$ClCH_2CH_2OCH_3$	(a)	3.1-4.1	3.6
		(b)	3.3-3.7	3.6
		(c)	3.3-3.7	3.4
(d)	$ClCH_2C(CH_3)_2$	(a)	3.1-4.1	(not available)
	Cl	(b)	0.9-1.8	(not available)
(e)	$ClCH_2CH_2CH_2Cl$	(a)	3.1-4.1	3.7
		(b)	0.9-1.8	2.25
(f)	$Cl_2CHCH_2CCH_2CH_3$	(a)	3.1-4.1	(not available)
		(b)	2.1-2.5	(not available)
		(c)	2.1-2.5	(not available)
		(d)	0.9-1.8	(not available)
(g)	$CH_3OCH_2CH_2OCH_3$	(a)	3.3-3.7	(not available)
		(b)	3.3-3.7	(not available)

9.21 From Table 9.1 it can be seen that infrared absorption at 1710 cm^{-1} is characteristic of a carbonyl group. Only the compound in (f), **1,1-dichloro-3-pentanone**, contains a carbonyl group.

9.22 Since each compound exhibits only a single peak in its ^1H nmr spectrum, all the hydrogen substituents are equivalent in each one. Structures are assigned on the basis of their molecular formulas and chemical shifts.

(a) This compound has the molecular formula C_8H_{18} and so must be an alkane. The 18 hydrogens are contributed by six equivalent methyl groups.

$(CH_3)_3CC(CH_3)_3$ 　　　　2,2,3,3-Tetramethylbutane (δ = 0.9 ppm)

(b) A hydrocarbon with the molecular formula C_5H_{10} has a SODAR of 1 and so is either a cycloalkane or an alkene. Since all 10 hydrogens are equivalent, this compound must be cyclopentane.

Cyclopentane 　　(δ = 1.5 ppm)

(c)   The chemical shift of the eight equivalent hydrogens in C_8H_8 is $\delta = 5.8$ ppm, which is consistent with protons attached to a carbon-carbon double bond.

1,3,5,7-Cyclooctatetraene   ($\delta = 5.8$ ppm)

(d)   The compound C_4H_9Br has no rings or double bonds. The nine hydrogens belong to three equivalent methyl groups.

$(CH_3)_3CBr$ *tert*–Butyl bromide   ($\delta = 1.8$ ppm)

9.23   (a)   A five-proton signal at $\delta = 7.1$ ppm indicates a monosubstituted aromatic ring. With a SODAR of 4, C_8H_{10} contains this monosubstituted aromatic ring and no other rings or multiple bonds. The triplet-quartet pattern at high field suggests an ethyl group.

—CH_2CH_3 Ethylbenzene

quartet   triplet
($\delta = 2.6$ ppm)   ($\delta = 1.2$ ppm)

(b)   The SODAR of 4 and the five-proton multiplet at $\delta = 7.0$ to 7.5 ppm are accommodated by a monosubstituted aromatic ring. The remaining four carbons and nine hydrogens are most reasonably a *tert*-butyl group since all nine hydrogens are equivalent.

—$C(CH_3)_3$ *tert*-Butylbenzene

singlet ($\delta = 1.3$ ppm)

(c)   Its molecular formula requires that C_6H_{14} be an alkane. The doublet-heptet pattern is consistent with an isopropyl group, and the total number of protons requires that two of these groups be present.

$(CH_3)_2CHCH(CH_3)_2$ 2,3-Dimethylbutane

doublet   doublet
($\delta = 0.8$ ppm)   heptet heptet   ($\delta = 0.8$ ppm)
($\delta = 1.4$ ppm)

(d)   The compound $C_4H_6Cl_4$ contains no double bonds or rings. There are no high-field peaks ($\delta = 0.5$-1.5 ppm), so there are no methyl groups. Therefore at least one chlorine substituent must be at each end of the chain. The most likely structure has the four chlorines divided in two two groups of two.

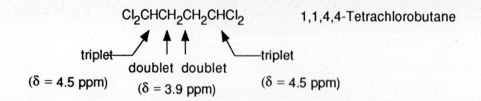

Cl$_2$CHCH$_2$CH$_2$CHCl$_2$　　　1,1,4,4-Tetrachlorobutane

triplet　　　doublet　doublet　　triplet

(δ = 4.5 ppm)　　　　　　　　(δ = 4.5 ppm)

(δ = 3.9 ppm)

(e)　The molecular formula C$_4$H$_6$Cl$_2$ indicates the presence of one double bond or ring. A signal at δ = 5.7 ppm is consistent with a proton attached to a double bonded carbon. The following structural units are present:

　　　　H　(δ = 5.7 ppm; triplet)

C=C　　　　　　　　　　　C=CCH$_3$　(δ = 2.2 ppm; allylic)

　　　CH$_2$

In order for the methyl group to appear as a singlet and the methylene group to appear as a doublet, the chlorine substituents must be distributed as shown below. (The stereochemistry of the double bond (*E* or *Z*) is not revealed by the ^{1}H nmr spectrum.)

singlet　CH$_3$

　　　　　C=CHCH$_2$Cl　　1,3-Dichloro-2-butene

Cl

triplet　doublet　(δ = 4.1 ppm)

9.24　The compounds of molecular formula C$_4$H$_9$Cl are the isomeric chlorides: butyl, isobutyl, *sec*-butyl and *tert*-butyl chloride.

(a)　All nine methyl protons of *tert*-butyl chloride (CH$_3$)$_3$CCl are equivalent; its ^{1}H nmr spectrum has only one peak.

(b)　A doublet at δ = 3.4 ppm indicates a —CH$_2$Cl group attached to a carbon that bears a single proton. This spectrum corresponds to isobutyl chloride (CH$_3$)$_2$CHCH$_2$Cl.

(c)　A triplet at δ = 3.5 ppm indicates that the —CH$_2$Cl group is attached to a methylene group (CH$_2$). This compound is butyl chloride CH$_3$CH$_2$CH$_2$CH$_2$Cl.

(d)　There are two nonequivalent methyl groups in this compound. It is *sec*-butyl chloride.

CH$_3$CHCH$_2$CH$_3$

　　　Cl

(δ = 1.5 ppm)　doublet　　　　triplet　(δ = 1.0 ppm)

9.25　(a)　The five-proton signal at δ = 7.2 ppm suggests the presence of a monosubstituted aromatic ring. The molecular formula C$_{10}$H$_{12}$O indicates a SODAR of 5, and the aromatic ring accounts for four of them. The remaining unsaturation is a carbonyl group, as indicated by the infrared absorption at 1710 cm^{-1}. The three-proton triplet and the two-proton quartet suggest the presence of an ethyl group. The compound most consistent with these data is **ethyl benzyl ketone**.

quartet (δ = 2.4 ppm)

$$\text{C}_6\text{H}_5-\text{CH}_2\overset{\overset{\displaystyle O}{\|}}{\text{C}}\text{CH}_2\text{CH}_3 \longleftarrow \text{triplet} \quad (\delta = 1.0 \text{ ppm})$$

singlet   singlet (δ = 3.6 ppm)

(δ = 7.2 ppm)

(b)   From its molecular formula, this compound contains no rings or double bonds, and an infrared absorption at 3400 cm^{-1} suggests the presence of a hydroxyl group. From the ^{1}H nmr data, the compound must be a diol, as the OH protons are best explained in terms of the two-proton singlet at δ = 2.0 ppm. The twelve-proton singlet at δ = 2.0 ppm is due to four equivalent methyl groups. The compound which best fits these data is **2,3-dimethyl-2,3-butanediol**.

$$\underset{\underset{\displaystyle \text{HO}}{|}}{\overset{\overset{\displaystyle \text{CH}_3}{|}}{\text{CH}_3\text{C}}}-\underset{\underset{\displaystyle \text{OH}}{|}}{\overset{\overset{\displaystyle \text{CH}_3}{|}}{\text{CCH}_3}} \qquad \text{2,3-Dimethyl-2,3-butanediol}$$

(c)   The compound contains a hydroxyl group, as evidenced by the infrared absorption at 3400 cm^{-1}. The hydroxyl proton appears in the nmr spectrum at δ = 3.7 ppm. The other nmr signal is explained by two equivalent methyl groups. All the atoms in the molecular formula except for one carbon and the nitrogen have been explained. A cyano group (CN), would account for these atoms as well as the infrared absorption at 2240 cm^{-1}. Putting these pieces together gives

OH $\longleftarrow$ singlet (δ = 3.7 ppm)

$$\text{CH}_3-\underset{\underset{\displaystyle \text{CH}_3}{|}}{\overset{\overset{\displaystyle \text{OH}}{|}}{\text{C}}}-\text{C}\equiv\text{N}$$

singlet

(δ = 1.65 ppm)

9.26   Since Compound A has a five-proton signal at δ = 7.2 ppm and a SODAR of 4, we conclude that six of its eight carbons belong to a monosubstituted aromatic ring. The infrared spectrum exhibits absorption at 3300 cm^{-1}, indicating the presence of a hydroxyl group. Compound A is an alcohol. A three-proton doublet at δ = 1.4 ppm, along with a one-proton quartet at δ = 4.7 ppm, signals the presence of a CH$_3$CH unit. Compound A is 1-phenylethanol.

quartet (δ = 4.7 ppm)

$$\text{C}_6\text{H}_5-\underset{\underset{\displaystyle \text{OH}}{|}}{\overset{\overset{\displaystyle \text{H}}{|}}{\text{C}}}-\text{CH}_3 \qquad \text{1-Phenylethanol (Compound A)}$$

doublet (δ = 1.4 ppm)

singlet

(δ = 1.9 ppm)

183

9.27 The most prominent peak in the infrared spectrum of compound B is at 1725 cm^{-1}, a region characteristic of C=O stretching vibrations.

The ^{1}H nmr spectrum shows only two sets of signals, a triplet at δ = 1.1 ppm and a quartet at δ = 2.4 ppm. Thus all the protons of compound B are part of an ethyl (CH$_3$CH$_2$) group. The molecular formula C$_5$H$_{10}$O is explained by two ethyl groups and the carbonyl. Thus compound B is 3-pentanone.

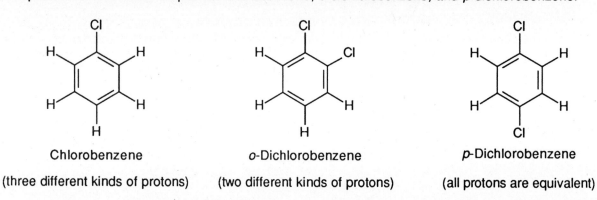

9.28 (a) All the hydrogens are equivalent in *p*-dichlorobenzene, therefore it has the simplest ^{1}H nmr spectrum of the three compounds chlorobenzene, *o*-dichlorobenzene, and *p*-dichlorobenzene.

Chlorobenzene

(three different kinds of protons)

o-Dichlorobenzene

(two different kinds of protons)

p-Dichlorobenzene

(all protons are equivalent)

(b) through (d) In addition to give the simplest ^{1}H nmr spectrum, *p*-dichlorobenzene gives the simplest ^{13}C nmr spectrum. It has two peaks in its ^{13}C nmr spectrum, chlorobenzene has four, and *o*-dichlorobenzene has three.

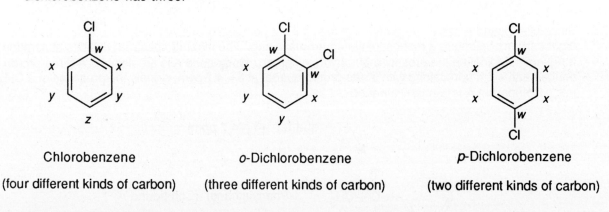

Chlorobenzene

(four different kinds of carbon)

o-Dichlorobenzene

(three different kinds of carbon)

p-Dichlorobenzene

(two different kinds of carbon)

9.29 The molecular formula corresponds to a SODAR of 0, and so there are no double bonds or rings. The absence of infrared absorption between 3000 and 3500 cm^{-1}, along with the strong peak at 1150 cm^{-1}, suggests the compound is an ether.

The singlet at δ = 3.4 ppm is due to a CH$_3$O— group in the molecule. The six-proton doublet at δ = 1.0 ppm suggests a (CH$_3$)$_2$CH— is present. The best structure to explain the data is

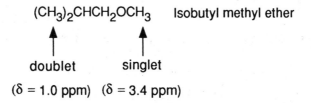

(CH₃)₂CHCH₂OCH₃   Isobutyl methyl ether

doublet   singlet

(δ = 1.0 ppm)   (δ = 3.4 ppm)

As predicted by the presence of four signals in the ^{13}C nmr spectrum, the proposed structure has four different carbon atoms.

9.30   Each different carbon atom in a molecule gives a distinct signal in a ^{13}C nmr spectrum. Thus the seven different carbon atoms in the compound shown will produce a ^{13}C nmr spectrum having seven signals.

$$Cl-\underset{1}{\overset{2\quad 3}{\underset{2\quad 3}{\bigcirc}}}-\overset{\overset{O}{\|}}{\underset{5}{C}}-\underset{6}{CH_2}-\underset{7}{CH_3}$$

9.31   Among the possible molecular formulas that correspond to the molecular ion at m/z = 134, $C_{10}H_{14}$ correlates best with the information from the ^1H nmr spectrum. What is evident is that there is a signal due to aromatic protons, as well as a triplet-quartet pattern of an ethyl group. A molecular formula of $C_{10}H_{14}$ suggest a benzene ring that bears two ethyl groups. Since the signal for the aryl protons is so sharp, they are probably equivalent. Compound C is *p*-diethylbenzene.

(δ = 1.2 ppm)   (δ = 1.2 ppm)

triplet   triplet

CH₃CH₂—⟨benzene ring⟩—CH₂CH₃

quartet   quartet

(δ = 2.6 ppm)   (δ = 2.6 ppm)

185

CHAPTER 10

ALCOHOLS, ETHERS, AND PHENOLS

GLOSSARY OF TERMS

Alcohol A compound of the type ROH in which the hydroxyl functional group is attached to sp^3-hybridized carbon.

Alcohol dehydrogenase An enzyme in the liver which catalyzes the oxidation of alcohols to aldehydes and ketones.

Aldehyde A family of compounds containing the CH=O functional group as in

$$\underset{\text{RCH}}{\overset{\text{O}}{\|}} \quad \text{or} \quad \underset{\text{ArCH}}{\overset{\text{O}}{\|}}$$

Coenzyme Q A naturally occurring group of related quinones involved in the chemistry of cellular respiration. Also known as *ubiquinone*.

Collins' reagent A chromium trioxide-pyridine complex which, when used in anhydrous media such as dichloromethane, oxidizes primary alcohols to aldehydes and no further. Secondary alcohols are oxidized to ketones.

Condensation reaction A reaction in which two molecules combine to give a product accompanied by the expulsion of some small stable molecule (such as water). An example is acid-catalyzed ether formation.

$$\text{2 ROH} \xrightarrow{\text{H}_2\text{SO}_4} \text{ROR} + \text{H}_2\text{O}$$

Epoxidation The conversion of an alkene to an epoxide by treatment with a peroxy acid.

Epoxide A compound of the type

Ether Compounds of the type shown which contain the C—O—C functional group.

ROR   ArOR   ArOAr

When the two groups bonded to oxygen are the same, the ether is described as a *symmetrical ether*. When the groups are different, it is called a *mixed ether*.

Grain alcohol A nonsystematic name for ethanol (CH_3CH_2OH).

Intramolecular An intramolecular reaction is one which occurs between two atoms or groups within the same molecule.

Ketone A family of compounds in which both atoms attached to a carbonyl group (C=O) are carbon as in

$$\underset{\text{RCR}}{\overset{\text{O}}{\|}} \quad \underset{\text{RCAr}}{\overset{\text{O}}{\|}} \quad \underset{\text{ArCAr}}{\overset{\text{O}}{\|}}$$

Lithium aluminum hydride   The compound $LiAlH_4$. It is an extremely powerful reducing agent.

Mercaptan   An old name for the class of compounds now known as *thiols* RSH.

Oxidation   A loss in the number of electrons associated with an atom. In organic chemistry, oxidation of carbon corresponds to an increase in the number of bonds to oxygen or a decrease in the number of bonds to hydrogen.

Phenol   A family of compounds characterized by a hydroxyl substituent on an aromatic ring as in ArOH. Phenol is also the name of the parent compound C_6H_5OH.

Polyether   A molecule that contains many ether linkages. Polyethers occur naturally in a number of antibiotic substances.

Quinone   The product of oxidation of an ortho or para dihydroxybenzene derivative. Examples include:

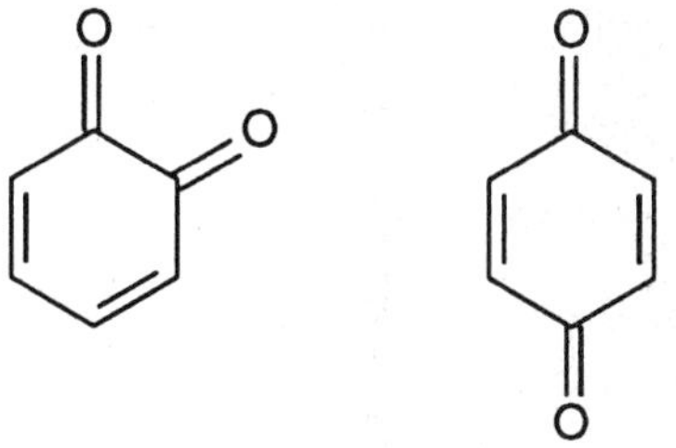

Reduction   A gain in the number of electrons associated with an atom. In organic chemistry, reduction of carbon corresponds to an increase in the number of bonds to hydrogen or a decrease in the number of bonds to oxygen.

Rubbing alcohol   A nonsystematic name for 2-propanol, $(CH_3)_2CHOH$.

Sodium borohydride   The compound $NaBH_4$. A good reducing agent for converting aldehydes and ketones to alcohols. It is not as powerful a reducing agent as lithium aluminum hydride.

Thiol   A compound of the type RSH or ArSH.

Vicinal   Substituents on adjacent atoms.

Williamson ether synthesis   A method for the preparation of ethers involving an S_N2 reaction between an alkoxide ion and a primary alkyl halide.

$$RONa + R'CH_2Br \longrightarrow R'CH_2OR + NaBr$$

Wood alcohol   A nonsystematic name for methanol CH_3OH.

SOLUTIONS TO TEXT PROBLEMS

10.1   The systematic name of tetrahydrolinalool is **3,7-dimethyl-3-octanol.** Tetrahydrolinalool is a tertiary alcohol.

10.2   The change from ethanol to acetic acid increases the oxygen content of the molecule, and so it is an oxidation. A reagent that brings about an oxidation is called an oxidizing agent.

10.3   The two primary alcohols 1-butanol (*n*-butyl alcohol) and 2-methyl-1-propanol (isobutyl alcohol) can be prepared by hydrogenation of the corresponding aldehydes:

$$\underset{\text{Butanal}}{CH_3CH_2CH_2\overset{\overset{\text{O}}{\|}}{C}H} \xrightarrow{\text{H}_2,\ \text{Ni}} \underset{\text{1-Butanol}}{CH_3CH_2CH_2CH_2OH}$$

187

$$(CH_3)_2CHCH \overset{O}{\underset{}{||}} \quad \xrightarrow{\text{H}_2,\ \text{Ni}} \quad (CH_3)_2CHCH_2OH$$

2-Methylpropanal   2-Methyl-1-propanol

The secondary alcohol 2-butanol (*sec*-butyl alcohol) arises by hydrogenation of a ketone.

$$CH_3\overset{O}{\overset{||}{C}}CH_2CH_3 \quad \xrightarrow{\text{H}_2,\ \text{Ni}} \quad CH_3\underset{OH}{\underset{|}{C}H}CH_2CH_3$$

2-Butanone   2-Butanol

Tertiary alcohols such as 2-methyl-2-propanol $(CH_3)_3COH$ (*tert*-butyl alcohol) cannot be prepared by hydrogenation of a carbonyl compound.

10.4   (b)   Catalytic hydrogenation of an aldehyde yields a primary alcohol.

$$(CH_3)_2CHCH_2CH \overset{O}{\underset{}{||}} \quad \xrightarrow[\text{ethanol}]{\text{H}_2,\ \text{Pt}} \quad (CH_3)_2CHCH_2CH_2OH$$

3-Methylbutanal   3-Methyl-1-butanol

(c)   Sodium borohydride reduces aldehydes to primary alcohols but does not reduce isolated double bonds. The product is the unsaturated alcohol, 3-cyclohexenylmethanol.

(d)   Catalytic hydrogenation reduces the carbon-carbon double bond and the carbon-oxygen double bond. The product of this reaction is the saturated alcohol, cyclohexylmethanol.

10.5   One method for preparation of a sodium alkoxide is reaction of the appropriate alcohol with sodium metal.

$$2\ CH_3OH \quad + \quad 2\ Na \quad \longrightarrow \quad 2\ CH_3O^-\ Na^+ \quad + \quad H_2$$

Methanol   Sodium   Sodium methoxide   Hydrogen

A second method involves reaction of the alcohol with a strong base such as sodium hydride.

$$CH_3OH \quad + \quad NaH \quad \longrightarrow \quad CH_3O^-\ Na^+ \quad + \quad H_2$$

Methanol   Sodium hydride   Sodium methoxide   Hydrogen

10.6 Heating ethanol in the presence of an acid produces diethyl ether by a condensation reaction.

$$2 \; CH_3CH_2OH \xrightarrow[\text{heat}]{H^+} CH_3CH_2OCH_2CH_3 \; + \; H_2O$$

Ethanol   Diethyl ether   Water

The acid is indicated above the arrow and does not appear as a reactant. It is a catalyst and is not consumed in the reaction.

10.7 (b) The substrate is a secondary alcohol and so gives a ketone on oxidation with sodium dichromate. 2-Octanone has been prepared in 92 to 96 percent yield under these reaction conditions.

$$CH_3CH(CH_2)_5CH_3 \; \underset{OH}{} \xrightarrow[\text{H}_2\text{SO}_4,\; \text{heat}]{Na_2Cr_2O_7} CH_3C(CH_2)_5CH_3$$

2-Octanol   2-Octanone

(c) The alcohol is primary, so oxidation can produce either an aldehyde or a carboxylic acid depending on the reaction conditions. Here the oxidation is carried out under anhydrous conditions with Collins' reagent. The product, the corresponding aldehyde, has been obtained in 70 to 84 percent yield.

$$CH_3CH_2CH_2CH_2CH_2CH_2CH_2OH \xrightarrow[\text{CH}_2\text{Cl}_2]{(C_5H_5N)_2CrO_3} CH_3CH_2CH_2CH_2CH_2CH_2CH$$

1-Heptanol   Heptanal

(d) The alcohol is a cyclic secondary alcohol and yields a cyclic ketone on oxidation.

Menthol   Menthone

10.8 (b) Oxirane is the IUPAC name for ethylene oxide. A chloromethyl group ($ClCH_2$-) is attached to position 2 of the ring in 2-(chloromethyl)oxirane.

Oxirane   2-(Chloromethyl)oxirane

(c) Epoxides may be named by adding the prefix *epoxy* to the IUPAC name of a parent compound, specifying both atoms to which the oxygen is attached by number.

$$CH_2=CHCH_2CH_3 \qquad CH_2=CHCH \!-\! CH_2$$

$$\qquad\qquad\qquad\qquad\qquad \underset{O}{\diagdown\diagup}$$

1-Butene   3,4-Epoxy-1-butene

10.9 (b) Methyl propyl ether can be prepared by the reaction of sodium methoxide with a propyl halide:

$$CH_3ONa \quad + \quad CH_3CH_2CH_2Br \longrightarrow CH_3OCH_2CH_2CH_3 \quad + \quad NaBr$$

Sodium methoxide   1-Bromopropane   Methyl propyl ether   Sodium bromide

It can also be made by the reaction of sodium propoxide with a methyl halide:

$$CH_3Br \quad + \quad CH_3CH_2CH_2ONa \longrightarrow CH_3OCH_2CH_2CH_3 \quad + \quad NaBr$$

Bromomethane   Sodium propoxide   Methyl propyl ether   Sodium bromide

(c) The two routes to benzyl ethyl ether are:

$$C_6H_5CH_2ONa \quad + \quad CH_3CH_2Br \longrightarrow C_6H_5CH_2OCH_2CH_3 \quad + \quad NaBr$$

Sodium benzyloxide   Bromoethane   Benzyl ethyl ether   Sodium bromide

$$C_6H_5CH_2Br \quad + \quad CH_3CH_2ONa \longrightarrow C_6H_5CH_2OCH_2CH_3 \quad + \quad NaBr$$

Benzyl bromide   Sodium ethoxide   Benzyl ethyl ether   Sodium bromide

10.10 The alkyl groups in disparlure are cis, thus the cis or *Z*-alkene is used in the epoxidation reaction. The desired alkene is (*Z*)-2-methyl-7-octadecene.

(*Z*)-2-Methyl-7-octadecene   Disparlure

10.11 By analogy with the reaction of ethylene oxide and ethanol shown in the text, the reaction of ethylene oxide and 1-butanol can be written:

$$\underset{\diagdown O \diagup}{H_2C \!-\! CH_2} \quad + \quad CH_3CH_2CH_2CH_2OH \longrightarrow CH_3CH_2CH_2CH_2OCH_2CH_2OH$$

Ethylene oxide   1-Butanol   "Butyl cellosolve"

10.12 Thiols contain the -SH functional group. The structure of 3-methyl-1-butanethiol is $(CH_3)_2CHCH_2CH_2SH$.

10.13 (b) Picric acid (2,4,6-trinitrophenol) is a phenol derivative with nitro groups (-NO$_2$) at positions 2, 4, and 6.

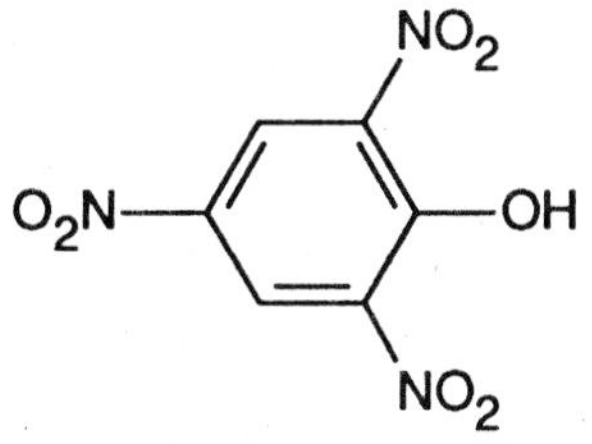

(c) 2, 4, 5-Trichlorophenol has the structure:

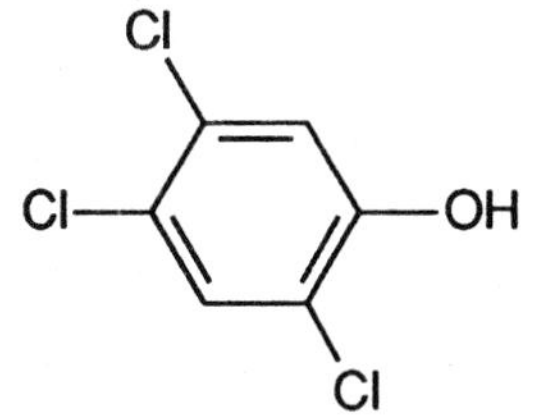

10.14 (a) The appropriate alkene for the preparation of 1-butanol by a hydroboration-oxidation sequence is 1-butene. Remember, hydroboration-oxidation leads to hydration of alkenes with a regioselectivity opposite to that seen in acid-catalyzed hydration.

$$CH_3CH_2CH=CH_2 \xrightarrow[\text{2. } H_2O_2,\ HO^-]{\text{1. } B_2H_6} CH_3CH_2CH_2CH_2OH$$

1-Butene   1-Butanol

(b) Lithium aluminum hydride reduces aldehydes to alcohols.

$$CH_3CH_2CH_2\overset{\overset{\displaystyle O}{\|}}{C}H \xrightarrow[\text{2. } H_2O]{\text{1. } LiAlH_4} CH_3CH_2CH_2CH_2OH$$

Butanal   1-Butanol

(c) Hydrogenation in the presence of a nickel or platinum catalyst also reduces aldehydes to primary alcohols.

$$CH_3CH_2CH_2\overset{\overset{\displaystyle O}{\|}}{C}H \xrightarrow{H_2,\ Ni} CH_3CH_2CH_2CH_2OH$$

Butanal   1-Butanol

(d) Reaction of 1-bromobutane with a source of hydroxide ion such as NaOH or KOH will produce 1-butanol by an S$_N$2 reaction.

$$CH_3CH_2CH_2CH_2Br \xrightarrow{NaOH} CH_3CH_2CH_2CH_2OH$$

1-Bromobutane   1-Butanol

10.15 (a) Hydroboration-oxidation of cyclohexene gives cyclohexanol. Regioselectivity is not a concern here because the alkene is symmetrically substituted.

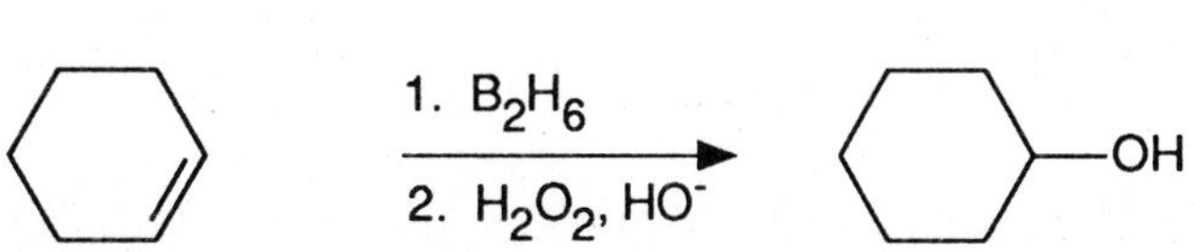

Cyclohexene   Cyclohexanol

(b)   Catalytic hydrogenation of the ketone cyclohexanone gives cyclohexanol.

Cyclohexanone   Cyclohexanol

$$\text{Cyclohexanone} \xrightarrow{\text{H}_2, \text{Ni (or Pt)}} \text{Cyclohexanol}$$

(c)   Lithium aluminum hydride reduces cyclohexanone to cyclohexanol.

$$\text{Cyclohexanone} \xrightarrow[\text{2. H}_2\text{O}]{\text{1. LiAlH}_4} \text{Cyclohexanol}$$

Cyclohexanone   Cyclohexanol

10.16 (a)   Alcohols react with thionyl chloride to give the corresponding alkyl chloride.

$$\text{CH}_3\text{CH}_2\text{CH}_2\text{CH}_2\text{OH} \xrightarrow{\text{SOCl}_2} \text{CH}_3\text{CH}_2\text{CH}_2\text{CH}_2\text{Cl}$$

1-Butanol   1-Chlorobutane

(b)   Oxidation of a primary alcohol with potassium permanganate gives a carboxylic acid having the same number of carbons.

$$\text{CH}_3\text{CH}_2\text{CH}_2\text{CH}_2\text{OH} \xrightarrow[\text{2. H}^+]{\text{1. KMnO}_4} \text{CH}_3\text{CH}_2\text{CH}_2\overset{\displaystyle O}{\overset{\displaystyle \|}{\text{C}}}\text{OH}$$

1-Butanol   Butanoic acid

(c)   Chromium trioxide-pyridine (Collins' reagent) is used to convert primary alcohols to the corresponding aldehyde.

$$\text{CH}_3\text{CH}_2\text{CH}_2\text{CH}_2\text{OH} \xrightarrow[\text{CH}_2\text{Cl}_2]{(\text{C}_5\text{H}_5\text{N})_2\text{CrO}_3} \text{CH}_3\text{CH}_2\text{CH}_2\overset{\displaystyle O}{\overset{\displaystyle \|}{\text{C}}}\text{H}$$

1-Butanol   Butanal

(d)   Oxidizing agents in aqueous acidic solution convert primary alcohols to the carboxylic acid.

$$\text{CH}_3\text{CH}_2\text{CH}_2\text{CH}_2\text{OH} \xrightarrow{\text{K}_2\text{Cr}_2\text{O}_7, \text{H}_2\text{SO}_4} \text{CH}_3\text{CH}_2\text{CH}_2\overset{\displaystyle O}{\overset{\displaystyle \|}{\text{C}}}\text{OH}$$

1-Butanol   Butanoic acid

(e)   Depending on the temperature at which the reaction is carried out, heating a primary alcohol in sulfuric acid can give either an alkene or an ether. Section 10.7 (p 284) shows the ether-forming reaction of 1-butanol.

$$CH_3CH_2CH_2CH_2OH \xrightarrow[\text{heat}]{H_2SO_4} CH_3CH_2CH_2CH_2OCH_2CH_2CH_2CH_3$$

1-Butanol   Dibutyl ether

(f) Sodium amide is a strong base, and removes the hydroxyl proton to form an alkoxide ion.

$$CH_3CH_2CH_2CH_2OH \xrightarrow{NaNH_2} CH_3CH_2CH_2CH_2O^- Na^+$$

1-Butanol   Sodium 1-butoxide

(g)   Potassium metal reacts with an alcohol to give the corresponding potassium alkoxide.

$$CH_3CH_2CH_2CH_2OH \xrightarrow{K} CH_3CH_2CH_2CH_2O^- K^+$$

1-Butanol   Potassium 1-butoxide

10.17 (a)   Thionyl chloride reacts with a secondary alcohol to give a secondary alkyl chloride.

$$\underset{\overset{|}{OH}}{CH_3CHCH_2CH_3} \xrightarrow{SOCl_2} \underset{\overset{|}{Cl}}{CH_3CHCH_2CH_3}$$

2-Butanol   2-Chlorobutane

(b-d)   Each of the oxidizing agents in the previous problem react with a secondary alcohol to give the corresponding ketone.

$$\underset{\overset{|}{OH}}{CH_3CHCH_2CH_3} \xrightarrow[\underset{K_2Cr_2O_7, H_2SO_4}{\overset{(C_5H_5N)_2CrO_3, CH_2Cl_2}{or}}]{\overset{KMnO_4}{or}} \overset{O}{\overset{||}{CH_3CCH_2CH_3}}$$

2-Butanol   2-Butanone

(e)   Secondary alcohols undergo dehydration on being heated with sulfuric acid. Dehydration occurs primarily in the direction that gives the more substituted alkene, as predicted by Zaitsev's rule. The major product is a mixture of *cis* - and *trans*-2-butene.

$$\underset{\overset{|}{OH}}{CH_3CHCH_2CH_3} \xrightarrow[\text{heat}]{H_2SO_4} CH_3CH=CHCH_3 \ + \ CH_2=CHCH_2CH_3$$

2-Butanol   2-Butene   1-Butene
                                   (mixture of cis and trans)

(f)   Reaction with sodium amide gives the sodium alkoxide.

$$CH_3CHCH_2CH_3 \xrightarrow{NaNH_2} CH_3CHCH_2CH_3$$

$$\underset{OH}{|} \qquad\qquad \underset{O^- Na^+}{|}$$

2-Butanol   Sodium 2-butoxide

(g)   Potassium metal reacts with 2-butanol to form the potassium alkoxide.

$$CH_3CHCH_2CH_3 \xrightarrow{K} CH_3CHCH_2CH_3$$

$$\underset{OH}{|} \qquad\qquad \underset{O^- K^+}{|}$$

2-Butanol   Potassium 2-butoxide

10.18 (a)   Dehydration of 1-hexanol by heating in strong acid gives 1-hexene. (Note: This is not a particularly effective synthesis. Dihexyl ether is also formed as well as, for reasons we have not discussed, alkenes which are isomeric with 1-hexene.)

$$CH_3CH_2CH_2CH_2CH_2CH_2OH \xrightarrow[heat]{H_2SO_4} CH_3CH_2CH_2CH_2CH=CH_2$$

1-Hexanol   1-Hexene

(b)   1-Chlorohexane is obtained by substitution of 1-hexanol using thionyl chloride. (Hydrogen chloride is not very effective with primary alcohols.)

$$CH_3CH_2CH_2CH_2CH_2CH_2OH \xrightarrow{SOCl_2} CH_3CH_2CH_2CH_2CH_2CH_2Cl$$

1-Hexanol   1-Chlorohexane

(c)   Oxidation of a primary alcohol to an aldehyde is achieved by using Collins' reagent.

$$CH_3CH_2CH_2CH_2CH_2CH_2OH \xrightarrow[CH_2Cl_2]{(C_5H_5N)_2CrO_3} CH_3CH_2CH_2CH_2CH_2\overset{\overset{O}{\|}}{C}H$$

1-Hexanol   Hexanal

(d)   Oxidation of 1-hexanol with either potassium permanganate or acidic dichromate will give the carboxylic acid, hexanoic acid.

$$CH_3CH_2CH_2CH_2CH_2CH_2OH \xrightarrow[CH_2Cl_2]{(C_5H_5N)_2CrO_3} CH_3CH_2CH_2CH_2CH_2\overset{\overset{O}{\|}}{C}OH$$

1-Hexanol   Hexanoic acid

10.19 Primary alcohols are formed by reduction of aldehydes; secondary alcohols by reduction of ketones. Therefore, the $C_5H_{12}O$ alcohols capable of being formed by sodium borohydride reduction are the isomeric primary and secondary alcohols shown.

Primary alcohols:

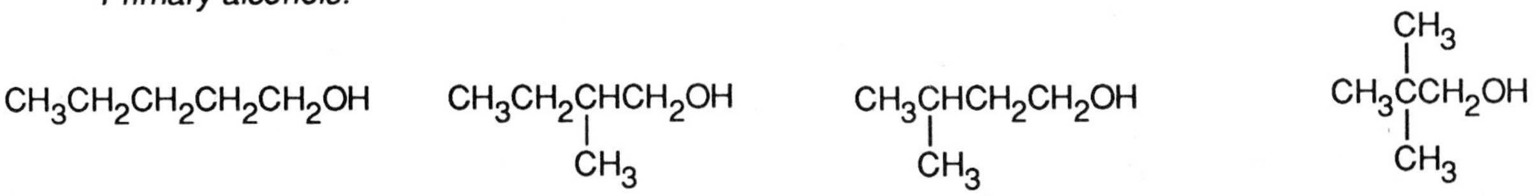

$CH_3CH_2CH_2CH_2CH_2OH$ $CH_3CH_2\underset{\underset{CH_3}{|}}{C}HCH_2OH$ $CH_3\underset{\underset{CH_3}{|}}{C}HCH_2CH_2OH$ $CH_3\underset{\underset{CH_3}{|}}{\overset{\overset{CH_3}{|}}{C}}CH_2OH$

  1-Pentanol   2-Methyl-1-butanol   3-Methyl-1-butanol   2,2-Dimethyl-1-propanol

Secondary alcohols:

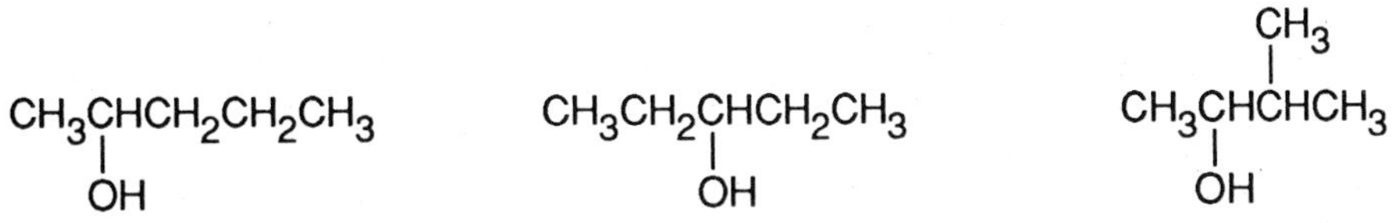

$CH_3\underset{\underset{OH}{|}}{C}HCH_2CH_2CH_3$ $CH_3CH_2\underset{\underset{OH}{|}}{C}HCH_2CH_3$ $CH_3\underset{\underset{OH}{|}}{\overset{\overset{CH_3}{|}}{C}}HCHCH_3$

      2-Pentanol   3-Pentanol   3-Methyl-2-butanol

The tertiary alcohol 2-methyl-2-butanol cannot be prepared by reduction of a carbonyl compound.

10.20 All the constitutionally isomeric ethers of molecular formula $C_5H_{12}O$ belong to one of two general groups, $C_4H_9OCH_3$ and $C_3H_7OCH_2CH_3$. Thus, we have:

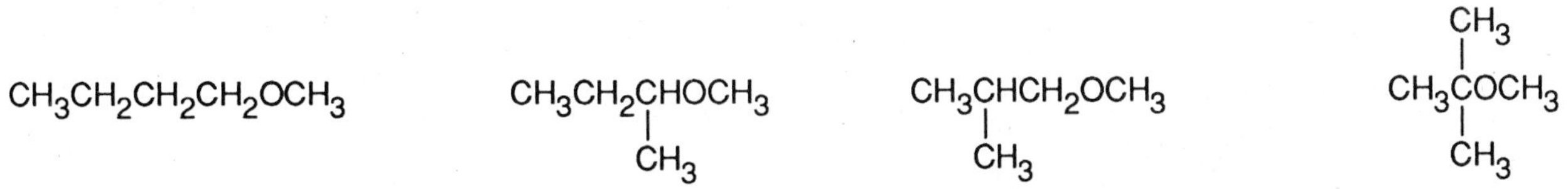

$CH_3CH_2CH_2CH_2OCH_3$ $CH_3CH_2\underset{\underset{CH_3}{|}}{C}HOCH_3$ $CH_3\underset{\underset{CH_3}{|}}{C}HCH_2OCH_3$ $CH_3\underset{\underset{CH_3}{|}}{\overset{\overset{CH_3}{|}}{C}}OCH_3$

  Butyl methyl ether   *sec*-Butyl methyl ether   Isobutyl methyl ether   *tert*-Butyl methyl ether

$CH_3CH_2CH_2OCH_2CH_3$ $CH_3\underset{\underset{CH_3}{|}}{C}HOCH_2CH_3$

    Ethyl propyl ether   Ethyl isopropyl ether

10.21

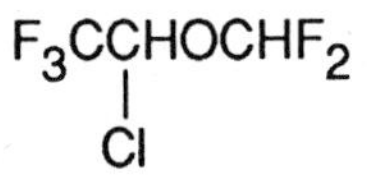

$F_3C\underset{\underset{Cl}{|}}{C}HOCHF_2$ Isoflurane: 1-Chloro-2,2,2-trifluoroethyl difluoromethyl ether

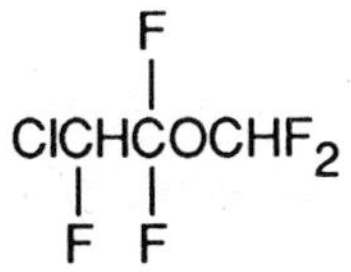

$Cl\underset{\underset{F}{|}}{C}H\underset{\underset{F}{|}}{\overset{\overset{F}{|}}{C}}OCHF_2$ Enflurane: 2-Chloro-1,1,2-trifluoroethyl difluromethyl ether

10.22 Methyl *tert*-butyl ether is $CH_3OC(CH_3)_3$. The Williamson ether synthesis must use a methyl halide and *tert*-butoxide as the reactants.

$$(CH_3)_3COH \xrightarrow{\text{Na}} (CH_3)_3CO^- \ Na^+ \xrightarrow{CH_3I} (CH_3)_3COCH_3$$

| *tert*-Butyl alcohol | Sodium *tert*-butoxide | *tert*-Butyl methyl ether |

The combination of a *tert*-butyl halide and sodium methoxide is incorrect. Tertiary halides undergo elimination, not substitution, on being treated with alkoxides.

10.23 (a) The ether linkage of cyclopentyl ethyl ether involves a primary carbon and a secondary one. Choose the alkyl halide corresponding to the primary alkyl group, leaving the secondary alkyl group to arise from the alkoxide nucleophile.

Sodium cyclopentanolate   Ethyl bromide   Cyclopentyl ethyl ether

The alternative combination, cyclopentyl bromide and sodium ethoxide, is not appropriate since elimination will be the principal reaction of secondary alkyl halides with alkoxides.

(b) A primary carbon and a secondary carbon are attached to the ether oxygen. The secondary carbon can only be derived from the alkoxide, as secondary alkyl halides cannot be used in the preparation of ethers by the Williamson method. The only effective route uses an allyl halide and sodium isopropoxide.

$$(CH_3)_2CHONa \ + \ CH_2=CHCH_2Br \longrightarrow CH_2=CHCH_2OCH(CH_3)_2$$

Sodium isopropoxide   Allyl bromide   Allyl isopropyl ether

Elimination will be the major reaction of an isopropyl halide with an alkoxide base.

(c) Here the ether is a mixed primary-tertiary one. The best combination is the one that uses the primary alkyl halide.

$$(CH_3)_3COK \ + \ C_6H_5CH_2Br \longrightarrow (CH_3)_3COCH_2C_6H_5$$

Potassium *tert*-butoxide   Benzyl bromide   Benzyl *tert*-butyl ether

The reaction between $(CH_3)_3CBr$ and $C_6H_5CH_2O^-$ will be one of elimination, not substitution.

10.24 (a) Acid-catalyzed hydration leads to the more substituted alcohol product.

2-Methyl-1-butene   2-Methyl-2-butanol

(b) Sodium hydride is a strong base and converts the alcohol to the corresponding alkoxide.

$$CH_3\overset{\overset{\displaystyle OH}{|}}{\underset{\underset{\displaystyle CH_3}{|}}{C}}CH_2CH_3 \quad \xrightarrow{\text{NaH}} \quad CH_3\overset{\overset{\displaystyle O^-\,Na^+}{|}}{\underset{\underset{\displaystyle CH_3}{|}}{C}}CH_2CH_3 \qquad \text{(Compound B)}$$

2-Methyl-2-butanol   Sodium 2-methyl-2-butanolate

(c) Phosphorus tribromide converts alcohols into alkyl bromides.

$$CH_3CH_2CH_2OH \quad \xrightarrow{\text{PBr}_3} \quad CH_3CH_2CH_2Br \qquad \text{(Compound C)}$$

1-Propanol   1-Bromopropane

(d) The alkoxide (Compound B) and alkyl bromide (Compound C) react to form an ether (Compound D) in a Williamson reaction.

$$CH_3\overset{\overset{\displaystyle O^-\,Na^+}{|}}{\underset{\underset{\displaystyle CH_3}{|}}{C}}CH_2CH_3 \quad + \quad CH_3CH_2CH_2Br \quad \longrightarrow \quad CH_3\overset{\overset{\displaystyle OCH_2CH_2CH_3}{|}}{\underset{\underset{\displaystyle CH_3}{|}}{C}}CH_2CH_3 \qquad \text{(Compound D)}$$

Sodium 2-methyl-2-butanolate   1-Bromopropane   1,1-Dimethylpropyl propyl ether

10.25 (a) The starting material is a primary allylic halide, a structural class known to be reactive in nucleophilic substitution reactions of the S$_N$2 type. The reaction produces an ether and is an example of the Williamson ether synthesis.

$$CH_3CH{=}CHCH_2Cl \quad \xrightarrow{\text{KOC(CH}_3)_3} \quad CH_3CH{=}CHCH_2OC(CH_3)_3$$

1-Chloro-2-butene   2-Butenyl *tert*-butyl ether

(b) The potassium alkoxide acts as a nucleophile toward iodoethane to yield an ethyl ether.

$$\underset{\underset{\displaystyle C_6H_5}{}}{\overset{\overset{\displaystyle CH_3}{}}{CH_3CH_2\!\!-\!\!C\!\!-\!\!O^-K^+}} \quad + \quad CH_3CH_2I \quad \longrightarrow \quad \underset{\underset{\displaystyle C_6H_5}{}}{\overset{\overset{\displaystyle CH_3}{}}{CH_3CH_2\!\!-\!\!C\!\!-\!\!OCH_2CH_3}}$$

Potassium (*R*)-2-phenyl-2-butanolate   Ethyl iodide   Ethyl (*R*)-1-methyl-1-phenylpropyl ether

The ether product has the same absolute configuration as the starting alkoxide because no bonds to the chiral center are made or broken in the reaction.

(c) Vicinal halohydrins are converted to epoxides on being treated with base.

$$CH_3CH_2\overset{}{\underset{\underset{\displaystyle OH}{|}}{C}}HCH_2Br \quad \xrightarrow{\text{NaOH}} \quad CH_3CH_2CH{-}CH_2{-}Br \quad \longrightarrow \quad CH_3CH_2CH{-}CH_2$$

1-Bromo-2-butanol   1,2-Epoxybutane

10.26 (a)   The parent compound is benzaldehyde. Vanillin bears a methoxy group (CH_3O) at C-3 and a hydroxy group (HO) at C-4.

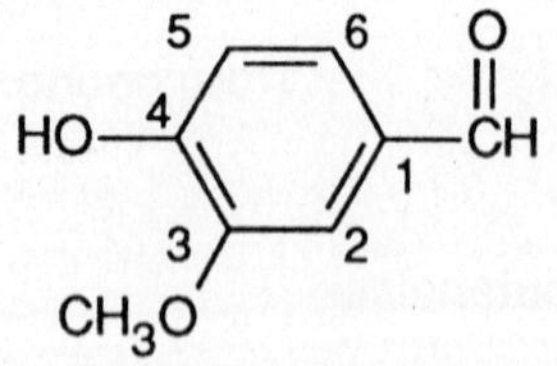

Vanillin   (4-hydroxy-3-methoxybenzaldehyde)

(b,c)   Thymol and carvacrol differ with respect to the position of the hydroxyl group.

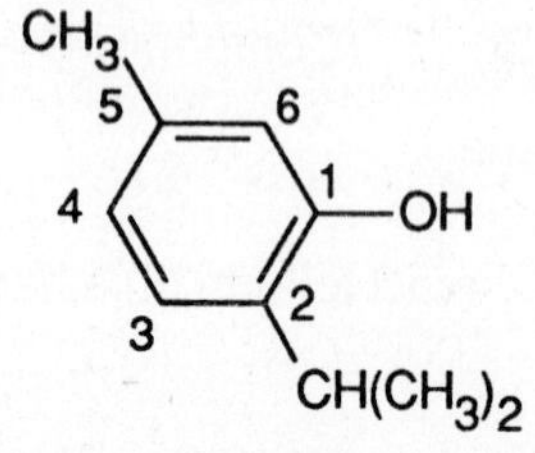

Thymol (2-isopropyl-5-methylphenol) 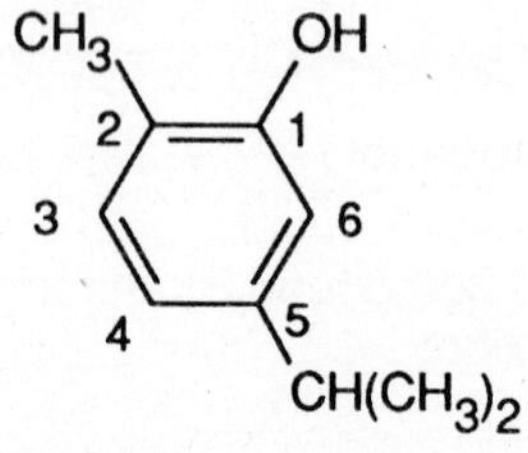   Carvacrol (5-isopropyl-2-methylphenol)

(d)   Benzyl alcohol is $C_6H_5CH_2OH$. Salicyl alcohol bears a hydroxyl group at the *ortho* position.

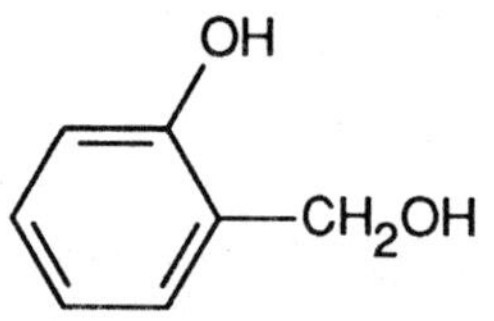

Salicyl alcohol (*o*-Hydroxybenzyl alcohol)

10.27 (a)   The compound is named as a derivative of phenol. The substituents (ethyl and nitro) are cited in alphabetical order with numbers assigned in the direction that gives the lowest number at the first point of difference.

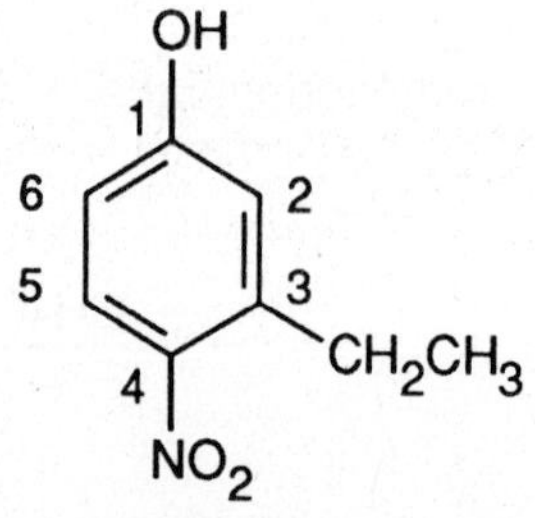

3-Ethyl-4-nitrophenol

(b)   An isomer of the above is 4-ethyl-3-nitrophenol.

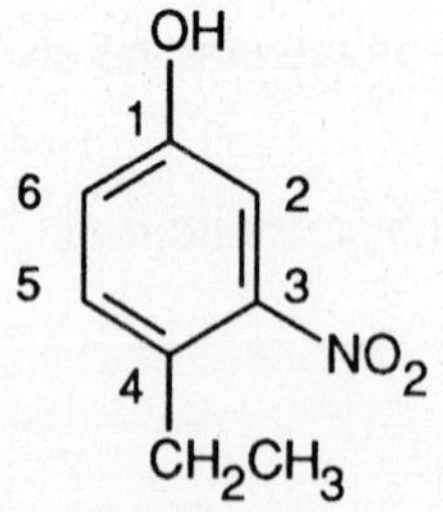

4-Ethyl-3-nitrophenol

(c) The parent compound is phenol. It bears, in alphabetical order, a benzyl group at C-4 and a chlorine at C-2.

4-Benzyl-2-chlorophenol

10.28 BHA may be named as a derivative of phenol (C_6H_5OH) or anisole ($C_6H_5OCH_3$).

2-*tert*-Butyl-4-methoxyphenol
or
3-*tert*-Butyl-4-hydroxyanisole

BHT may be named as a derivative of toluene or phenol. The name based on phenol as the parent compound is preferred.

2,6-di-*tert*-Butyl-4-methylphenol

10.29 (a) The reaction is an acid-base reaction. Phenol is the acid; sodium hydroxide is the base.

| Phenol | Sodium hydroxide | Sodium phenoxide | Water |
| (stronger acid) | (stronger base) | (weaker base) | (weaker acid) |

(b) Sodium phenoxide reacts with ethyl bromide to yield ethyl phenyl ether in a Williamson reaction. Phenoxide ion acts as a nucleophile.

$$C_6H_5ONa \quad + \quad CH_3CH_2Br \longrightarrow C_6H_5OCH_2CH_3 \quad + \quad NaBr$$

Sodium phenoxide   Ethyl bromide   Ethyl phenyl ether   Sodium bromide

(c) Phenols react as nucleophiles toward epoxides.

m-Cresol   Ethylene oxide   2-(*m*-Methylphenoxy)ethanol

The reaction as written conforms to the requirements of the problem that a balanced equation be written. Of course, the reaction will be much faster if catalyzed by acid or base, but the catalysts do not enter the equation representing the overall process.

10.30 Retrosynthetically, we can see that the cis carbon-carbon double bond is available by hydrogenation of the corresponding alkyne over the Lindlar catalyst.

$$CH_3CH_2CH{=}CHCH_2CH_2OH \implies CH_3CH_2C{\equiv}CCH_2CH_2OH$$

The $-CH_2CH_2OH$ unit can be appended to an alkynide anion by reaction with ethylene oxide.

$$CH_3CH_2C{\equiv}CCH_2CH_2OH \implies CH_3CH_2C{\equiv}\bar{C}{:} \quad + \quad \overset{H_2C{-}CH_2}{\underset{O}{\diagdown\diagup}}$$

The alkynide anion is derived from 1-butyne by alkylation of acetylene. This analysis suggests the following synthetic sequence:

$$HC{\equiv}CH \quad \xrightarrow[\text{2. }CH_3CH_2Br]{\text{1. }NaNH_2,\ NH_3} \quad CH_3CH_2C{\equiv}CH \quad \xrightarrow[\text{2. ethylene oxide}]{\text{1. }NaNH_2,\ NH_3} \quad CH_3CH_2C{\equiv}CCH_2CH_2OH$$

Acetylene   1-Butyne   3-Hexyn-1-ol

$$\Bigg\downarrow \begin{array}{l} H_2 \\ \text{Lindlar Pd} \end{array}$$

$$\underset{H}{\overset{CH_3CH_2}{\diagdown}}C{=}C\underset{H}{\overset{CH_2CH_2OH}{\diagup}}$$

cis-3-Hexen-1-ol

10.31 (a) On being heated in the presence of sulfuric acid, tertiary alcohols undergo elimination.

4-Methyl-1-phenylcyclohexanol   4-Methyl-1-phenylcyclohexene

(b) The substrate is a primary alcohol and is oxidized by Cr(VI) in the form of Collins' reagent to an aldehyde. The double bonds are unaffected.

$$CH_2{=}CHCH{=}CHCH_2CH_2CH_2OH \quad \xrightarrow[CH_2Cl_2]{(C_5H_5N)_2CrO_3} \quad CH_2{=}CHCH{=}CHCH_2CH_2\overset{\overset{\displaystyle O}{\|}}{C}H$$

4,6-Heptadien-1-ol   4,6-Heptadienal

200

(c)   Chromic acid oxidizes the secondary alcohol to the corresponding ketone but does not affect the triple bond.

$$CH_3CHC\equiv C(CH_2)_4CH_3 \quad\xrightarrow[\substack{H_2SO_4,\ H_2O \\ acetone}]{H_2CrO_4}\quad CH_3\overset{\overset{O}{\|}}{C}C\equiv C(CH_2)_4CH_3$$

3-Octyn-2-ol   3-Octyn-2-one

(e)   Oxidation of hydroquinone derivatives (*p*-dihydroxybenzenes) with Cr(VI) reagents is a method for preparing quinones.

2-Chloro-1,4-benzenediol   2-Chloro-1,4-benzoquinone

(f)   Silver oxide is a mild oxidizing agent, often used for the preparation of quinones from hydroquinones. It will not affect the allyl groups.

2,3-Diallylhydroquinone   2,3-Diallyl-1,4-benzoquinone

10.32 Glucose contains five hydroxyl groups and an aldehyde functional group. Its hydrogenation will not affect the hydroxyl groups but will reduce the aldehyde to a primary alcohol.

Glucose   Sorbitol

10.33 Even though we are given the structure of the starting material, it is still better to reason backward from the target molecule rather than forward from the starting material. The desired product contains a cyano (—CN) group. The only method we have seen so far for introducing such a function into a molecule is by nucleophilic substitution.

Therefore, the last step in the synthesis must be:

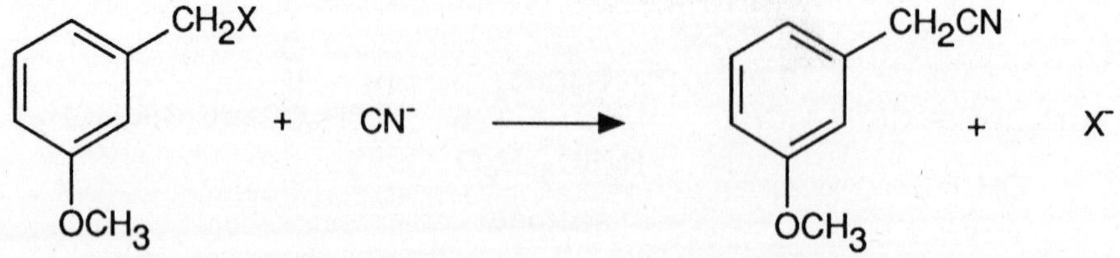

This step should work very well since the substrate is a primary benzylic halide, cannot undergo elimination, and is very reactive in S_N2 reactions.

The primary benzylic halide can be prepared from the corresponding alcohol by any of a number of methods.

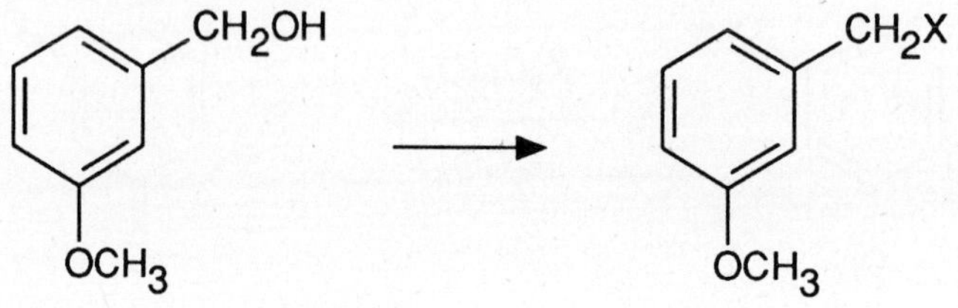

Suitable reagents include HBr, PBr$_3$, or SOCl$_2$.

There remains now only to prepare the primary alcohol from the given aldehyde, which is accomplished by reduction.

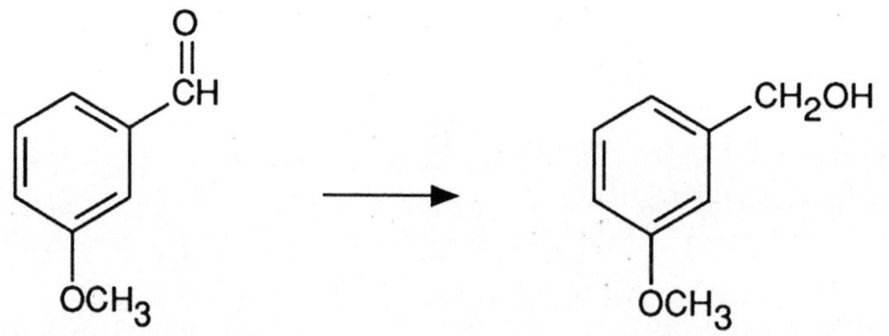

Reduction can be achieved by catalytic hydrogenation, with lithium aluminum hydride, or with sodium borohydride.

The actual sequence of reactions as carried out is as shown:

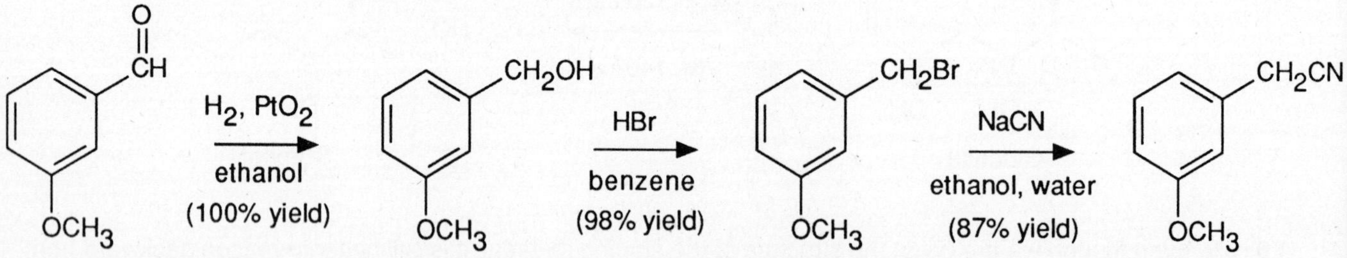

m-Methoxybenzaldehyde	*m*-Methoxybenzyl alcohol	*m*-Methoxybenzyl bromide	*m*-Methoxybenzyl cyanide

10.34 (a) Since all the peaks in the ^1H nmr spectrum of this ether are singlets, none of the protons can be vicinal to any other nonequivalent proton. The only $C_5H_{12}O$ ether that satisfies this requirement is **tert-butyl methyl ether**.

$$CH_3-O-\underset{\underset{CH_3}{|}}{\overset{\overset{CH_3}{|}}{C}}-CH_3$$

singlet at δ = 3.2 ppm

singlet at δ = 1.2 ppm

(b) A doublet-heptet pattern is characteristic of an isopropyl group. There are two isomeric $C_5H_{12}O$ ethers that contain an isopropyl group, ethyl isopropyl ether and isobutyl methyl ether.

$$(CH_3)_2CHOCH_2CH_3$$
Ethyl isopropyl ether

$$(CH_3)_2CHCH_2OCH_3$$
Isobutyl methyl ether

The methine proton in isobutyl methyl ether will be split into more than a heptet, however, because in addition to being split by two methyl groups, it is coupled to the two protons in the methylene group. Thus, isobutyl methyl ether does not have the correct splitting pattern to be the answer. The correct answer is **ethyl isopropyl ether.**

(c) The low-field signals are due to the protons on the carbon atoms of the C-O-C linkage. Since one is a doublet, it must be vicinal to only one other proton. Therefore, we can specify the partial structure:

$$C-O-\underset{\underset{H}{|}}{C}-\underset{\underset{H}{|}}{\overset{\overset{C}{|}}{C}}-C$$

low-field doublet

This partial structure contains all the carbon atoms in the molecule. Fill in the remaining valences with hydrogen atoms to reveal **isobutyl methyl ether** as the correct choice.

$$CH_3-O-CH_2-CH(CH_3)_2$$

low-field singlet

low-field doublet

(d) Here again, signals at low field arise from protons on the carbons of the C-O-C unit. One of these signals is a quartet and so corresponds to a proton on a carbon bearing a methyl group.

$$CH_3-\underset{\underset{H}{|}}{C}-O-C$$

quartet

The other carbon of the C-O-C unit has a hydrogen substituent that is split into a triplet. Therefore, this hydrogen must be attached to a carbon that bears a methylene group.

$$CH_3-\underset{\underset{H}{|}}{C}-O-\underset{\underset{H}{|}}{C}-CH_2-$$

quartet

triplet

These data permit us to complete the structure by adding an additional carbon and the requisite number of hydrogens in such a way that the protons attached to the carbons of the ether linkage are not split further. The correct structure is **ethyl propyl ether**.

$$CH_3CH_2OCH_2CH_2CH_3$$

quartet $\nearrow$ $\quad$ $\nwarrow$ — triplet

10.35 An intense, broad absorption at about 3400 cm^{-1} in its infrared spectrum indicates that compound is an alcohol, and the presence of peaks in the C—O stretching region (1025 to 1200 cm^{-1}) supports this conclusion. A signal at $\delta = 3.5$ ppm in the ^1H nmr spectrum, equivalent to one proton, can be assigned to the OH proton of an alcohol.

The presence of aromatic rings is indicated by the low hydrogen-to-carbon ratio ($C_{14}H_{14}O$) of its molecular formula and by the presence of weak absorbances in the 1670 to 2000 cm^{-1} region of the ir spectrum. The difference of the molecular formula ($C_{14}H_{14}O$) from C_nH_{2n+2} ($C_{14}H_{30}$) tells us that the sum of double bonds plus rings is eight. A SODAR equal to 8 can be accommodated by two benzene rings. That both benzene rings are monosubstituted is shown by the presence of 10 aromatic protons in the nmr spectrum at $\delta = 7$ to 7.5 ppm and by a peak in the ir spectrum at 700 cm^{-1}.

At this point we have identified three structural units accounting for the oxygen, 12 carbons, and 11 hydrogens.

and   OH

The three remaining hydrogens are part of a methyl group. These methyl protons appear as a singlet in the nmr at $\delta = 1.9$ ppm. The methyl group must be attached to a carbon that bears no hydrogens.

$$CH_3-\overset{|}{\underset{|}{C}}-$$

Putting these structural units together, we arrive at **1,1-diphenylethanol** as the correct structure of compound G.

CHAPTER 11

ALDEHYDES AND KETONES

GLOSSARY OF TERMS

Acetal   The product of the reaction of an aldehyde or ketone with two moles of an alcohol according to the equation:

$$
\underset{RCR'}{\overset{O}{\|}} + 2\,R''OH \xrightarrow{H^+} \underset{RCR'}{\overset{OR''}{\underset{OR''}{|}}} + H_2O
$$

Aldehyde   A compound of the type:

$$
\underset{RCH}{\overset{O}{\|}}
$$

Aldol addition   The nucleophilic addition of an aldehyde or ketone enolate to the carbonyl group of an aldehyde or ketone. The most typical case involves two molecules of an aldehyde, and is usually catalyzed by bases.

$$
2\ \underset{RCH_2CH}{\overset{O}{\|}} \longrightarrow \underset{RCH_2CHCHR}{\overset{OH}{\underset{CH=O}{|}}}
$$

Aldol condensation   When an aldol addition is carried out so that the β–hydroxy aldehyde dehydrates under the conditions of its formation, the product is described as arising by an aldol condensation.

$$
2\ \underset{RCH_2CH}{\overset{O}{\|}} \xrightarrow[\text{heat}]{HO^-} \underset{RCH_2CH=CR}{\underset{CH=O}{|}} + H_2O
$$

Cyanohydrin   A compound of the type:

$$
\underset{RCR'}{\overset{OH}{\underset{C\equiv N}{|}}}
$$

Cyanohydrins are formed by nucleophilic addition of HCN to the carbonyl group of an aldehyde or ketone.

Enol   A compound of the type:

$$
\underset{RC=CR_2}{\overset{OH}{|}}
$$

Enols are in equilibrium with an isomeric aldehyde or ketone, but are much less stable than aldehydes and ketones.

Enolate ion   The conjugate base of an enol. Enolate ions are stabilized by electron delocalization.

$$\text{RC}=\text{CR}_2 \quad \longleftrightarrow \quad \overset{\displaystyle O}{\overset{\|}{\text{RC}}}-\overset{\displaystyle -}{\text{CR}}_2$$

Geminal diol The hydrate of an aldehyde or ketone.

$$\overset{\displaystyle OH}{\underset{}{\text{R}_2\text{C}}}-\text{OH}$$

Grignard reagent An organomagnesium compound of the type RMgX formed by the reaction of magnesium with an alkyl or aryl halide.

Hemiacetal The product of nucleophilic addition of one molecule of an alcohol to an aldehyde or ketone. A compound of the type:

$$\overset{\displaystyle OH}{\underset{}{\text{R}_2\text{C}}}-\text{OR}'$$

Hydrazone A compound of the type $\text{R}_2\text{C}=\text{NNHR}'$ formed by the reaction of a substituted hydrazine (RNHNH_2) with an aldehyde or ketone.

Imine A compound of the type $\text{R}_2\text{C}=\text{NR}'$ formed by the reaction of an aldehyde or ketone with a primary amine ($\text{R}'\text{NH}_2$).

Ketone A compound of the type:

$$\overset{\displaystyle O}{\overset{\|}{\text{RCR}'}}$$

R and R' may be the same or different and may be alkyl or aryl.

Nucleophilic addition The characteristic reaction of an aldehyde or ketone. An atom possessing an unshared electron pair bonds to the carbon of the C=O group, and some other species (normally hydrogen) bonds to the oxygen.

$$\overset{\displaystyle O}{\overset{\|}{\text{RCR}'}} + \text{H}-\text{Y}: \quad \longrightarrow \quad \overset{\displaystyle OH}{\underset{\displaystyle R'}{\text{RC}}}-\text{Y}:$$

Retrosynthetic analysis A technique for synthetic planning based on reasoning backward from the target molecule to appropriate starting materials. An arrow of the type ⇨ designates a retrosynthetic step.

Tautomerism The process by which two isomers are interconverted by (formal) movement of an atom or group. Enolization is a form of tautomerism.

$$\overset{\displaystyle O}{\overset{\|}{\text{RC}}}-\text{CHR}_2 \quad \rightleftharpoons \quad \overset{\displaystyle OH}{\underset{}{\text{RC}}}=\text{CR}_2$$

Tollens' reagent A solution of silver nitrate in aqueous ammonia. It is a weak oxidizing agent capable of oxidizing aldehydes to carboxylic acids. When used for analysis, the deposition of a shiny silver "mirror" when an unknown is treated with Tollens' reagent is considered to be a positive test for an aldehyde.

SOLUTIONS TO TEXT PROBLEMS

11.1 **(b)** Both **trichloroethanal** and **trichloroacetaldehyde** are acceptable systematic names for $\text{Cl}_3\text{CCH}=\text{O}$ (*chloral*).

(c) The three-carbon parent chain of *cinnamaldehyde* has a double bond between C-2 and C-3 and a phenyl substituent at C-3. Its IUPAC name is **3-phenyl-2-propenal**.

$$\underset{3 \quad\quad 2 \quad 1}{\text{C}_6\text{H}_5\text{CH}=\text{CHCH}}\overset{\displaystyle O}{\overset{\|}{}}$$

(d) *Vanillin* is named as a derivative of benzaldehyde. The substituents are cited in alphabetical order in its IUPAC name **4-hydroxy-3-methoxybenzaldehyde**.

206

11.2 (b) *Acetophenone* is a common name accepted by the IUPAC. The systematic name is **1-phenylethanone**. Acetophenone may also be called **methyl phenyl ketone**.

$$\text{C}_6\text{H}_5\!-\!\overset{\displaystyle O}{\overset{\|}{\text{C}}}\text{CH}_3$$

(c) Write out the structure of *pinacolone* to reveal that the longest chain contains four carbon atoms. Pinacolone is **3,3-dimethyl-2-butanone**. The compound may also be named *tert* - **butyl methyl ketone**.

$$\underset{\underset{\displaystyle \text{CH}_3}{|}}{\overset{\overset{\displaystyle \text{H}_3\text{C}}{|}}{\text{CH}_3\text{C}}}\!-\!\overset{\displaystyle O}{\overset{\|}{\text{C}}}\text{CH}_3$$

11.3 Using both its oxygen and the hydroxyl proton, benzyl alcohol is able to hydrogen bond to water more effectively than benzaldehyde can. Benzaldehyde can only hydrogen bond using the carbonyl oxygen.

11.4 The equation describing the formation of chloral hydrate is analogous to the examples given in the text.

$$\overset{\displaystyle O}{\overset{\|}{\text{CCl}_3\text{CH}}} \quad + \quad \text{H}_2\text{O} \quad \rightleftharpoons \quad \underset{\underset{\displaystyle \text{OH}}{|}}{\overset{\overset{\displaystyle \text{OH}}{|}}{\text{CCl}_3\text{CH}}}$$

Chloral   Water   Chloral hydrate

11.5 (a) The hemiacetal intermediate corresponds to addition of methanol to the carbonyl group.

$$\overset{\displaystyle O}{\overset{\|}{\text{CH}_3\text{CH}}} \quad + \quad \text{CH}_3\text{OH} \quad \overset{\text{H}^+}{\rightleftharpoons} \quad \underset{}{\overset{\overset{\displaystyle \text{OCH}_3}{|}}{\text{CH}_3\text{CHOH}}}$$

Acetaldehyde   Methanol   Hemiacetal intermediate

(b) The hemiacetal is converted to a carbocation by protonation followed by loss of a water molecule from the protonated form.

$$\underset{}{\overset{\overset{\displaystyle \text{OCH}_3}{|}}{\text{CH}_3\text{CHOH}}} \quad \overset{\text{H}^+}{\rightleftharpoons} \quad \overset{\overset{\displaystyle \text{OCH}_3}{|}}{\text{CH}_3\text{CH}}\!-\!\overset{+}{\text{O}}\overset{\displaystyle \text{H}}{\diagdown}_{\diagdown \text{H}} \quad \overset{-\text{H}_2\text{O}}{\rightleftharpoons} \quad \text{CH}_3\overset{+}{\text{CH}}\text{OCH}_3$$

Hemiacetal   Carbocation

The two principal forms of this carbocation are:

$$CH_3\overset{+}{C}H\!\!-\!\!\overset{..}{\underset{..}{O}}CH_3 \longleftrightarrow CH_3CH\!\!=\!\!\overset{+}{\underset{..}{O}}CH_3$$

(c)　The dimethyl acetal of acetaldehyde is $CH_3CH(OCH_3)_2$.

11.6　Benzaldehyde cyanohydrin corresponds to the addition product of benzaldehyde (Section 11.1) and hydrogen cyanide.

$$\underset{\text{Benzaldehyde}}{C_6H_5\overset{\overset{\displaystyle O}{\|}}{C}H} \quad + \quad \underset{\text{Hydrogen cyanide}}{HCN} \quad \longrightarrow \quad \underset{\text{Benzaldehyde cyanohydrin}}{C_6H_5\overset{\overset{\displaystyle OH}{|}}{\underset{\underset{\displaystyle CN}{|}}{C}}H}$$

11.7　(b)　Allyl chloride is converted to allylmagnesium chloride on reaction with magnesium in diethyl ether.

$$\underset{\text{Allyl chloride}}{CH_2\!\!=\!\!CHCH_2Cl} \quad \xrightarrow[\text{diethyl ether}]{Mg} \quad \underset{\text{Allylmagnesium chloride}}{CH_2\!\!=\!\!CHCH_2MgCl}$$

(c)　The carbon-iodine bond of iodocyclobutane is replaced by a carbon-magnesium bond in the Grignard reagent.

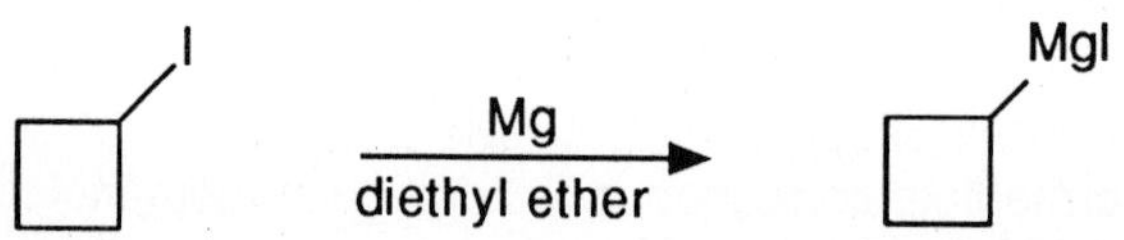

Iodocyclobutane　　　　　　　Cyclobutylmagnesium iodide

(d)　Bromine is attached to sp^3 hybridized carbon in 1-bromocyclohexene. The product of its reaction with magnesium has a carbon-magnesium bond in place of the carbon-bromine bond.

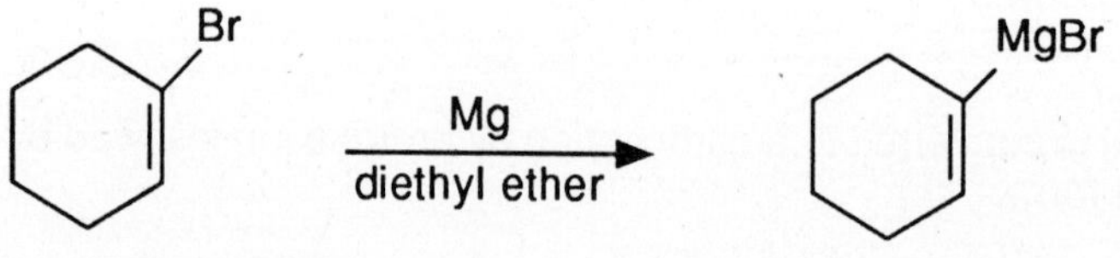

1-Bromocyclohexene　　　　　　　1-Cyclohexenylmagnesium bromide

11.8　(b)　Propyl magnesium bromide reacts with benzaldehyde by addition to the carbonyl group:

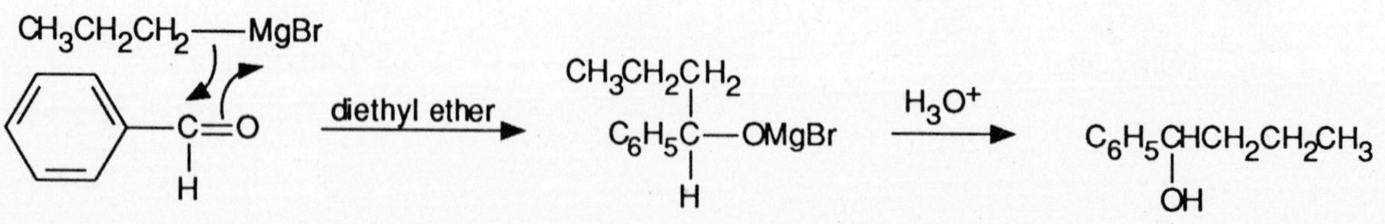

1-Phenyl-1-butanol

208

(c) Tertiary alcohols result from the reaction of Grignard reagents and ketones:

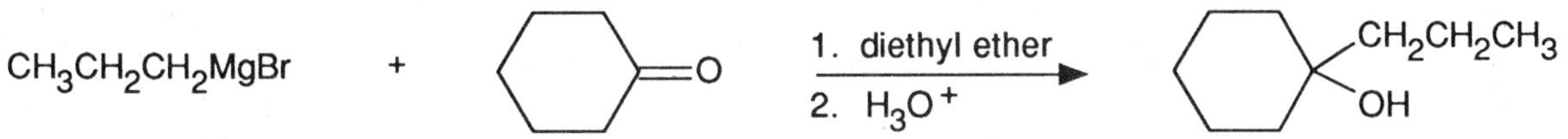

$CH_3CH_2CH_2MgBr$ + (cyclohexanone) $\xrightarrow[\text{2. } H_3O^+]{\text{1. diethyl ether}}$ (1-propylcyclohexanol structure) $CH_2CH_2CH_3$ / OH

The product is **1-propylcyclohexanol.**

(d) The starting material is a ketone and so reacts with a Grignard reagent to give a tertiary alcohol.

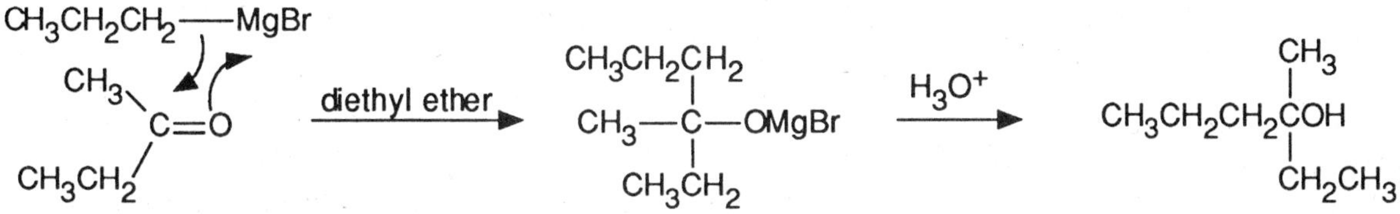

$CH_3CH_2CH_2$—$MgBr$

CH_3
$\diagdown$
$\quad\quad C=O$ $\xrightarrow{\text{diethyl ether}}$
$\diagup$
CH_3CH_2

$CH_3CH_2CH_2$
$|$
CH_3—C—$OMgBr$ $\xrightarrow{H_3O^+}$
$|$
CH_3CH_2

CH_3
$|$
$CH_3CH_2CH_2COH$
$|$
CH_2CH_3

3-Methyl-3-hexanol

11.9 (b) Since 2-hexanol is a secondary alcohol, it arises by the reaction of a Grignard reagent with an aldehyde. Disconnection of bonds to the hydroxyl-bearing carbon generates two pairs of structural fragments:

$CH_3CHCH_2CH_2CH_2CH_3$
$\quad\quad|$
$\quad\quad OH$ ⇨ CH_3^- $HCCH_2CH_2CH_2CH_3$
$\quad\quad\quad\quad\quad\quad\quad\quad\quad\quad\quad\quad\quad\quad\quad\quad||$
$\quad\quad\quad\quad\quad\quad\quad\quad\quad\quad\quad\quad\quad\quad\quad\quad O$

and $CH_3CHCH_2CH_2CH_2CH_3$
$\quad\quad\quad\quad|$
$\quad\quad\quad\quad OH$ ⇨ CH_3CH $^-CH_2CH_2CH_2CH_3$
$\quad\quad\quad\quad\quad\quad\quad\quad\quad\quad\quad\quad\quad||$
$\quad\quad\quad\quad\quad\quad\quad\quad\quad\quad\quad\quad\quad O$

Therefore, one of the routes involves the addition of a methyl Grignard reagent to a five-carbon aldehyde.

CH_3MgI + $CH_3CH_2CH_2CH_2CH$ (with O double bond) $\xrightarrow[\text{2. } H_3O^+]{\text{1. diethyl ether}}$ $CH_3CH_2CH_2CH_2CHCH_3$
$\quad\quad\quad\quad\quad\quad\quad\quad\quad\quad\quad\quad\quad\quad\quad\quad\quad\quad\quad\quad\quad\quad\quad\quad\quad\quad\quad\quad\quad|$
$\quad\quad\quad\quad\quad\quad\quad\quad\quad\quad\quad\quad\quad\quad\quad\quad\quad\quad\quad\quad\quad\quad\quad\quad\quad\quad\quad\quad\quad OH$

Methylmagnesium iodide   Pentanal   2-Hexanol

while the other route requires addition of a butylmagnesium halide to a two-carbon aldehyde.

$CH_3CH_2CH_2CH_2MgBr$ + CH_3CH (with O double bond) $\xrightarrow[\text{2. } H_3O^+]{\text{1. diethyl ether}}$ $CH_3CH_2CH_2CH_2CHCH_3$
$\quad\quad\quad\quad\quad\quad\quad\quad\quad\quad\quad\quad\quad\quad\quad\quad\quad\quad\quad\quad\quad\quad\quad\quad\quad\quad\quad\quad\quad\quad\quad|$
$\quad\quad\quad\quad\quad\quad\quad\quad\quad\quad\quad\quad\quad\quad\quad\quad\quad\quad\quad\quad\quad\quad\quad\quad\quad\quad\quad\quad\quad\quad\quad OH$

Butylmagnesium bromide   Acetaldehyde   2-Hexanol

(c) The target alcohol is secondary and so can be prepared by the reaction of a Grignard reagent with an aldehyde. The two retrosynthetic transformations are:

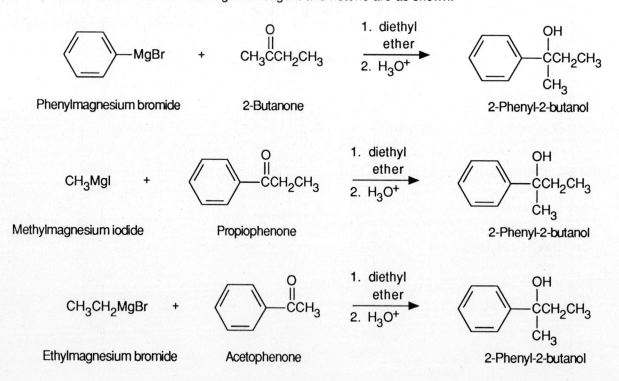

Therefore, two plausible syntheses are:

Phenylmagnesium bromide   +   Propanal   →   1-Phenyl-1-propanol

and   CH₃CH₂MgBr   +   Benzaldehyde   →   1-Phenyl-1-propanol

Ethylmagnesium bromide

11.10 The three combinations of Grignard reagent and ketone are as shown:

Phenylmagnesium bromide   +   2-Butanone   →   2-Phenyl-2-butanol

CH₃MgI   +   Propiophenone   →   2-Phenyl-2-butanol

Methylmagnesium iodide

CH₃CH₂MgBr   +   Acetophenone   →   2-Phenyl-2-butanol

Ethylmagnesium bromide

11.11 (b) There are no α hydrogen atoms in 2,2-dimethylpropanal, as the α carbon atom bears three methyl groups.

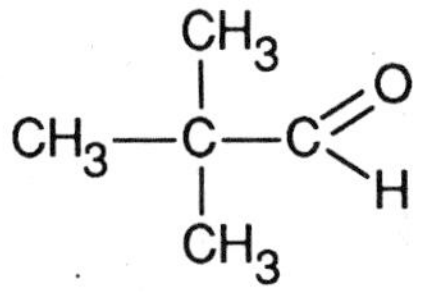

(c) All three protons of the methyl group as well as the two benzylic protons are α hydrogens in benzyl methyl ketone.

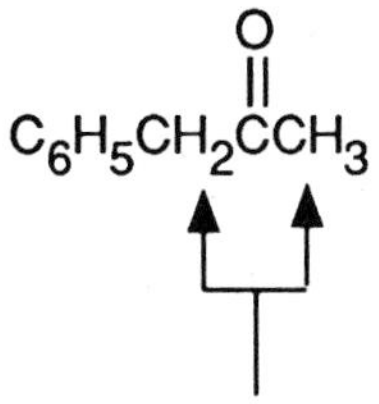

$$C_6H_5CH_2CCH_3$$

Benzyl methyl ketone

five α hydrogens

(d) Cyclohexanone has four equivalent α hydrogens. (The α hydrogens are the ones indicated in the structural formula.

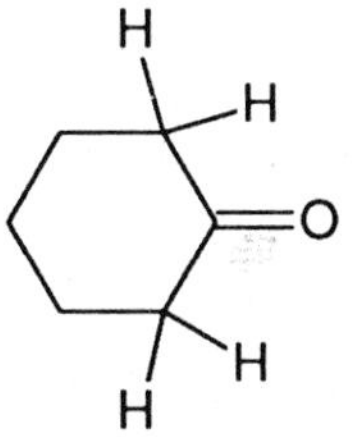

11.12 (b) Acetophenone can enolize only in the direction of the methyl group.

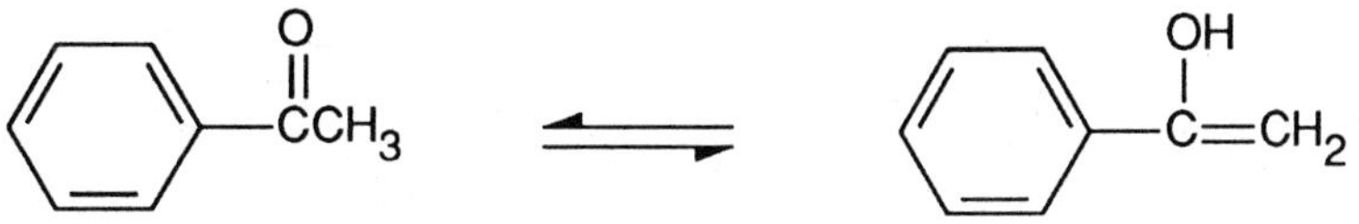

Acetophenone   Enol form of acetophenone

(c) The enol form of cyclohexanone is shown.

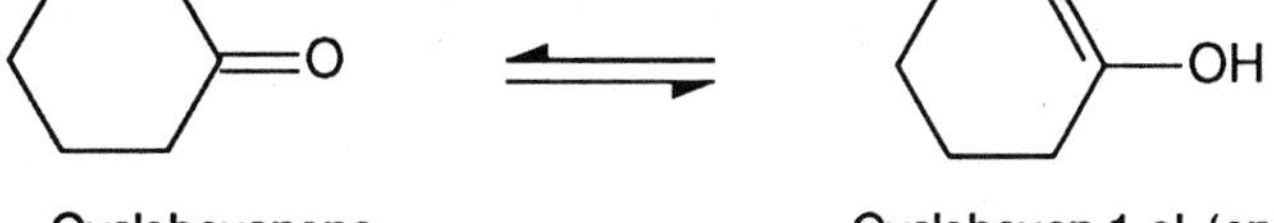

Cyclohexanone   Cyclohexen-1-ol (enol form)

(d) Enolization of 2-methylcyclohexanone can take place in two different directions.

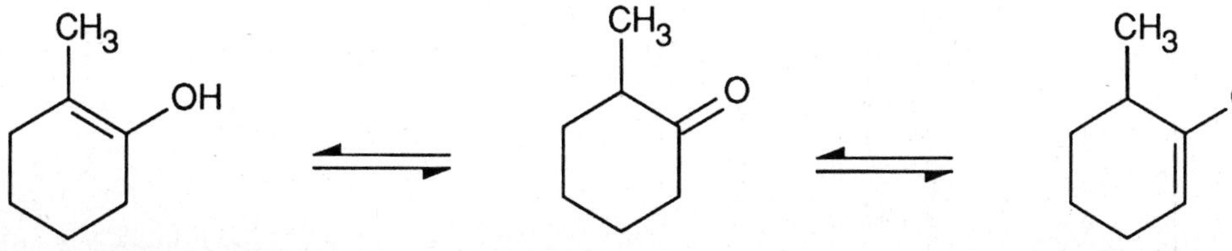

2-Methyl-1-cyclohexenol   2-Methylcyclohexanol   6-Methyl-1-cyclohexenol

211

11.13 (b) Approaching this problem in the same way as that of (a), write the structure of the enolate ion from 2-methylbutanal.

$$\underset{\substack{| \\ CH_3 \\ \text{2-Methylbutanal}}}{CH_3CH_2CHCH} \overset{O}{\overset{\|}{}} \quad + \quad HO^- \quad \rightleftharpoons \quad \underset{\substack{| \\ CH_3}}{CH_3CH_2CCH}\overset{O}{\overset{\|}{}} \quad \longleftrightarrow \quad \underset{\substack{| \\ CH_3}}{CH_3CH_2C=CH}\overset{O^-}{}$$

Enolate of 2-methylbutanal

This enolate adds to the carbonyl group of the aldehyde.

2-Methylbutanal   Enolate of 2-methylbutanal $\longrightarrow$ $CH_3CH_2CHCH-CCH_2CH_3$ (O⁻, CH₃, CH₃, HC=O)

The alkoxide ion formed in this step accepts a proton from solvent to yield the product of aldol addition.

$$\underset{\substack{| \quad\quad | \\ CH_3 \quad HC=O}}{CH_3CH_2CHCH-CCH_2CH_3}\overset{O^- \quad CH_3}{} \quad + \quad H_2O \quad \longrightarrow \quad \underset{\substack{| \quad\quad | \\ CH_3 \quad HC=O}}{CH_3CH_2CHCH-CCH_2CH_3}\overset{HO \quad CH_3}{} \quad + \quad HO^-$$

2-Ethyl-2,4-dimethyl-3-hydroxyhexanal

(c) The aldol addition product of 3-methylbutanal can be identified through the same mechanistic approach.

$$\underset{\text{3-Methylbutanal}}{(CH_3)_2CHCH_2CH}\overset{O}{\overset{\|}{}} \quad + \quad HO^- \quad \rightleftharpoons \quad \underset{\text{Enolate of 3-methylbutanal}}{(CH_3)_2CHCHCH}\overset{O}{\overset{\|}{}} \quad + \quad H_2O$$

$$(CH_3)_2CHCH_2CH\overset{O}{\overset{\|}{}} \quad + \quad \underset{\substack{| \\ HC=O}}{:CHCH(CH_3)_2}^- \quad \longrightarrow \quad \underset{\substack{| \\ HC=O}}{(CH_3)_2CHCH_2CH-CHCH(CH_3)_2}\overset{O^-}{}$$

3-Methylbutanal   Enolate of
                        3-methylbutanal

$\Big\downarrow H_2O$

$$\underset{\substack{| \\ HC=O}}{(CH_3)_2CHCH_2CH-CHCH(CH_3)_2}\overset{OH}{} \quad + \quad OH^-$$

3-Hydroxy-2-isopropyl-5-methylhexanal

11.14 Dehydration of the aldol addition product involves loss of a proton from the α carbon atom and hydroxide from the β carbon atom.

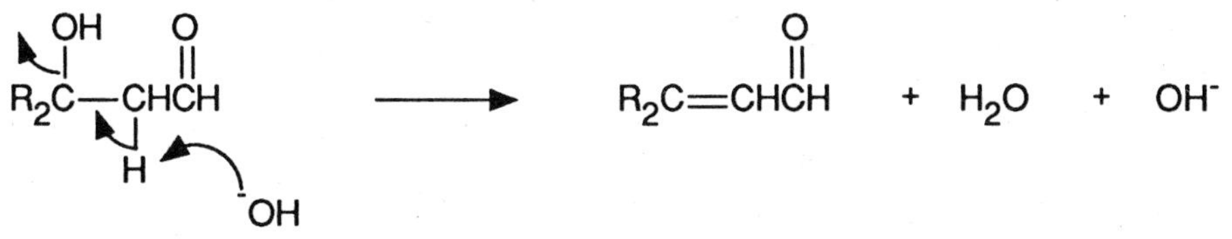

$$R_2C-CHCH \longrightarrow R_2C=CHCH + H_2O + OH^-$$

(with OH and O groups, H and OH shown in mechanism)

(b) The product of aldol addition of 3-methylbutanal has no α hydrogens. It cannot dehydrate to an aldol condensation product.

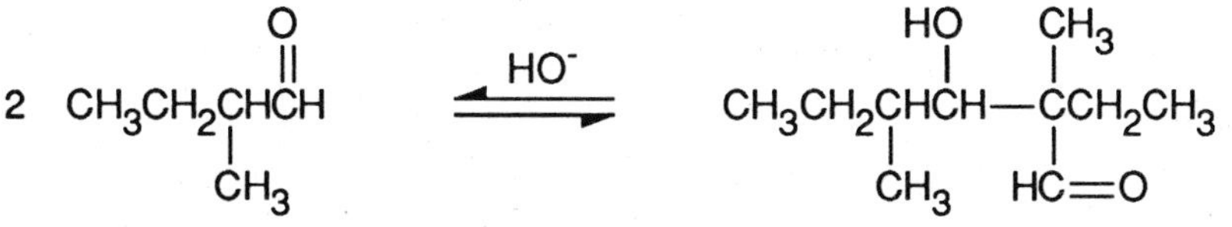

$$2 \ CH_3CH_2CHCH \ \underset{}{\overset{HO^-}{\rightleftharpoons}} \ CH_3CH_2CHCH-CCH_2CH_3$$

2-Methylbutanal

(no protons on α carbon atom)

(c) Aldol condensation is possible with 3-methylbutanal.

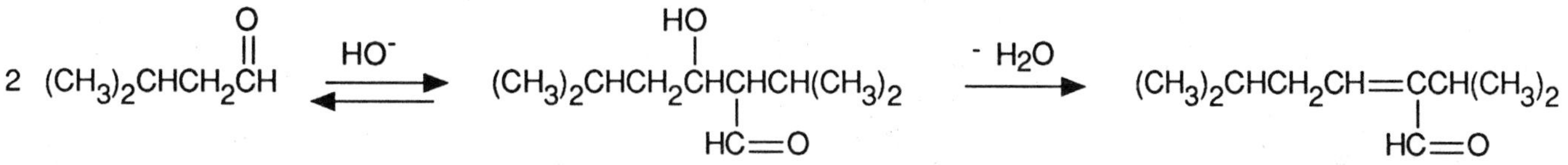

$$2 \ (CH_3)_2CHCH_2CH \ \underset{}{\overset{HO^-}{\rightleftharpoons}} \ (CH_3)_2CHCH_2CHCHCH(CH_3)_2 \ \overset{-H_2O}{\longrightarrow} \ (CH_3)_2CHCH_2CH=CCH(CH_3)_2$$

3-Methylbutanal

2-Isopropyl-5-methyl-2-hexenal

11.15 (a) Pivaldehyde has two methyl groups attached to C-2 of propanal.

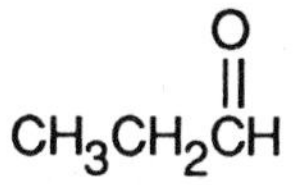

$$CH_3CH_2CH$$

Propanal

$$CH_3C-CH$$

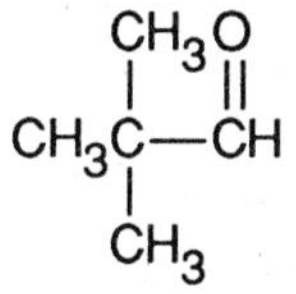

2,2-Dimethylpropanal
(pivaldehyde)

(b) Acrolein has a double bond between C-2 and C-3 of a three-carbon aldehyde.

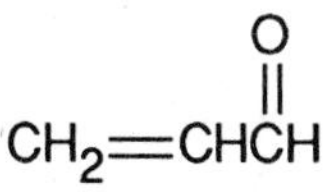

$$CH_2=CHCH$$

2-Propenal (acrolein)

(c) Deoxybenzoin (benzyl phenyl ketone) has a benzyl group and a phenyl group attached to a carbonyl group.

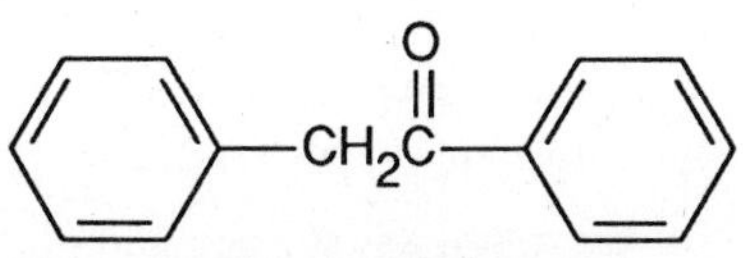

$$-CH_2C-$$

Benzyl phenyl ketone

(d) Diacetone alcohol is

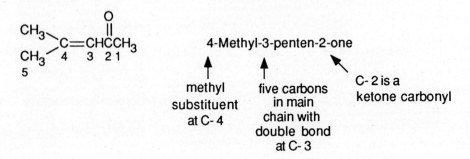

4-Hydroxy-4-methyl-2-pentanone

(e) The systematic name for mesityl oxide tells us that the correct structure is as shown.

CH₃
   ⟩C=CHCCH₃   4-Methyl-3-penten-2-one
CH₃ 4   3 21
5

↑ methyl substituent at C-4

↑ five carbons in main chain with double bond at C-3

↖ C-2 is a ketone carbonyl

(f) Citral has two double bonds, one between C-2 and C-3 and the other between C-6 and C-7. The one at C-2 has the *E* configuration. There are methyl substituents at C-3 and C-7.

(E)-3,7-Dimethyl-2,6-octadienal
(citral)

11.16 (a) The carbonyl group is at C-3 of a seven carbon chain in this ketone.

$$\text{(CH}_3)_3\text{CCHCH}_2\text{CCH}_2\text{CH}_3$$

5-Chloro-6,6-dimethyl-3-heptanone

(b) The compound is a trimethyl derivative of cyclohexanone. The position of the carbonyl group is assumed to be C-1, and is not numbered.

2,2,3-Trimethylcyclohexanone

214

(c) This aldehyde is named as a methyl-substituted pentenal.

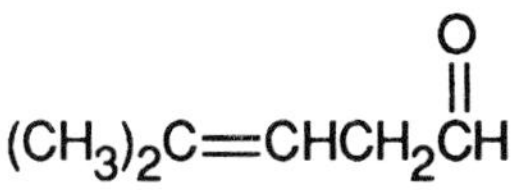

$$(CH_3)_2C=CHCH_2\overset{\overset{\displaystyle O}{\|}}{C}H$$

4-Methyl-3-pentenal

(d) The longest chain containing the carbonyl group has six carbons.

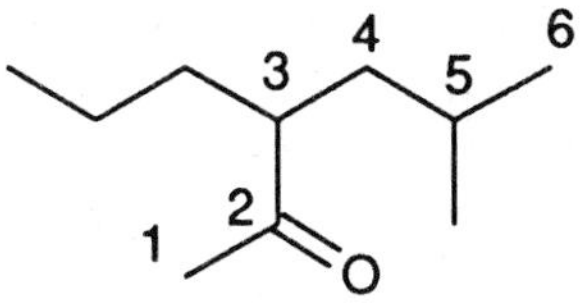

5-Methyl-3-propyl-2-hexanone

11.17 (a) First write the structure corresponding to the name. Dibenzyl ketone has two benzyl groups attached to a carbonyl.

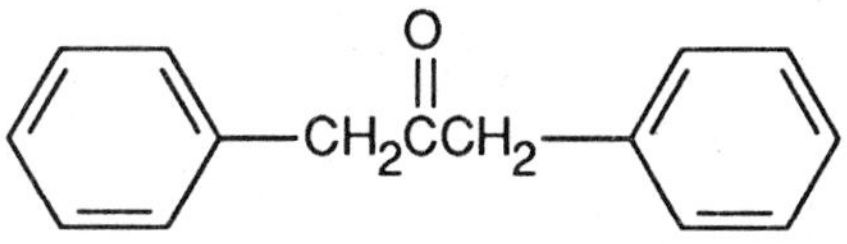

Dibenzyl ketone

The longest continuous chain contains three carbons, and C-2 is the carbon of the carbonyl group. A systematic IUPAC name for this ketone is **1,3-diphenyl-2-propanone.**

(b) The structure that corresponds to benzyl *tert*- butyl ketone is

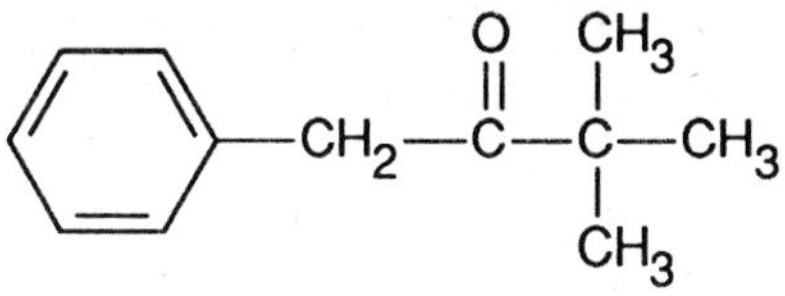

The longest carbon chain contains four carbon atoms; the compound is therefore named as a derivative of butane. An acceptable IUPAC name is **3,3-dimethyl-1-phenyl-2-butanone**.

(c) First write the structure form the name given. Ethyl isopropyl ketone has an ethyl group and an isopropyl group bonded to a carbonyl group.

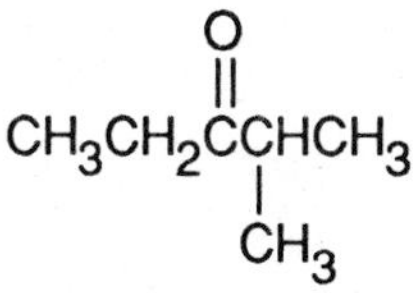

$$CH_3CH_2\overset{\overset{\displaystyle O}{\|}}{C}\underset{\underset{\displaystyle CH_3}{|}}{C}HCH_3$$

Ethyl isopropyl ketone may be alternatively named **2-methyl-3-pentanone**. Its longest continuous chain has five carbons. The carbonyl carbon is C-3 irrespective of the direction in which the chain is numbered, so we choose the one that gives the lowest number to the position that bears the methyl group.

(d)   Isobutyl methyl ketone has an isobutyl group [—CH2CH(CH3)2] and a methyl group bonded to a carbonyl group.

$$
\underset{\displaystyle CH_3\overset{\displaystyle O}{\overset{\|}{C}}CH_2\overset{\displaystyle CH_3}{\overset{|}{C}}HCH_3}{}
$$

The longest continuous chain has five carbons and the carbonyl carbon is C-2. The ketone may be named **4-methyl-2-pentanone**.

(e)   The structure corresponding to allyl methyl ketone is

$$
CH_3\overset{\displaystyle O}{\overset{\|}{C}}CH_2CH{=}CH_2
$$

Since the carbonyl group is given the lowest possible number in the chain, an appropriate name is **4-penten-2-one**.

11.18 (a)   The *-al* ending in the name suggests the compound is an aldehyde, however the carbonyl carbon of an aldehyde must bear at least one hydrogen, and therefore cannot be part of a ring.

(b)   Write the structure of the compound described by the name.

$$
CH_3\overset{\displaystyle CH_3}{\overset{|}{\underset{\underset{\displaystyle CH_3}{|}}{C}}}CH_2\overset{\displaystyle O}{\overset{\|}{C}}CH_2CH_3
$$

The chain should be numbered from the end that gives the carbonyl group the lower number. The correct name is **5,5-dimethyl-3-hexanone**.

(c)   Writing the structure reveals that the correct name of the compound is **3-chloropropanal**.

$$
ClCH_2CH_2\overset{\displaystyle O}{\overset{\|}{C}}H
$$

3-Chloropropanal

(d)   Trimethylacetaldehyde is an acceptable name for the compound shown. The preferred IUPAC name is **2,2-dimethylpropanal**.

$$
CH_3{-}\overset{\displaystyle H_3C}{\overset{|}{\underset{\underset{\displaystyle CH_3}{|}}{C}}}{-}\overset{\displaystyle O}{\overset{\|}{C}}H
$$

11.19 Hydrogen bonding forces present in alcohols give these compounds higher boiling points than compounds of similar molecular weight in which hydrogen bonds are absent. Attractive dipole forces give aldehydes a higher boiling point than hydrocarbons of similar molecular weight. The compounds in the problem and their boiling points are:

1-Pentanol (138°C);   pentanal (103.4°C);   hexane (68.8°C)

216

11.20 (a)   Lithium aluminum hydride reduces aldehydes to primary alcohols.

$$CH_3CH_2\overset{\overset{\displaystyle O}{\|}}{C}H \quad \xrightarrow[\text{2. } H_2O]{\text{1. } LiAlH_4} \quad CH_3CH_2CH_2OH$$

Propanal   1-Propanol

(b)   Sodium borohydride reduces aldehydes to primary alcohols.

$$CH_3CH_2\overset{\overset{\displaystyle O}{\|}}{C}H \quad \xrightarrow[CH_3OH]{NaBH_4} \quad CH_3CH_2CH_2OH$$

Propanal   1-Propanol

(c)   Aldehydes can be reduced to primary alcohols by catalytic hydrogenation.

$$CH_3CH_2\overset{\overset{\displaystyle O}{\|}}{C}H \quad \xrightarrow[Ni]{H_2} \quad CH_3CH_2CH_2OH$$

Propanal   1-Propanol

(d)   Aldehydes react with Grignard reagents to form secondary alcohols.

$$CH_3CH_2\overset{\overset{\displaystyle O}{\|}}{C}H \quad \xrightarrow[\text{2. } H_3O^+]{\substack{\text{1. } CH_3MgI \\ \text{diethyl ether}}} \quad CH_3CH_2\overset{\overset{\displaystyle OH}{|}}{C}HCH_3$$

Propanal   2-Butanol

(e)   Aldehydes are converted to acetals on reaction with alcohols in the presence of an acid catalyst.

$$CH_3CH_2\overset{\overset{\displaystyle O}{\|}}{C}H \quad + \quad 2\ CH_3OH \quad \xrightarrow{HCl} \quad CH_3CH_2CH(OCH_3)_2$$

Propanal   Methanol   Propanal dimethyl acetal

(f)   Aldehydes react with primary amines to yield imines.

$$CH_3CH_2\overset{\overset{\displaystyle O}{\|}}{C}H \quad + \quad C_6H_5NH_2 \quad \xrightarrow{-H_2O} \quad CH_3CH_2CH{=}NC_6H_5$$

Propanal   Aniline   N-Propylideneaniline

(g)   Reaction of an aldehyde with p-nitrophenylhydrazine is analogous to that with hydrazine.

$$CH_3CH_2\overset{\overset{\displaystyle O}{\|}}{C}H \quad + \quad O_2N{-}\!\!\bigcirc\!\!{-}NHNH_2 \quad \longrightarrow$$

Propanal   p-Nitrophenylhydrazine

$$CH_3CH_2CH{=}NNH{-}\!\!\bigcirc\!\!{-}NO_2 \quad + \quad H_2O$$

(h)   Acidification of solutions of sodium cyanide generates HCN, which reacts with aldehydes to form cyanohydrins.

$$CH_3CH_2\overset{\displaystyle O}{\overset{\|}{C}}H \quad + \quad HCN \quad \longrightarrow \quad CH_3CH_2\overset{\displaystyle OH}{\underset{|}{C}}HCN$$

Propanal   Hydrogen cyanide   1-Cyano-1-propanol

(i)   Silver oxide oxidizes aldehydes to carboxylic acids with formation of metallic silver.

$$CH_3CH_2\overset{\displaystyle O}{\overset{\|}{C}}H \quad \xrightarrow{Ag_2O} \quad CH_3CH_2CO_2H$$

Propanal   Propanoic acid

(j)   Chromic acid oxidizes aldehydes to carboxylic acids.

$$CH_3CH_2\overset{\displaystyle O}{\overset{\|}{C}}H \quad \xrightarrow{H_2CrO_4} \quad CH_3CH_2CO_2H$$

Propanal   Propanoic acid

11.21 (a)   Lithium aluminum hydride reduces ketones to secondary alcohols.

Cyclopentanone   Cyclopentanol

(b)   Sodium borohydride converts ketones to secondary alcohols.

Cyclopentanone   Cyclopentanol

(c)   Catalytic hydrogenation of ketones yields secondary alcohols.

Cyclopentanone   Cyclopentanol

218

(d) Grignard reagents react with ketones to form tertiary alcohols.

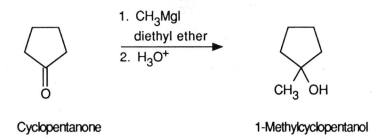

Cyclopentanone   1-Methylcyclopentanol

(e) The equilibrium constant for acetal formation from ketones is generally unfavorable.

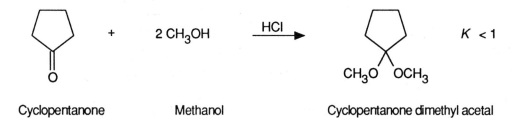

Cyclopentanone   Methanol   Cyclopentanone dimethyl acetal

(f) Ketones react with primary amines to form imines.

Cyclopentanone   Aniline   N-Cyclopentylideneaniline

(g) A p- nitrophenylhydrazone is formed.

Cyclopentanone   p-Nitrophenylhydrazine   Cyclopentanone p-nitrophenylhydrazone

(h) Cyanohydrin formation takes place.

Cyclopentanone   1-Cyanocyclopentanol

(i) No reaction occurs between silver oxide and cyclopentanone.

219

11.22 (a)   First consider all the isomeric aldehydes of molecular formula $C_5H_{10}O$.

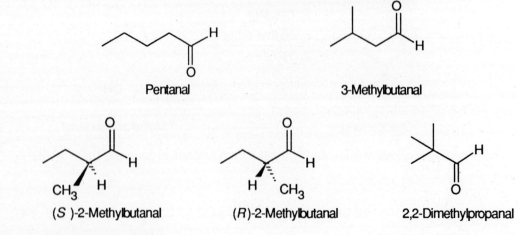

Pentanal

3-Methylbutanal

(S)-2-Methylbutanal

(R)-2-Methylbutanal

2,2-Dimethylpropanal

There are three isomeric ketones.

2-Pentanone

3-Pentanone

3-Methyl-2-butanone

(b)   Reduction of an aldehyde to a primary alcohol does not introduce a chiral center into the molecule. Therefore the only aldehydes that yield chiral alcohols on reduction are those that already contain a chiral center.

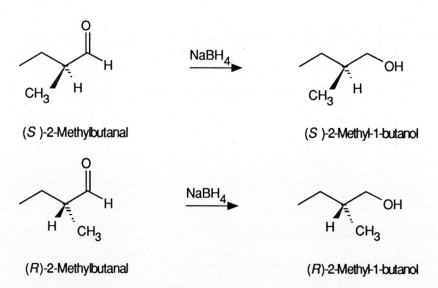

(S)-2-Methylbutanal

$\xrightarrow{\text{NaBH}_4}$

(S)-2-Methyl-1-butanol

(R)-2-Methylbutanal

$\xrightarrow{\text{NaBH}_4}$

(R)-2-Methyl-1-butanol

Among the ketones, 2-pentanone and 3-methyl-2-butanone are reduced to chiral alcohols.

2-Pentanone

$\xrightarrow{\text{NaBH}_4}$

2-Pentanol (chiral but racemic)

220

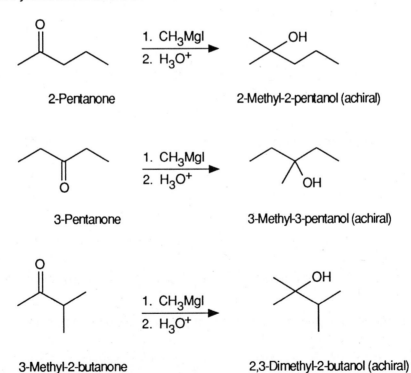

3-Pentanone → 3-Pentanol (achiral)

$NaBH_4$

3-Methyl-2-butanone → 3-Methyl-2-butanol (chiral but racemic)

$NaBH_4$

(c) All the aldehydes yield chiral alcohols on reaction with methylmagnesium iodide. Thus

$$C_4H_9\overset{\overset{O}{\|}}{C}H \quad \xrightarrow[2.\ H_3O^+]{1.\ CH_3MgI} \quad C_4H_9\overset{\overset{H}{|}}{\underset{OH}{C}}CH_3$$

A chiral center is introduced in each case.

None of the ketones yields chiral alcohols.

2-Pentanone

$\xrightarrow[2.\ H_3O^+]{1.\ CH_3MgI}$

2-Methyl-2-pentanol (achiral)

3-Pentanone

$\xrightarrow[2.\ H_3O^+]{1.\ CH_3MgI}$

3-Methyl-3-pentanol (achiral)

3-Methyl-2-butanone

$\xrightarrow[2.\ H_3O^+]{1.\ CH_3MgI}$

2,3-Dimethyl-2-butanol (achiral)

221

11.23 (a)   Grignard reagents are prepared by reaction of an alkyl halide with magnesium.

$$\underset{\substack{| \\ \text{I}}}{CH_3CHCH_3} \quad + \quad Mg \quad \xrightarrow{\text{diethyl ether}} \quad \underset{\substack{| \\ MgI}}{CH_3CHCH_3}$$

2-Iodopropane   Magnesium   Isopropylmagnesium iodide

(b)   Grignard reagents react with formaldehyde to give primary alcohols.

$$\underset{\substack{| \\ MgI}}{CH_3CHCH_3} \quad + \quad H_2C=O \quad \xrightarrow[\text{2. } H_3O^+]{\text{1. diethyl ether}} \quad \underset{\substack{| \\ CH_2OH}}{CH_3CHCH_3}$$

Isopropylmagnesium iodide   Formaldehyde   2-Methyl-1-propanol

(c)   Grignard reagents react with aldehydes (other than formaldehyde) to give secondary alcohols.

$$\underset{\substack{| \\ MgI}}{CH_3CHCH_3} \quad + \quad \underset{\substack{|| \\ C_6H_5CH}}{\overset{O}{}} \quad \xrightarrow[\text{2. } H_3O^+]{\text{1. diethyl ether}} \quad \underset{\substack{| \\ C_6H_5CHCH(CH_3)_2}}{\overset{OH}{}}$$

Isopropylmagnesium iodide   Benzaldehyde   2-Methyl-1-phenyl-1-propanol

(d)   Grignard reagents react with ketones to give tertiary alcohols.

$$\underset{\substack{| \\ MgI}}{CH_3CHCH_3} \quad + \quad \text{cyclopentanone} \quad \xrightarrow[\text{2. } H_3O^+]{\text{1. diethyl ether}} \quad \text{HO} \quad CH(CH_3)_2$$

Isopropylmagnesium iodide   Cyclopentanone   1-Isopropylcyclopentanol

11.24 In these problems the principles of retrosynthetic analysis are applied. The alkyl groups attached to the carbon that bears the hydroxyl group are mentally disconnected to reveal the Grignard reagent and carbonyl compound.

(a)   The two combinations that give the alcohol 5-methyl-3-hexanol

$$\underset{\substack{| \\ OH}}{CH_3CH_2CHCH_2CH(CH_3)_2}$$

are:

$$\underset{\substack{|| \\ O}}{CH_3CH_2CH} \quad + \quad XMgCH_2CH(CH_3)_2$$

Propanal   Isobutylmagnesium halide

222

and

$$CH_3CH_2MgX \quad + \quad \underset{\underset{O}{\|}}{HCCH_2CH(CH_3)_2}$$

Ethylmagnesium   3-Methylbutanal
halide

(b)   There is only one combination of Grignard reagent and carbonyl compound that gives 2,2-dimethyl-1-propanol $(CH_3)_3CCH_2OH$.

$$(CH_3)_3CMgX \qquad + \qquad \underset{\underset{H\overset{\|}{C}H}{}}{\overset{O}{}}$$

tert-Butylmagnesium halide   Formaldehyde

(c)   The tertiary alcohol 4-ethyl-4-octanol may be prepared by three different combinations of Grignard reagent and ketone.

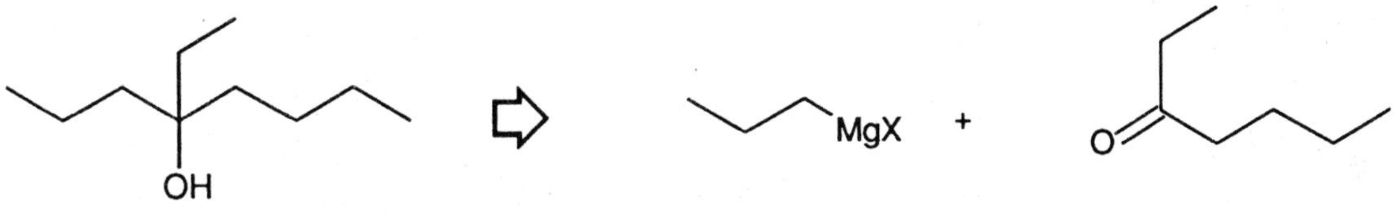

4-Ethyl-4-octanol   Propylmagnesium halide   3-Heptanone

or

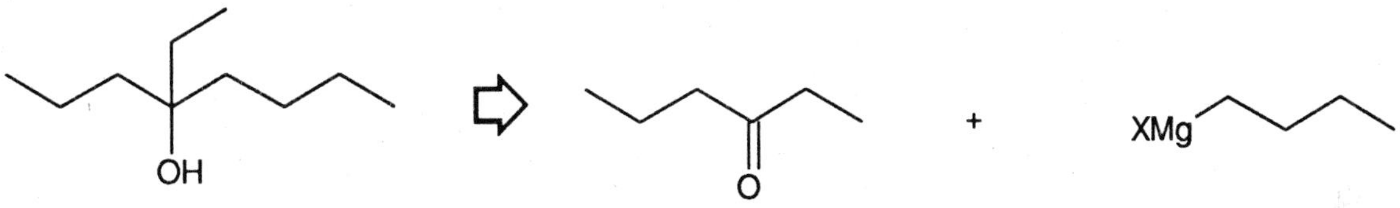

4-Ethyl-4-octanol   3-Hexanone   Butylmagnesium halide

or

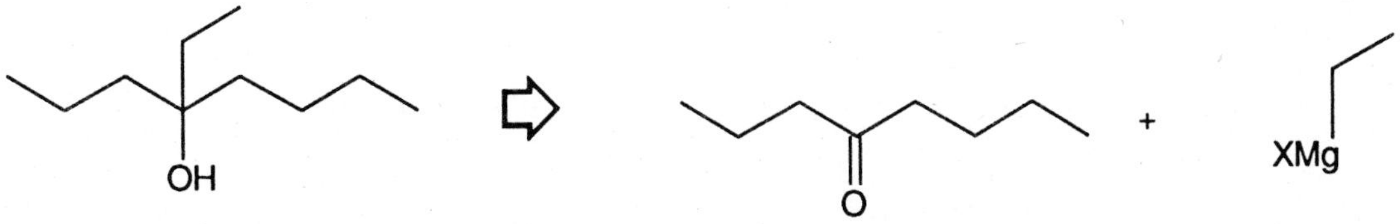

4-Ethyl-4-octanol   4-Octanone   Ethylmagnesium halide

11.25 Primary and secondary alcohols can be prepared by reduction of the corresponding aldehyde or ketone. The alcohols in (a) and (b) of the preceding problem can be prepared by reduction.

(a)

$$CH_3CH_2\overset{\overset{\displaystyle O}{\|}}{C}CH_2CH(CH_3)_2 \quad \xrightarrow[\text{2. } H_2O]{\text{1. LiAlH}_4} \quad CH_3CH_2\overset{\overset{\displaystyle OH}{|}}{C}HCH_2CH(CH_3)_2$$

5-Methyl-3-hexanone 5-Methyl-3-hexanol

(b)

$$(CH_3)_3\overset{\overset{\displaystyle O}{\|}}{C}CH \quad \xrightarrow[\text{2. } H_2O]{\text{1. LiAlH}_4} \quad (CH_3)_3CCH_2OH$$

2,2-Dimethylpropanal 2,2-Dimethyl-1-propanol

11.26 (a) The product is an acetal, formed by reaction of ethanol with acetaldehyde.

$$CH_3\overset{\overset{\displaystyle O}{\|}}{C}H \quad + \quad 2\ CH_3CH_2OH \quad \xrightarrow{H^+} \quad CH_3CH(OCH_2CH_3)_2$$

Acetaldehyde Ethanol Acetaldehyde diethyl acetal

(b) The product is an imine, formed from the reaction of acetophenone with *tert*-butylamine.

Acetophenone *tert*-Butylamine *N*-1-Phenylethylidene-*tert*-butylamine

(c) The tertiary alcohol is formed by reaction of phenylmagnesium bromide with benzyl methyl ketone.

$$C_6H_5MgBr \quad + \quad CH_3\overset{\overset{\displaystyle O}{\|}}{C}CH_2C_6H_5 \quad \xrightarrow[\text{2. } H_3O^+]{\text{1. diethyl ether}} \quad C_6H_5\overset{\overset{\displaystyle OH}{|}}{\underset{\underset{\displaystyle CH_3}{|}}{C}}CH_2C_6H_5$$

Phenylmagnesium bromide Benzyl methyl ketone 1,2-Diphenyl-2-propanol

11.27 (a) Only $(CH_3)_2CHCH=O$ can enolize, as the α carbon (shown in bold) in $(CH_3)_3\mathbf{C}CH=O$ has no hydrogens.

$$(CH_3)_2CH\overset{\overset{\displaystyle O}{\|}}{C}H \quad \rightleftharpoons \quad (CH_3)_2C=\overset{\overset{\displaystyle OH}{|}}{C}H$$

2-Methylpropanal Enol form

(b) Benzophenone has no α hydrogens; it cannot form an enol.

Enolization is impossible

224

Dibenzyl ketone enolizes slightly to form a small amount of enol.

$$C_6H_5CH_2\overset{\overset{\displaystyle O}{\|}}{C}CH_2C_6H_5 \quad \rightleftharpoons \quad C_6H_5CH=\overset{\overset{\displaystyle OH}{|}}{C}CH_2C_6H_5$$

Dibenzyl ketone   Enol form

11.28 (a) Aldehydes undergo aldol addition on treatment with base in cold ethanol.

$$2\ CH_3CH_2\overset{\overset{\displaystyle O}{\|}}{C}H \quad \xrightarrow[\text{cold ethanol}]{\text{NaOH}} \quad CH_3CH_2\overset{\overset{\displaystyle OH}{|}}{C}H\underset{\underset{\displaystyle CH_3}{|}}{C}H\overset{\overset{\displaystyle O}{\|}}{C}H$$

Propanal   3-Hydroxy-2-methylpentanal

(b) Dehydration of the aldol addition product occurs when the reaction is carried out at elevated temperature.

$$2\ CH_3CH_2\overset{\overset{\displaystyle O}{\|}}{C}H \quad \xrightarrow[\text{hot ethanol}]{\text{NaOH}} \quad CH_3CH_2CH=\underset{\underset{\displaystyle CH_3}{|}}{C}\overset{\overset{\displaystyle O}{\|}}{C}H$$

Propanal   2-Methyl-2-pentenal

(c) Sodium borohydride reduces the aldehyde function to the corresponding primary alcohol.

$$CH_3CH_2CH=\underset{\underset{\displaystyle CH_3}{|}}{C}\overset{\overset{\displaystyle O}{\|}}{C}H \quad \xrightarrow[\text{ethanol}]{\text{NaBH}_4} \quad CH_3CH_2CH=\underset{\underset{\displaystyle CH_3}{|}}{C}CH_2OH$$

2-Methyl-2-pentenal   2-Methyl-2-penten-1-ol

11.29 The product of aldol addition is the β–hydroxy aldehyde, and the final product after reduction is a diol.

$$CH_3CH_2CH_2\overset{\overset{\displaystyle O}{\|}}{C}H \quad \xrightarrow{\text{NaOH}} \quad CH_3CH_2CH_2\overset{\overset{\displaystyle OH}{|}}{C}H\underset{\underset{\displaystyle CH_2CH_3}{|}}{C}H\overset{\overset{\displaystyle O}{\|}}{C}H$$

Butanal   2-Ethyl-3-hydroxyhexanal

$$CH_3CH_2CH_2\overset{\overset{\displaystyle OH}{|}}{C}H\underset{\underset{\displaystyle CH_2CH_3}{|}}{C}H\overset{\overset{\displaystyle O}{\|}}{C}H \quad \xrightarrow[\text{catalyst}]{\text{H}_2} \quad CH_3CH_2CH_2\overset{\overset{\displaystyle OH}{|}}{C}H\underset{\underset{\displaystyle CH_2CH_3}{|}}{C}HCH_2OH$$

2-Ethyl-3-hydroxyhexanal   2-Ethyl-1, 3-hexanediol   (6-12)

11.30 Diphepanol is a tertiary alcohol and so may be prepared by reaction of a Grignard or organolithium reagent with a ketone. Retrosynthetically, two possibilities seem reasonable.

$$(C_6H_5)_2\underset{\underset{OH}{|}}{\overset{\overset{CH_3}{|}}{C}}CH-N\bigcirc \Rightarrow C_6H_5^- + C_6H_5\underset{\underset{O}{||}}{\overset{\overset{CH_3}{|}}{C}}CH-N\bigcirc$$

and

$$(C_6H_5)_2\underset{\underset{OH}{|}}{\overset{\overset{CH_3}{|}}{C}}CH-N\bigcirc \Rightarrow (C_6H_5)_2C{=}O + {}^-CH-N\bigcirc \overset{CH_3}{}$$

In principle either strategy is acceptable; in practice the one involving phenylmagnesium bromide is used.

$$C_6H_5MgBr + C_6H_5\underset{\underset{O}{||}}{\overset{\overset{CH_3}{|}}{C}}CH-N\bigcirc \xrightarrow[\text{2. }H_3O^+]{\text{1. ether}} (C_6H_5)_2\underset{\underset{CH_3}{|}}{\overset{\overset{CH_3}{|}}{C}}CH-N\bigcirc$$

11.31 (a)   1-Pentanol is a primary alcohol having one more carbon atom than 1-bromobutane. This suggests the reaction of a Grignard reagent with formaldehyde.

$$CH_3CH_2CH_2CH_2Br \xrightarrow[\text{diethyl ether}]{Mg} CH_3CH_2CH_2CH_2MgBr \xrightarrow[\text{2. }H_3O^+]{\text{1. }HCH\ (\overset{O}{||})} CH_3CH_2CH_2CH_2CH_2OH$$

| 1-Bromobutane | Butylmagnesium bromide | 1-Pentanol |

(b)   2-Hexanol is a secondary alcohol having two more carbon atoms than 1-bromobutane. It may be prepared by reaction of ethanal with butylmagnesium bromide.

$$CH_3CH_2CH_2CH_2Br \xrightarrow[\text{diethyl ether}]{Mg} CH_3CH_2CH_2CH_2MgBr \xrightarrow[\text{2. }H_3O^+]{\text{1. }CH_3CH\ (\overset{O}{||})} CH_3CH_2CH_2CH_2\underset{\underset{OH}{|}}{CH}CH_3$$

| 1-Bromobutane | Butylmagnesium bromide | 2-Hexanol |

(c)   1-Phenyl-1-pentanol is a secondary alcohol. This suggests that it can be prepared from butylmagnesium bromide and an aldehyde; benzaldehyde is the appropriate one.

$$CH_3CH_2CH_2CH_2MgBr + \bigcirc\!\!-\overset{\overset{O}{||}}{CH} \xrightarrow[\text{2. }H_3O^+]{\text{1. ether}} \bigcirc\!\!-\underset{\underset{OH}{|}}{CH}CH_2CH_2CH_2CH_3$$

| Butylmagnesium bromide | Benzaldehyde | 1-Phenyl-1-pentanol |

11.32 Since the target molecule is an eight-carbon secondary alcohol and the problem restricts our choices of starting materials to alcohols of five carbons or less, we are led to consider building up the carbon chain by a Grignard reaction.

$$CH_3CH_2CH-\!\!\!\!\!-CHCH_2CH_2CH_3 \quad \Rightarrow \quad CH_3CH_2CH \quad + \quad ^-CHCH_2CH_2CH_3$$

with CH_3 groups and OH and O as shown.

4-Methyl-3-heptanol

The disconnection shown leads to a three-carbon aldehyde and a five carbon Grignard reagent. Starting with the corresponding alcohols, the following synthetic scheme seems reasonable. First propanal is prepared.

$$CH_3CH_2CH_2OH \xrightarrow[\text{pyridine}]{(C_5H_5N)_2CrO_3} CH_3CH_2CH$$

1-Propanol   Propanal

After converting 2-pentanol to its bromo derivative, a solution of the Grignard reagent is prepared.

$$CH_3CHCH_2CH_2CH_3 \xrightarrow{PBr_3} CH_3CHCH_2CH_2CH_3 \xrightarrow[\text{diethyl ether}]{Mg} CH_3CHCH_2CH_2CH_3$$
$$OH \qquad\qquad\qquad\qquad Br \qquad\qquad\qquad\qquad MgBr$$

2-Pentanol   2-Bromopentane   1-Methylbutylmagnesium bromide

Reaction of the Grignard reagent with the aldehyde yields the desired 4-methyl-3-heptanol.

$$CH_3CHCH_2CH_2CH_3 \quad + \quad CH_3CH_2CH \xrightarrow[\text{2. } H_3O^+]{\text{1. diethyl ether}} CH_3CHCH_2CH_2CH_3$$
$$MgBr \qquad\qquad\qquad\qquad\qquad\qquad\qquad\qquad HOCHCH_2CH_3$$

1-Methylbutylmagnesium bromide   Propanal   4-Methyl-3-heptanol

11.33   Phenylmagnesium bromide reacts with 4-*tert*-butylcyclohexanone as shown:

$$\xrightarrow[\text{2. } H_3O^+]{\text{1. } C_6H_5MgBr, \text{ ether}}$$

4-*tert*-Butylcyclohexanone   4-*tert*-Butyl-1-phenylcyclohexanol

The phenyl substituent can be introduced either cis or trans to the *tert*-butyl group. Therefore the two alcohols are stereoisomers (diastereomers).

Dehydration of either alcohol yields 4-*tert*-butyl-1-phenylcyclohexene.

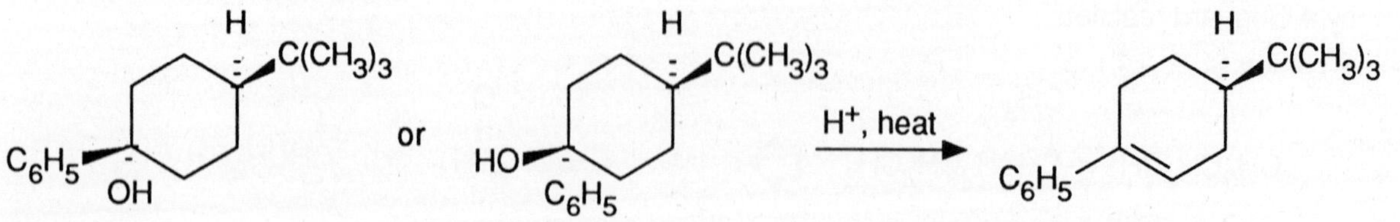

11.34 Cyclic hemiacetals are formed by intramolecular nucleophilic addition of a hydroxyl group to a carbonyl.

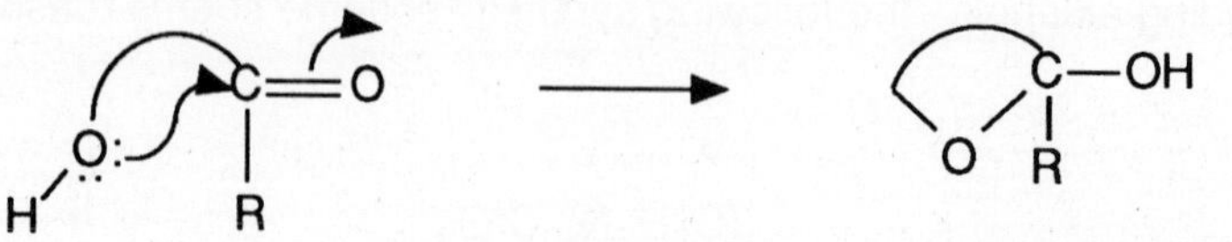

Cyclic hemiacetal

The ring oxygen is derived from the hydroxyl group; the carbonyl oxygen becomes the hydroxyl oxygen of the hemiacetal.

(a)   This compound is the cyclic hemiacetal of 5-hydroxypentanal.

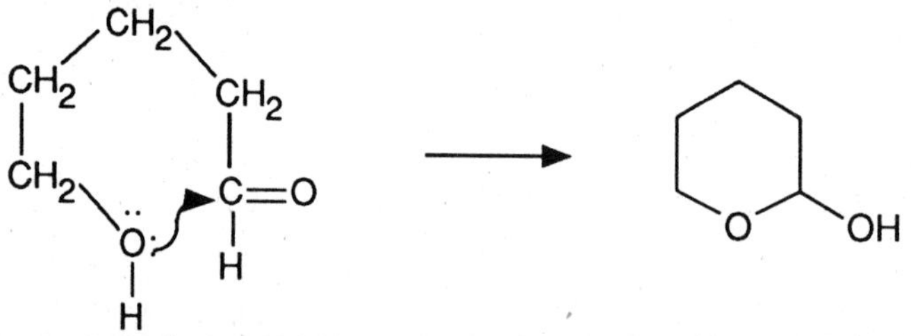

Indeed, 5-hydroxypentanal seems to exist entirely as the cyclic hemiacetal. Its infrared spectrum is devoid of absorption in the carbonyl region.

(b)   The carbon connected to two oxygens is the one that is derived from the carbonyl group. Using retrosynthetic symbolism, disconnect the ring oxygen from this carbon.

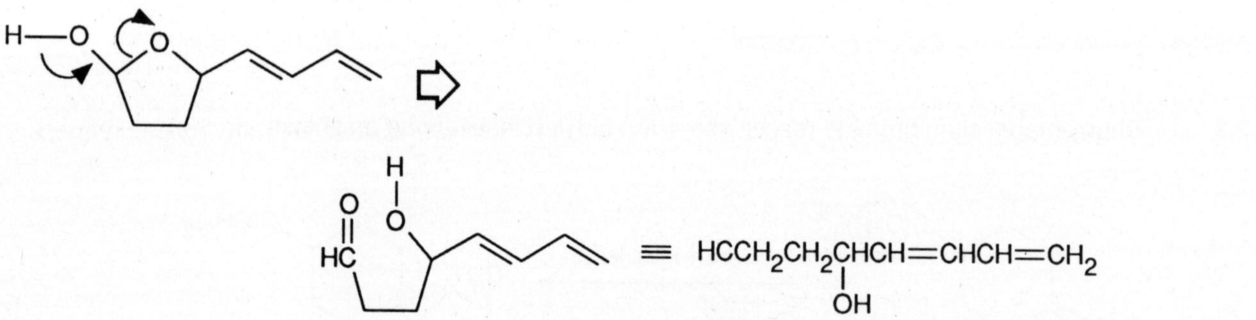

$$HCCH_2CH_2CHCH=CHCH=CH_2$$
$$OH$$

4-Hydroxy-5,7-octadienal

(c)   This compound is a cyclic acetal. The original carbonyl group is identifiable as the one that bears two oxygen substituents, which originate as hydroxyl oxygens of a diol.

228

Frontalin   6-Methyl-6,7-dihydroxy-2-heptanone

11.35 Compound C gives propanal and acetone upon ozonolysis, therefore compound C is **2-methyl-2-pentene**.

2-Methyl-2-pentene   Acetone   Propanal
(Compound C)

Compound A gives a positive Tollens test, and is therefore an aldehyde. Reaction with ethylmagnesium bromide followed by treatment with sulfuric acid gives compound C. The second reaction is a dehydration, as indicated by the difference in molecular formulas of compounds B and C. Compounds A and B are therefore **2-methylpropanal** and **2-methyl-2-pentanol**, respectively.

Compound A   Compound B

11.36 Compound F is achiral and gave 2-pentene as the *only* product upon dehydration, therefore compound F must be **3-pentanol**. Dehydration of 2-pentanol would have given a mixture of 1-pentene and 2-pentene.

Compound F   2-Pentene

Compound F was formed by reduction of compound D, so compound D must be the corresponding ketone, **3-pentanone**. This is consistent with compound D's failure to react with the Tollens reagent.

Compound D   Compound F

Compound E is the cyanohydrin formed by reaction of hydrogen cyanide with compound D.

Compound D   Compound E

229

11.37 Both the ketone 2-pentanone and the aldehyde pentanal would exhibit a strong infrared absorption in the region characteristic of carbonyl groups (1700 - 1750 cm^{-1}). The aldehyde would exhibit an additional absorption near 2750 cm^{-1} due to the formyl hydrogen, —CH=O. The alcohol 1- pentanol would exhibit a strong absorption due to the hydroxyl group in the region 3200 to 3600 cm^{-1}, and of course the carbonyl absorptions would be absent.

$$
\begin{array}{ccc}
& O & \\
& \parallel & \\
CH_3CH_2CH_2CCH_3 & & \\
\text{2-Pentanone} & &
\end{array}
$$

$$
\begin{array}{c}
O \\
\parallel \\
CH_3CH_2CH_2CH_2CH \\
\text{Pentanal}
\end{array}
$$

$$
CH_3CH_2CH_2CH_2CH_2OH
$$
1-Pentanol

11.38 A carbonyl group is evident in the strong infrared absorption at 1710 cm^{-1}. Since all the ^{1}H nmr signals are singlets, there are no nonequivalent hydrogens in a vicinal or "three bond" relationship. The three proton signal at δ = 2.1 ppm and the two proton signal at δ =2.4 ppm can be understood as arising from a

$$
\begin{array}{c}
O \\
\parallel \\
CH_2CCH_3
\end{array}
$$
unit. The intense nine proton singlet at δ =1.0 ppm is due to the three equivalent methyl groups of a $(CH_3)_3C$ unit. The compound is **4,4-dimethyl-2-pentanone**.

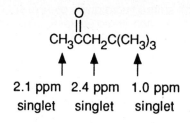

$$
\begin{array}{c}
O \\
\parallel \\
CH_3CCH_2C(CH_3)_3
\end{array}
$$

2.1 ppm   2.4 ppm   1.0 ppm
singlet   singlet   singlet

CHAPTER 12

CARBOXYLIC ACIDS

GLOSSARY OF TERMS

Carboxylate ion   The conjugate base of a carboxylic acid. An ion of the type:

$$\underset{\underset{RCO^-}{\overset{\parallel}{}}}{\overset{O}{}} \quad \text{also written as} \quad RCO_2^-$$

Carboxylation   In the preparation of a carboxylic acid, the reaction of a carbanion with carbon dioxide. Typically, the carbanion source is a Grignard reagent.

$$RMgX \xrightarrow[\text{2. } H_3O^+]{\text{1. } CO_2} RCO_2H$$

Carboxylic acid   A compound of the type:

$$\underset{\underset{RCOH}{\overset{\parallel}{}}}{\overset{O}{}} \quad \text{also written as} \quad RCO_2H$$

Detergent   Substances that clean by micellar action. While the term usually refers to a synthetic detergent, soaps are also detergents.

Fatty acid   Carboxylic acids, usually possessing unbranched carbon chains of 12-18 carbons, obtained by the hydrolysis of fats.

Hydrophilic   Literally, "water-loving." A term applied to substances which are soluble in water, usually because of their ability to form hydrogen bonds with water.

Hydrophobic   Literally, "water-hating." A term applied to substances which are not soluble in water, but are soluble in nonpolar hydrocarbon-like media.

Lipophilic   Literally, "fat-loving." Synonymous in practice with *hydrophobic*.

Micelle   A spherical aggregate of species such as carboxylate salts of fatty acids which contain a lipophilic end and a hydrophilic end. Micelles containing 50-100 carboxylate salts of fatty acids are soaps.

Nitrile   A compound of the type $RC\equiv N$. "R" may be alkyl or aryl. Also known as alkyl or aryl cyanides.

Soap   Cleansing substances obtained by the hydrolysis of fats in aqueous base. Soaps are sodium or potassium salts of unbranched carboxylic acids having 12-18 carbon atoms.

SOLUTIONS TO TEXT PROBLEMS

12.1 (b)   The carbon chain is four carbons long, with a methyl substituent at C-3.

$$(CH_3)_2CHCH_2CO_2H$$

3-Methylbutanoic acid

(c)   The longest continuous chain contains three carbons, and therefore it is named as a derivative of propanoic acid.

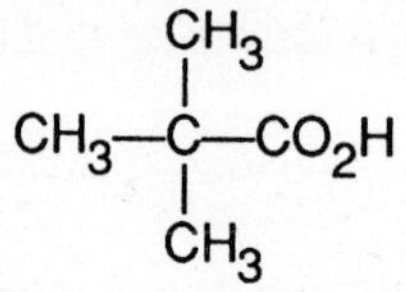

$$CH_3-\underset{\underset{CH_3}{|}}{\overset{\overset{CH_3}{|}}{C}}-CO_2H$$

2,2-Dimethylpropanoic acid

(d)   The longest chain is three carbons long, and contains a double bond as well as the carboxyl group. The systematic name for methacrylic acid is 2-methylpropenoic acid

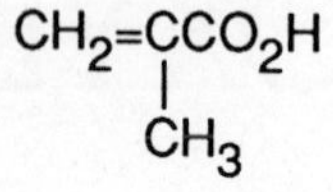

$$CH_2=\underset{\underset{CH_3}{|}}{C}CO_2H$$

2-Methylpropenoic acid

12.2 Carboxylic acids form hydrogen bonds with water using both the carbonyl oxygen and the hydroxyl proton.

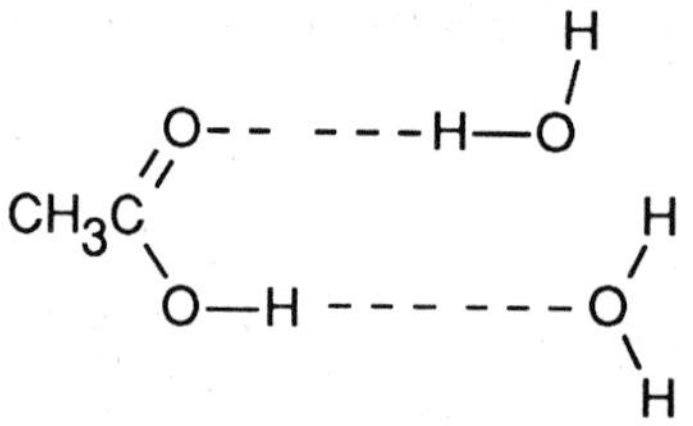

12.3 The equation for calculating pK_a is given in Section 3.7.

$$pK_a = -\log K_a$$
$$pK_a = -\log (3.3 \times 10^{-4})$$
$$pK_a = -(-3.48)$$
$$pK_a = 3.48$$

12.4 (b)   Propanoic acid is similar to acetic acid in its acidity. A hydroxyl group at C-2 is electron-withdrawing and stabilizes the carboxylate ion of lactic acid.

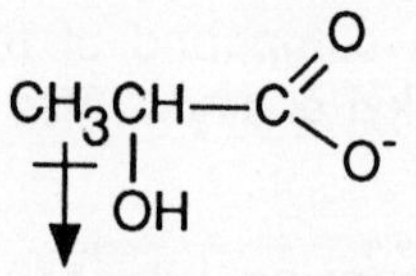

$$CH_3\underset{\underset{OH}{|}}{CH}-C\overset{\overset{O}{\parallel}}{\underset{}{\diagdown}}O^-$$

Hydroxyl group stabilizes negative charge by attracting electrons.

Lactic acid is more acidic than propanoic acid. The measured ionization constants are:

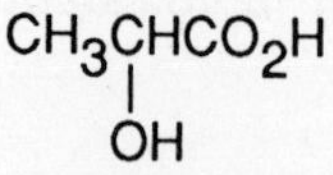

$$CH_3\underset{\underset{OH}{|}}{CH}CO_2H$$

Lactic acid

$K_a = 1.4 \times 10^{-4}$
$pK_a = 3.8$

$$CH_3CH_2CO_2H$$

Propanoic acid

$K_a = 1.3 \times 10^{-5}$
$pK_a = 4.9$

(c) Changing the hybridization of C-2 from sp^3 to sp^2 increases its *s* character and makes it more electron-withdrawing. Also, the electronegative carbonyl oxygen withdraws electrons and adds to the stabilization of the carboxylate ion. Pyruvic acid is a stronger acid than 2-methylpropanoic acid.

$$\underset{\text{O}}{\overset{\text{O}}{\|}}$$
$$CH_3CCO_2H \qquad\qquad CH_3CHCO_2H$$

Pyruvic acid
$K_a = 5.1 \times 10^{-4}$
$pK_a = 3.3$

2-Methylpropanoic acid
$K_a = 1.3 \times 10^{-5}$
$pK_a = 4.9$

12.5 (b) The acid-base reaction between acetic acid and *tert*-butoxide ion is represented by the equation:

$$CH_3CO_2H \;+\; (CH_3)_3CO^- \;\rightleftharpoons\; CH_3CO_2^- \;+\; (CH_3)_3COH$$

| Acetic acid | *tert*-Butoxide | Acetate ion | *tert*-Butyl alcohol |
| (stronger acid) | (stronger base) | (weaker base) | (weaker acid) |

Alcohols are weaker acids than carboxylic acids, so the equilibrium lies to the right.

(c) Bromide ion is the conjugate base of hydrogen bromide, a strong acid.

$$CH_3CO_2H \;+\; Br^- \;\rightleftharpoons\; CH_3CO_2^- \;+\; HBr$$

| Acetic acid | Bromide ion | Acetate ion | Hydrogen bromide |
| (weaker acid) | (weaker base) | (stronger base) | (stronger acid) |

In this case, the position of equilibrium favors the starting materials because acetic acid is a weaker acid than hydrogen bromide.

(d) Acetylide ion is a rather strong base and acetylene, with a K_a of 10^{-26}, is a much weaker acid than acetic acid. The position of equilibrium favors the formation of products.

$$CH_3CO_2H \;+\; HC\equiv C:^- \;\rightleftharpoons\; CH_3CO_2^- \;+\; HC\equiv CH$$

| Acetic acid | Acetylide ion | Acetate ion | Acetylene |
| (stronger acid) | (stronger base) | (weaker base) | (weaker acid) |

(e) Amide ion is a very strong base. The position of equilibrium lies to the right.

$$CH_3CO_2H \;+\; H_2N^- \;\rightleftharpoons\; CH_3CO_2^- \;+\; NH_3$$

| Acetic acid | Amide ion | Acetate ion | Ammonia |
| (stronger acid) | (stronger base) | (weaker base) | (weaker acid) |

12.6 (b) Reaction of 2-bromopropane with magnesium metal gives the corresponding Grignard reagent. Reaction with carbon dioxide followed by acid hydrolysis leads to the desired carboxylic acid.

$$(CH_3)_2CHBr \quad + \quad Mg \quad \xrightarrow{\text{ether}} \quad (CH_3)_2CHMgBr$$

2-Bromopropane   Magnesium   Isopropylmagnesium
                                                          bromide

$$(CH_3)_2CHMgBr \quad \xrightarrow[\text{2. } H_3O^+]{\text{1. } CO_2} \quad (CH_3)_2CHCO_2H$$

2-Methylpropanoic acid

(c) Benzoic acid is obtained by reaction of phenylmagnesium bromide with carbon dioxide, followed by hydrolysis.

Bromobenzene   Phenylmagnesium   Benzoic acid
                                      bromide

12.7 (b) Reaction of 2-bromopropane with sodium cyanide followed by hydrolysis leads to 2-methylpropanoic acid.

$$(CH_3)_2CHBr \quad \xrightarrow{\text{NaCN}} \quad (CH_3)_2CHCN \quad \xrightarrow[\text{heat}]{H_2O, H_2SO_4} \quad (CH_3)_2CHCO_2H$$

Isopropyl bromide   2-Methylpropanenitrile   2-Methylpropanoic acid

(c) The reaction between sodium cyanide and bromobenzene fails, thus benzoic acid cannot be prepared by this method. (Most aryl halides do not react with nucleophiles.)

12.8 (b) Acyl chlorides are formed by the reaction of a carboxylic acid and thionyl chloride.

$$\underset{\text{Phenylacetic acid}}{C_6H_5CH_2\overset{\overset{O}{\|}}{C}OH} + \underset{\text{Thionyl chloride}}{SOCl_2} \longrightarrow \underset{\text{Phenylacetyl chloride}}{C_6H_5CH_2\overset{\overset{O}{\|}}{C}Cl} + SO_2 + HCl$$

(c) Lithium aluminum hydride reduces carboxylic acids to the corresponding primary alcohol.

$$\underset{\text{Phenylacetic acid}}{C_6H_5CH_2\overset{\overset{O}{\|}}{C}OH} \quad \xrightarrow[\text{2. } H_2O]{\text{1. } LiAlH_4} \quad \underset{\text{2-Phenylethanol}}{C_6H_5CH_2CH_2OH}$$

(d) Sodium bicarbonate is a weak base, and reacts with phenylacetic acid to give the carboxylate salt.

$$C_6H_5CH_2\overset{\overset{\displaystyle O}{\|}}{C}OH \ + \ NaHCO_3 \ \longrightarrow \ C_6H_5CH_2\overset{\overset{\displaystyle O}{\|}}{C}O^- \ Na^+ \ + \ CO_2 \ + \ H_2O$$

Phenylacetic acid   Sodium bicarbonate   Sodium phenylacetate

12.9 (a) The parent hydrocarbon is tetradecane, an alkane with an unbranched chain of 14 carbons. The terminal methyl group is transformed to a carboxyl function in tetradecanoic acid.

$$CH_3(CH_2)_{12}\overset{\overset{\displaystyle O}{\|}}{C}OH$$
Tetradecanoic acid
(myristic acid)

(b) Undecane is an unbranched alkane with 11 carbon atoms, undecanoic acid is the corresponding carboxylic acid, and undecenoic acid is an 11-carbon carboxylic acid that contains a double bond. Since the carbon chain is numbered beginning with the carboxyl group, 10-undecenoic acid has its double bond at the opposite end of the chain from the carboxyl group.

$$CH_2{=}CH(CH_2)_8CO_2H$$
10-Undecenoic acid
(undecylenic acid)

(c) Mevalonic acid has a five-carbon chain with hydroxyl groups at C-3 and C-5, along with a methyl group at C-3.

$$HOCH_2CH_2\overset{\overset{\displaystyle CH_3}{|}}{\underset{\underset{\displaystyle OH}{|}}{C}}CH_2CO_2H$$
3,5-Dihydroxy-3-methylpentanoic acid
(mevalonic acid)

(d) The constitution represented by the systematic name 2-methyl-2-butenoic acid gives rise to two stereoisomers.

$$CH_3CH{=}\overset{\overset{\displaystyle\ }{}}{\underset{\underset{\displaystyle CH_3}{|}}{C}}CO_2H$$
2-Methyl-2-butenoic acid

Tiglic acid is the *E* isomer, and the *Z* isomer is known as angelic acid. The higher-priority substituents, methyl and carboxyl, are placed on opposite sides of the double bond in tiglic acid and on the same side in angelic acid.

(E)-2-Methyl-2-butenoic acid   (Z)-2-Methyl-2-butenoic acid

(tiglic acid)   (angelic acid)

12.10 (a) The longest chain of isobutyric acid is three carbons. Therefore, the correct systematic name is **2-methylpropanoic acid.**

$$CH_3\overset{\overset{\displaystyle CH_3}{|}}{C}HCO_2H$$
2-Methylpropanoic acid

(b) γ-Methylcaprilic acid is a methyl-substituted octanoic acid.

$$CH_3CH_2CH_2CH_2CHCH_2CH_2CO_2H$$
$$|$$
$$CH_3$$
                                              4-Methyloctanoic acid

(c) The correct systematic name is **2-chloropropanoic acid**.

$$CH_3CHCO_2H$$
$$|$$
$$Cl$$
                    2-Chloropropanoic acid

12.11 (a) The carboxylic acid contains an unbranched chain of eight carbon atoms. The parent alkane is octane, so the systematic IUPAC name of $CH_3(CH_2)_6CO_2H$ is **octanoic acid**.

(b) The compound shown is the potassium salt of octanoic acid. It is **potassium octanoate**.

(c) The presence of a double bond in $CH_2=CH(CH_2)_5CO_2H$ is indicated by the ending -*enoic acid*. Numbering of the chain begins with the carboxylic acid, so the double bond is between C-7 and C-8. The compound is **7-octenoic acid**.

12.12 All of the compounds in this problem are salts of the corresponding carboxylic acids.

(a) $C_6H_5CO_2^-$ Na^+ Sodium benzoate

(b) $[CH_3CH_2CH_2CO_2^-]_2$ Ca^{2+} Calcium propionate

(c) $CH_3CH=CHCH=CHCO_2^-$ K^+ Potassium sorbate

12.13 Dimethyl ether has the weakest intermolecular forces of these compounds, dipole association. The hydrogen bonding present in ethanol raises its boiling point relative to dimethyl ether. The strongest intermolecular forces are the hydrogen bonds present in acetic acid.

Dimethyl ether
(dipole association)

Ethanol
(hydrogen bonding)

Acetic acid
(hydrogen bonding)

12.14 (a) The order of decreasing acidity is:

		K_a	pK_a
Acetic acid	CH_3CO_2H	1.8×10^{-5}	4.7
Ethanol	CH_3CH_2OH	10^{-16}	16
Ethane	CH_3CH_3	$\sim 10^{-46}$	~46

(b)   Here again, the carboxylic acid is the strongest acid and the hydrocarbon is the weakest. Phenols are stronger acids than alcohols.

		K_a	pK_a
Benzoic acid	$C_6H_5CO_2H$	6.7×10^{-5}	4.2
Phenol	C_6H_5OH	10^{-10}	10
Benzyl alcohol	$C_6H_5CH_2OH$	$10^{-16} - 10^{-18}$	16 - 18
Benzene	C_6H_6	$\sim10^{-43}$	~43

(c)   Fluorine substituents increase the acidity of carboxylic acids and alcohols relative to their nonfluorinated analogs, but not enough to make fluorinated alcohols as acidic as carboxylic acids.

		K_a	pK_a
Trifluoroacetic acid	CF_3CO_2H	5.9×10^{-1}	0.2
Acetic acid	CH_3CO_2OH	1.8×10^{-5}	4.7
2,2,2-Trifluoroethanol	CF_3CH_2OH	4.2×10^{-13}	12.4
Ethanol	CH_3CH_2OH	$\sim10^{-16}$	~16

12.15 Carboxylate salts are prepared by reaction of the appropriate carboxylic acid with a base.

(a) $C_6H_5CO_2H + NaOH \longrightarrow C_6H_5CO_2^- Na^+ + H_2O$

(b) $2\ CH_3CH_2CO_2H + Ca(OH)_2 \longrightarrow [CH_3CH_2CO_2^-]_2\ Ca_2^+ + 2\ H_2O$

(c) $CH_3CH=CHCH=CHCO_2H + KOH \longrightarrow CH_3CH=CHCH=CHCO_2^- K^+ + H_2O$

12.16 (a)   The conversion of 1-butanol to butanoic acid is simply the oxidation of a primary alcohol to a carboxylic acid. Suitable oxidizing agents include potassium permanganate and chromic acid.

$$CH_3CH_2CH_2CH_2OH \xrightarrow[\text{2. } H^+]{\text{1. } KMnO_4} CH_3CH_2CH_2CO_2H$$

1-Butanol   Butanoic acid

(b)   Aldehydes may be oxidized to carboxylic acids by any of the oxidizing agents that convert primary alcohols to carboxylic acids or by silver oxide, a mild oxidizing agent that oxidizes aldehydes to carboxylic acids.

$$\underset{\text{Butanal}}{CH_3CH_2CH_2\overset{\displaystyle O}{\overset{\displaystyle \|}{C}}H} \xrightarrow[H_2SO_4]{K_2Cr_2O_7} \underset{\text{Butanoic acid}}{CH_3CH_2CH_2\overset{\displaystyle O}{\overset{\displaystyle \|}{C}}OH}$$

(c)   The starting material has the same number of carbon atoms as does butanoic acid, so all that is required is a series of functional group transformations. Carboxylic acids may be obtained by oxidation of the corresponding primary alcohol. The alcohol is available from the designated starting material, 1-butene.

$$CH_3CH_2CH_2CO_2H \implies CH_3CH_2CH_2CH_2OH \implies CH_3CH_2CH=CH_2$$

Hydroboration-oxidation of 1-butene yields 1-butanol, which can then be oxidized to butanoic acid as in (a).

$$CH_3CH_2CH=CH_2 \xrightarrow[\text{2. } H_2O_2, HO^-]{\text{1. } B_2H_6} CH_3CH_2CH_2CH_2OH \xrightarrow[\text{2. } H^+]{\text{1. } KMnO_4} CH_3CH_2CH_2CO_2H$$

1-Butene   1-Butanol   Butanoic acid

(d)   The starting material has one less carbon atom than the product. Chain extensions can be carried out by either of two methods, substitution with cyanide followed by hydrolysis or carboxylation of a Grignard reagent.

$$CH_3CH_2CH_2Cl \xrightarrow{NaCN} CH_3CH_2CH_2C{\equiv}N \xrightarrow[\text{heat}]{H_2O, H^+} CH_3CH_2CH_2CO_2H$$

1-Chloropropane   Butanenitrile   Butanoic acid

or

$$CH_3CH_2CH_2Cl \xrightarrow[\text{diethyl ether}]{Mg} CH_3CH_2CH_2MgCl \xrightarrow[\text{2. } H_3O^+]{\text{1. } CO_2} CH_3CH_2CH_2CO_2H$$

1-Chloropropane   Propylmagnesium chloride   Butanoic acid

12.17 The product has one more carbon atom than the starting material, the carbon of the carboxyl group.

$$CH_3CH_2CH_2CHCH_2{-}\{{-}CO_2H \quad \Rightarrow \quad CH_3CH_2CH_2CHCH_2OH \quad + \quad \text{"C"}$$
$$\qquad\qquad\quad | \qquad\qquad\qquad\qquad\qquad\qquad\qquad\quad |$$
$$\qquad\qquad\quad CH_3 \qquad\qquad\qquad\qquad\qquad\qquad\quad CH_3$$

The chain extension can be carried out by either carboxylation of a Grignard reagent, or hydrolysis of a nitrile. In either case, the alcohol must first be converted to the corresponding alkyl halide.

$$CH_3CH_2CH_2CHCH_2OH \xrightarrow{PBr_3} CH_3CH_2CH_2CHCH_2Br$$
$$\qquad\qquad\quad | \qquad\qquad\qquad\qquad\qquad\qquad\quad |$$
$$\qquad\qquad\quad CH_3 \qquad\qquad\qquad\qquad\qquad\qquad CH_3$$

2-Methyl-1-pentanol   1-Bromo-2-methylpentane

$$CH_3CH_2CH_2CHCH_2Br \xrightarrow[\substack{\text{2. } CO_2 \\ \text{3. } H_3O^+}]{\text{1. Mg, diethyl ether}} CH_3CH_2CH_2CHCH_2CO_2H$$
$$\qquad\qquad\quad | \qquad\qquad\qquad\qquad\qquad\qquad\qquad\qquad\quad |$$
$$\qquad\qquad\quad CH_3 \qquad\qquad\qquad\qquad\qquad\qquad\qquad\qquad CH_3$$

1-Bromo-2-methylpentane   3-Methylhexanoic acid

$$CH_3CH_2CH_2CHCH_2Br \xrightarrow[\text{2. } H_3O^+, \text{heat}]{\text{1. NaCN}} CH_3CH_2CH_2CHCH_2CO_2H$$
$$\qquad\qquad\quad | \qquad\qquad\qquad\qquad\qquad\qquad\qquad\quad |$$
$$\qquad\qquad\quad CH_3 \qquad\qquad\qquad\qquad\qquad\qquad\qquad CH_3$$

1-Bromo-2-methylpentane   3-Methylhexanoic acid

12.18 (a)   An acid-base reaction takes place when pentanoic acid is combined with sodium hydroxide.

$$CH_3CH_2CH_2CH_2CO_2H \;+\; NaOH \longrightarrow CH_3CH_2CH_2CH_2CO_2Na \;+\; H_2O$$

Pentanoic acid   Sodium hydroxide   Sodium pentanoate   Water

(b)   Carboxylic acids react with sodium bicarbonate to give carbonic acid, which dissociates to carbon dioxide and water, so the actual reaction that takes place is:

$$CH_3CH_2CH_2CH_2CO_2H \;+\; NaHCO_3 \longrightarrow CH_3CH_2CH_2CH_2CO_2Na \;+\; CO_2 \;+\; H_2O$$

Pentanoic acid   Sodium bicarbonate   Sodium pentanoate   Carbon dioxide   Water

(c)   Thionyl chloride is a reagent that converts carboxylic acids to the corresponding acyl chlorides.

$$CH_3CH_2CH_2CH_2CO_2H \;+\; SOCl_2 \longrightarrow CH_3CH_2CH_2CH_2\overset{\displaystyle O}{\overset{\displaystyle \|}{C}}Cl \;+\; SO_2 \;+\; HCl$$

Pentanoic acid   Thionyl chloride   Pentanoyl chloride   Sulfur dioxide   Hydrogen chloride

(d)   Lithium aluminum hydride is a powerful reducing agent and reduces carboxylic acids to primary alcohols.

$$CH_3CH_2CH_2CH_2CO_2H \xrightarrow[\text{2. } H_2O]{\text{1. } LiAlH_4} CH_3CH_2CH_2CH_2CH_2OH$$

Pentanoic acid   1-Pentanol

12.19 (a)   Conversion of butanoic acid to 1-butanol is a reduction and requires lithium aluminum hydride as the reducing agent.

$$CH_3CH_2CH_2CO_2H \xrightarrow[\text{2. } H_2O]{\text{1. } LiAlH_4} CH_3CH_2CH_2CH_2OH$$

Butanoic acid   1-Butanol

(b)   Alkyl halides are usually prepared from alcohols. Therefore, convert the 1-butanol in part (a) to 1-bromobutane by treating with phosphorus tribromide or heating with hydrogen bromide.

$$CH_3CH_2CH_2CH_2OH \xrightarrow[\text{or HBr, heat}]{PBr_3} CH_3CH_2CH_2CH_2Br$$

1-Butanol   1-Bromobutane

(c)   Pentanoic acid has one more carbon than butanoic acid. Chain extension can be achieved by carboxylation of a Grignard reagent or hydrolysis of a nitrile. Both these methods use 1-bromobutane, prepared in (b).

$$CH_3CH_2CH_2CH_2Br \xrightarrow[\substack{\text{2. } CO_2 \\ \text{3. } H_3O^+}]{\text{1. } Mg} CH_3CH_2CH_2CH_2\overset{\displaystyle O}{\overset{\displaystyle \|}{C}}OH$$

1-Bromobutane   Pentanoic acid

$$CH_3CH_2CH_2CH_2Br \xrightarrow[\text{2. } H_3O^+, \text{ heat}]{\text{1. NaCN}} CH_3CH_2CH_2CH_2\overset{\displaystyle O}{\overset{\|}{C}}OH$$

1-Bromobutane   Pentanoic acid

(d)   Use thionyl chloride to convert a carboxylic acid to the corresponding acyl chloride.

$$CH_3CH_2CH_2CO_2H \xrightarrow{\text{SOCl}_2} CH_3CH_2CH_2\overset{\displaystyle O}{\overset{\|}{C}}Cl$$

Butanoic acid   Butanoyl chloride

(e) Aromatic ketones are frequently prepared by Friedel-Crafts acylation of the appropriate acyl chloride and benzene. Butanoyl chloride, prepared in (d), can be used to acylate benzene in a Friedel-Crafts reaction.

Butanoyl chloride   Benzene   Phenyl propyl ketone

12.20 (a) Oxidation of an aldehyde gives the corresponding carboxylic acid.

$$CH_3CH_2CH_2CH_2\overset{\displaystyle O}{\overset{\|}{C}}H \xrightarrow[\text{H}_2\text{SO}_4]{\text{Na}_2\text{Cr}_2\text{O}_7} CH_3CH_2CH_2CH_2\overset{\displaystyle O}{\overset{\|}{C}}OH$$

Pentanal   Pentanoic acid

(b) Hexanoic acid has one more carbon than pentanoic acid. Chain extension can be achieved by first preparing the alkyl halide and then carboxylating the Grignard reagent. Hydrolysis of the corresponding nitrile would also be effective.

$$CH_3CH_2CH_2CH_2\overset{\displaystyle O}{\overset{\|}{C}}H \xrightarrow[\text{2. } H_2O]{\text{1. LiAlH}_4} CH_3CH_2CH_2CH_2CH_2OH \xrightarrow[\text{or HBr, heat}]{\text{PBr}_3} CH_3CH_2CH_2CH_2CH_2Br$$

Pentanal   1-Pentanol   1-Bromopentane

$$CH_3CH_2CH_2CH_2CH_2Br \xrightarrow[\substack{\text{2. CO}_2 \\ \text{3. } H_3O^+}]{\text{1. Mg}} CH_3CH_2CH_2CH_2CH_2CO_2H$$

1-Bromopentane   Hexanoic acid

(c) Hydroxycarboxylic acids can be prepared by hydrolysis of the appropriate cyanohydrin.

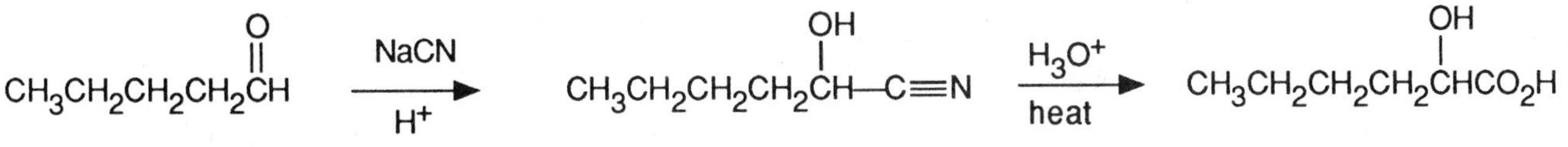

$$CH_3CH_2CH_2CH_2\overset{\overset{\displaystyle O}{\|}}{C}H \xrightarrow[H^+]{NaCN} CH_3CH_2CH_2CH_2\overset{\overset{\displaystyle OH}{|}}{C}H-C\equiv N \xrightarrow[\text{heat}]{H_3O^+} CH_3CH_2CH_2CH_2\overset{\overset{\displaystyle OH}{|}}{C}HCO_2H$$

| Pentanal | 2-Hydroxyhexanenitrile | 2-Hydroxyhexanoic acid |

12.21 The desired chain extension can be achieved by first reducing octanoic acid to 1-octanol, then preparing 1-bromooctane. The desired carboxylic acid is then obtained by carboxylation of the corresponding Grignard reagent. Nitrile hydrolysis is also effective for the last conversion.

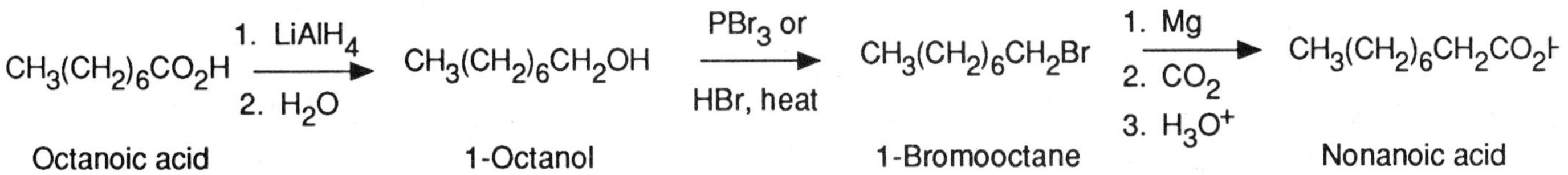

$$CH_3(CH_2)_6CO_2H \xrightarrow[\text{2. }H_2O]{\text{1. LiAlH}_4} CH_3(CH_2)_6CH_2OH \xrightarrow[\text{HBr, heat}]{PBr_3 \text{ or}} CH_3(CH_2)_6CH_2Br \xrightarrow[\substack{\text{2. }CO_2 \\ \text{3. }H_3O^+}]{\text{1. Mg}} CH_3(CH_2)_6CH_2CO_2H$$

| Octanoic acid | 1-Octanol | 1-Bromooctane | Nonanoic acid |

12.22 Hydration of the double bond can occur in two different directions.

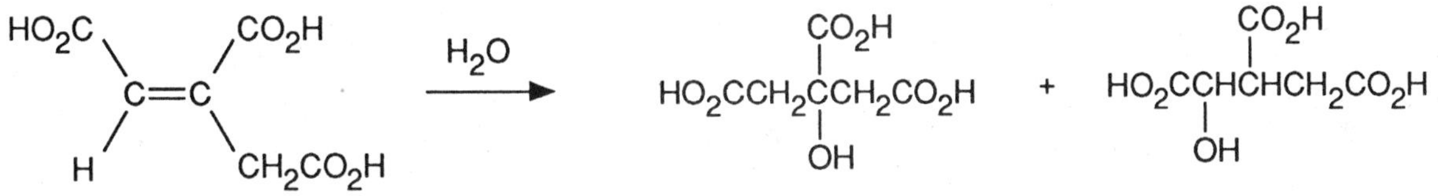

$$\underset{H}{\overset{HO_2C}{\diagdown}}C=C\underset{CH_2CO_2H}{\overset{CO_2H}{\diagup}} \xrightarrow{H_2O} HO_2CCH_2\overset{\overset{\displaystyle CO_2H}{|}}{\underset{\underset{\displaystyle OH}{|}}{C}}CH_2CO_2H \quad + \quad HO_2CCH\overset{\overset{\displaystyle CO_2H}{|}}{\underset{\underset{\displaystyle OH}{|}}{C}}HCH_2CO_2H$$

(a) The achiral isomer is citric acid. The structure shown contains no chiral centers.

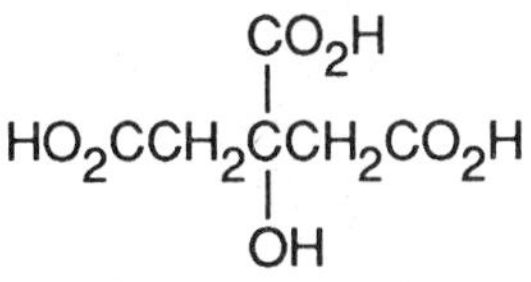

$$HO_2CCH_2\overset{\overset{\displaystyle CO_2H}{|}}{\underset{\underset{\displaystyle OH}{|}}{C}}CH_2CO_2H$$

(b) The other isomer, isocitric acid, has two chiral centers. Isocitric acid has the constitution shown below where the chiral centers are indicated by asterisks.

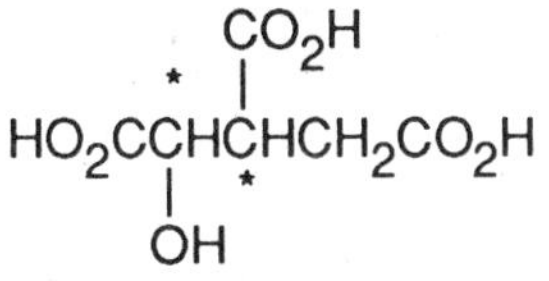

$$HO_2C\overset{*}{C}H\overset{\overset{\displaystyle CO_2H}{|}}{\underset{\underset{\displaystyle OH}{|}}{\overset{*}{C}}}HCH_2CO_2H$$

With two chiral centers, there are 2^2, or four stereoisomers represented by this constitution. The one that is actually formed in this enzyme catalyzed reaction is the 2*R*,3*S* isomer.

12.23 The molecular formula of compound A corresponds to a SODAR of 1. This can be accounted for by noting the strong infrared absorption in the carbonyl stretching region (1700 - 1750 cm^{-1}). The broad absorption from 2500 - 3400 cm^{-1} is typical of the hydroxyl band exhibited by carboxylic acids.

The ^{1}H nmr spectrum has three signals, all singlets. The signal farthest downfield at δ = 11.8 ppm is the proton of the carboxyl group. A methylene (-CH$_2$-) group adjacent to the carbonyl explains the 2H singlet at δ = 2.2 ppm. The 9H singlet at δ = 1.1 ppm is due to three equivalent methyl groups. Compound A is **3,3-dimethylbutanoic acid**.

$$\delta = 1.1 \text{ ppm} \longrightarrow \quad CH_3-\overset{\displaystyle \overset{CH_3}{\mid}}{\underset{\displaystyle \underset{CH_3}{\mid}}{C}}-CH_2-\overset{\displaystyle \overset{O}{\parallel}}{C}OH \quad \longleftarrow \delta = 11.8 \text{ ppm}$$

$$\delta = 2.2 \text{ ppm}$$

12.24 Carboxylic acid protons give signals in the range $\delta = 10$ to 12 ppm. A signal in this region suggests the presence of a carboxyl group but tells little about its environment. Thus, in assigning structures to compounds I, J, and K the data that will be most useful are the chemical shifts of the protons other than the carboxyl protons. Compare the three structures:

$$\overset{\displaystyle \overset{O}{\parallel}}{H}COH$$

$$\underset{HO_2C}{\overset{H}{\diagdown}}C=C\underset{CO_2H}{\overset{H}{\diagup}}$$

$$HO_2CCH_2CO_2H$$

Formic acid   **Maleic acid**   **Malonic acid**

The proton that is diagnostic of structure in formic acid is bonded to a carbonyl group; it is an aldehyde proton. Typical chemical shifts of aldehyde protons are 8 to 10 ppm, and therefore formic acid is compound K.

$$\text{Compound K} \quad H-\overset{\displaystyle \overset{O}{\parallel}}{C}-O-H \quad \longleftarrow \delta = 11.4 \text{ ppm}$$

$$\delta = 8.0 \text{ ppm}$$

The critical signals in maleic acid are those of the vinyl protons, which are normally found in the range $\partial = 5$ to 7 ppm. Maleic acid is compound J.

$$\text{Compound J} \quad \underset{HO_2C}{\overset{H}{\diagdown}}C=C\underset{CO_2H}{\overset{H}{\diagup}}$$

$\longleftarrow \delta = 6.3 \text{ ppm}$

$\longleftarrow \delta = 12.4 \text{ ppm}$

Compound I is malonic acid. Here we have a methylene group bearing two carbonyl substituents. These methylene protons are more shielded than the aldehyde proton of formic acid or the vinyl protons of maleic acid.

$$\text{Compound I} \quad HO_2CCH_2CO_2H \quad \longleftarrow \delta = 12.1 \text{ ppm}$$

$$\delta = 3.2 \text{ ppm}$$

CHAPTER 13

CARBOXYLIC ACID DERIVATIVES

GLOSSARY OF TERMS

Acid anhydride   A compound of the type

$$
\underset{\text{RCOCR}}{\overset{\text{O O}}{\underset{\|\;\|}{}}}
$$

Both R groups are usually the same, although they need not always be.

Acyl chloride   A compound of the type $\overset{\text{O}}{\overset{\|}{\text{RCCl}}}$. R may be alkyl or aryl.

Acyl group   The group $\overset{\text{O}}{\overset{\|}{\text{RC}}}$— .   R may be alkyl or aryl.

Acyl transfer   A reaction in which one type of carboxylic acid derivative is converted to another.

Amide   A compound of the type

$$
\underset{\text{RCNR'}_2}{\overset{\text{O}}{\underset{\|}{}}}
$$

The groups designated as R' may be any combination of H, alkyl, or aryl.

Carboxylic acid derivative   A compound which yields a carboxylic acid on hydrolysis. Carboxylic acid derivatives include acyl chlorides, anhydrides, esters, and amides.

Ester   A compound of the type $\overset{\text{O}}{\overset{\|}{\text{RCOR'}}}$.

Fatty acid   A carboxylic acid obtained by the hydrolysis of a fat or oil. Fatty acids typically have 12-18 carbon atoms and unbranched carbon chains. They may contain carbon-carbon double bonds.

Fischer esterification   Acid-catalyzed ester formation between an alcohol and a carboxylic acid.

$$
\overset{\text{O}}{\overset{\|}{\text{RCOH}}} + \text{R'OH} \; \overset{\text{H}^+}{\underset{}{\rightleftharpoons}} \; \overset{\text{O}}{\overset{\|}{\text{RCOR'}}} + \text{H}_2\text{O}
$$

Nucleophilic acyl substitution   Nucleophilic substitution at the carbon atom of an acyl group. An acyl transfer reaction.

Polyamide   A polymer in which individual structural units are joined by amide bonds. Nylon is a synthetic polyamide; proteins are naturally occurring polyamides.

Polyester   A polymer in which individual structural units are joined by ester bonds.

Saponification   Hydrolysis of esters in basic solution. The products are an alcohol and a carboxylate salt.

Tetrahedral intermediate   The key intermediate in nucleophilic acyl substitution. Formed by nucleophilic addition to the carbonyl group of a carboxylic acid derivative.

SOLUTIONS TO TEXT PROBLEMS

13.1 (b) Carboxylic acid anhydrides bear two acyl groups on oxygen, as in $\underset{\text{O}}{\overset{\text{O}}{\text{RCOCR}}}$. They are named as derivatives of carboxylic acids.

$$\underset{\underset{\text{C}_6\text{H}_5}{|}}{\text{CH}_3\text{CH}_2\text{CHCOH}}$$

$$\underset{\underset{\text{C}_6\text{H}_5}{|} \quad \underset{\text{C}_6\text{H}_5}{|}}{\text{CH}_3\text{CH}_2\text{CHCOCCHCH}_2\text{CH}_3}$$

2-Phenylbutanoic acid   2-Phenylbutanoic anhydride

(c) Butyl 2-phenylbutanoate is the butyl ester of 2-phenylbutanoic acid

$$\underset{\underset{\text{C}_6\text{H}_5}{|}}{\text{CH}_3\text{CH}_2\text{CHCOCH}_2\text{CH}_2\text{CH}_2\text{CH}_3}$$ Butyl 2-phenylbutanoate

(d) In 2-phenylbutyl butanoate, the 2-phenylbutyl group is an alkyl group bonded to oxygen of the ester. It is not involved in the acyl group of the molecule

$$\underset{\underset{\text{C}_6\text{H}_5}{|}}{\text{CH}_3\text{CH}_2\text{CH}_2\text{COCH}_2\text{CHCH}_2\text{CH}_3}$$ 2-Phenylbutyl butanoate

(e) The ending -*amide* reveals this to be a compound of the type $\underset{\text{O}}{\overset{\text{O}}{\text{RCNH}_2}}$.

$$\underset{\underset{\text{C}_6\text{H}_5}{|}}{\text{CH}_3\text{CH}_2\text{CHCNH}_2}$$ 2-Phenylbutanamide

13.2 Benzoyl chloride reacts rapidly with water in an exothermic reaction to yield benzoic acid

$$\underset{}{\overset{\text{O}}{\text{C}_6\text{H}_5\text{CCl}}} + \text{H}_2\text{O} \longrightarrow \underset{}{\overset{\text{O}}{\text{C}_6\text{H}_5\text{COH}}} + \text{HCl}$$

13.3 Alcohols react with carboxylic acids in the presence of an acid catalyst to give an ester and water.

(b)

$$\text{C}_6\text{H}_5\text{CH}_2\text{OH} + (\text{CH}_3)_2\text{CHCH}_2\overset{\text{O}}{\text{COH}} \xrightarrow{\text{H}^+} (\text{CH}_3)_2\text{CHCH}_2\overset{\text{O}}{\text{COCH}_2\text{C}_6\text{H}_5} + \text{H}_2\text{O}$$

Benzyl alcohol   3-Methylbutanoic Acid   Benzyl 3-methylbutanoate   Water

(c) The relationship of the molecular formula of the ester ($C_{10}H_{10}O_4$) to that of the starting dicarboxylic acid ($C_8H_6O_4$) indicates that the diacid reacted with two moles of methanol to form a diester.

$$2CH_3OH \quad + \quad HOC\!\!-\!\!\bigcirc\!\!-\!\!COH \quad \xrightarrow{H^+} \quad CH_3OC\!\!-\!\!\bigcirc\!\!-\!\!COCH_3$$

Methanol   1,4-Benzenedicarboxylic acid   Dimethyl 1,4-benzenedicarboxylate

13.4 (b) The structure of the tetrahedral intermediate corresponds to addition of the alcohol to the carbonyl group of the carboxylic acid.

$$(CH_3)_2CHCH_2COH \; + \; HOCH_2C_6H_5 \; \longrightarrow \; (CH_3)_2CHCH_2\!\!\underset{OH}{\overset{OH}{C}}\!\!OCH_2C_6H_5 \; \longrightarrow \; \text{ester}$$

3-Methylbutanoic acid   Benzyl alcohol   Tetrahedral intermediate

(c) Each carboxyl group reacts with a molecule of methanol, thus two tetrahedral intermediates form in this reaction.

$$HOC\!\!-\!\!\bigcirc\!\!-\!\!COH \; \xrightarrow{CH_3OH} \; HOC\!\!-\!\!\bigcirc\!\!-\!\!\underset{OH}{\overset{OH}{C}}OCH_3 \; \longrightarrow \; \text{monoester}$$

First tetrahedral intermediate

$$\text{monoester} \; \xrightarrow{CH_3OH} \; CH_3O\underset{OH}{\overset{OH}{C}}\!\!-\!\!\bigcirc\!\!-\!\!COCH_3 \; \longrightarrow \; \text{diester}$$

Second tetrahedral intermediate

13.5 (b) The name *benzyl acetate* reveals that the acyl portion is derived from acetic acid, and the alkyl group from benzyl alcohol.

$$CH_3COH \; \xrightarrow{SOCl_2} \; CH_3CCl \; \xrightarrow[\text{pyridine}]{C_6H_5CH_2OH} \; CH_3COCH_2C_6H_5$$

Acetic acid   Acetyl chloride   Benzyl acetate

(c) The acyl portion of this ester is derived from 2-methylpropanoic acid, the alkyl portion from isopropyl alcohol.

$$(CH_3)_2CHCOH \; \xrightarrow{SOCl_2} \; (CH_3)_2CHCCl \; \xrightarrow[\text{pyridine}]{(CH_3)_2CHOH} \; (CH_3)_2CHCOCH(CH_3)_2$$

2-Methylpropanoic acid   2-Methylpropanoyl chloride   Isopropyl 2-methylpropanoate

13.6 The acetate ester is formed by reaction of the phenolic hydroxyl and acetic anhydride.

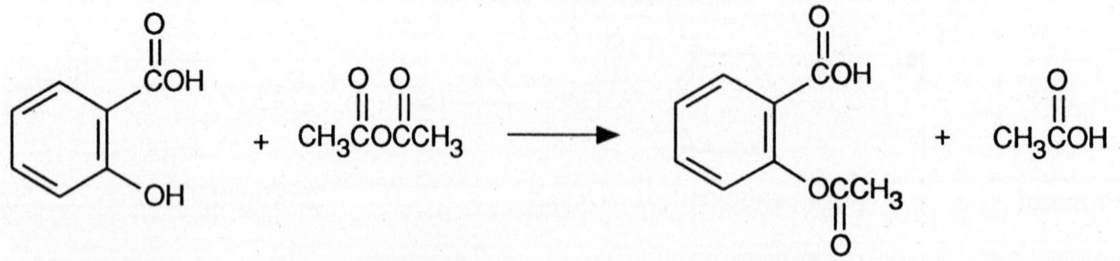

Salicylic acid | Acetic anhydride | Aspirin | Acetic acid

13.7 The products of hydrolysis of this component of beeswax are hexadecanoic acid and 1-triacontanol.

$$CH_3(CH_2)_{14}\overset{O}{\overset{\|}{C}}OCH_2(CH_2)_{28}CH_3 \xrightarrow{H_3O^+} CH_3(CH_2)_{14}CO_2H + HOCH_2(CH_2)_{28}CH_3$$

Triacontyl hexadecanoate | Hexadecanoic Acid | 1-Triacontanol

13.8 Based on trimyristin's molecular formula $C_{45}H_{86}O_6$ and on the fact that its hydrolysis gives only glycerol and tetradecanoic acid $CH_3(CH_2)_{12}CO_2H$, it must have the structure shown.

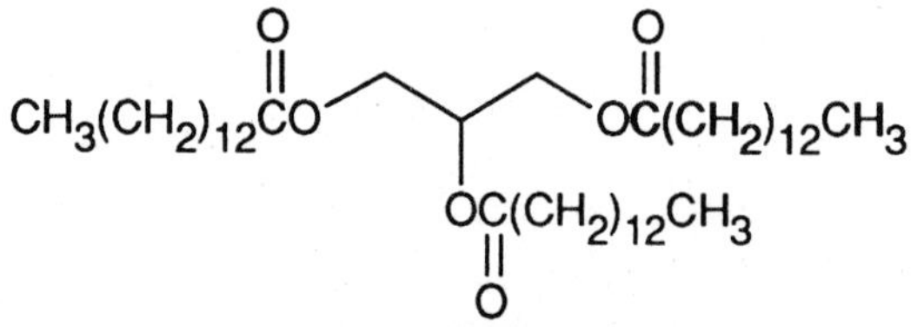

Trimyristin
($C_{45}H_{86}O_6$)

13.9 **Step 1:** Nucleophilic addition of hydroxide ion to the carbonyl group

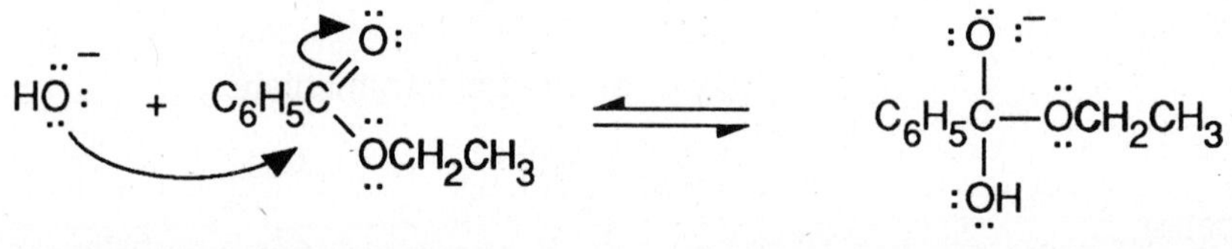

Hydroxide Ion | Ethyl benzoate | Anionic form of tetrahedral intermediate

Step 2: Proton transfer from water to give the neutral form of the tetrahedral intermediate

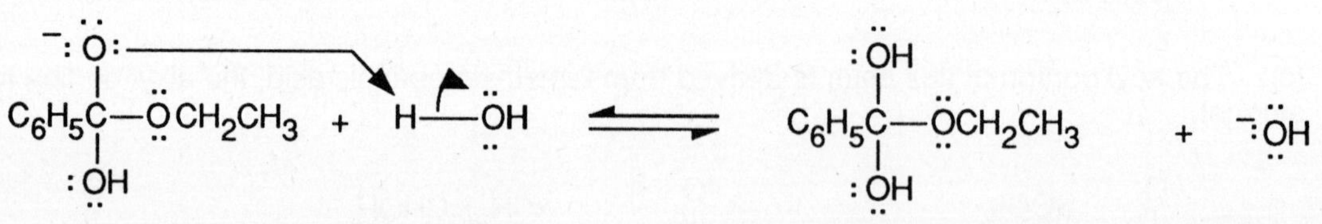

Anionic form of tetrahedral intermediate | Water | Tetrahedral intermediate | Hydroxide ion

Step 3: Hydroxide ion-promoted dissociation of tetrahedral intermediate

| Hydroxide ion | Tetrahedral intermediate | | Water | Benzoic acid | Ethoxide ion |

Step 4: Proton abstraction from benzoic acid

| Benzoic acid | Hydroxide ion | | Benzoate ion | Water |

13.10 (b)   The preparation of 6-methyl-6-undecanol has been described in the chemical literature. It was achieved in 75 percent yield by the reaction of pentylmagnesium bromide with ethyl acetate.

$$2CH_3CH_2CH_2CH_2CH_2MgBr \ + \ CH_3\overset{O}{\overset{||}{C}}OCH_2CH_3 \ \xrightarrow[\text{2. } H_3O^+]{\text{1. diethyl ether}} \ CH_3\overset{OH}{\overset{|}{C}}(CH_2CH_2CH_2CH_2CH_3)_2$$

| Pentylmagnesium bromide | Ethyl acetate | 6-Methyl-6-undecanol |

(c)   The two phenyl substituents arise by addition of a phenyl Grignard reagent to an ester of formic acid, HCO_2H.

$$2C_6H_5MgBr \ + \ H\overset{O}{\overset{||}{C}}OCH_2CH_3 \ \xrightarrow[\text{2. } H_3O^+]{\text{1. diethyl ether}} \ (C_6H_5)_2CHOH$$

| Phenylmagnesium bromide | Ethyl formate | Diphenylmethanol |

(d)   All the substituents bonded to the hydroxyl-bearing carbon in this compound are phenyl groups. Therefore use an ester benzoic acid and phenylmagnesium bromide.

| Phenylmagnesium bromide | Ethyl benzoate | Triphenylmethanol | Ethanol |

13.11 Each of the phenolic hydroxyls of bisphenol-A reacts with a molecule of phosgene. The "free" acyl chloride end groups react with additional molecules of bisphenol-A, giving rise to a polymer chain.

$$
\underset{\text{ClCCl}}{\overset{\overset{\displaystyle O}{\|}}{}} + HO-\!\!\left\langle\bigcirc\right\rangle\!\!-\underset{\overset{\displaystyle CH_3}{\underset{\displaystyle CH_3}{|}}}{\overset{|}{C}}-\!\!\left\langle\bigcirc\right\rangle\!\!-OH \longrightarrow \left[\underset{CO}{\overset{\overset{\displaystyle O}{\|}}{}}-O-\!\!\left\langle\bigcirc\right\rangle\!\!-\underset{\overset{\displaystyle CH_3}{\underset{\displaystyle CH_3}{|}}}{\overset{|}{C}}-\!\!\left\langle\bigcirc\right\rangle\!\!-O\underset{OCO}{\overset{\overset{\displaystyle O}{\|}}{}}\right]_n
$$

13.12 (b) *N*-Methylacetamide may be prepared by the reaction of methylamine with acetyl chloride.

$$
\underset{\text{Acetyl chloride}}{\overset{\overset{\displaystyle O}{\|}}{CH_3CCl}} + \underset{\text{Methylamine}}{2\ CH_3NH_2} \longrightarrow \underset{\text{\textit{N}-Methylacetamide}}{\overset{\overset{\displaystyle O}{\|}}{CH_3CNHCH_3}} + \underset{\text{Methylammonium chloride}}{CH_3NH_3^+\ Cl^-}
$$

(c) The acyl group is derived from benzoic acid, $C_6H_5CO_2H$. Reaction of benzoyl chloride with dimethylamine yields the desired amide.

$$
\underset{\text{Benzoyl chloride}}{\overset{\overset{\displaystyle O}{\|}}{C_6H_5CCl}} + \underset{\text{Dimethylamine}}{2\ HN(CH_3)_2} \longrightarrow \underset{\text{\textit{N,N}-Dimethylbenzamide}}{\overset{\overset{\displaystyle O}{\|}}{C_6H_5CN(CH_3)_2}} + \underset{\text{Dimethylammonium chloride}}{(CH_3)_2NH_2^+\ Cl^-}
$$

13.13 The ester acts as an acyl transfer agent, and forms an amide on reaction with methylamine.

$$
\underset{\text{Methylamine}}{CH_3NH_2} + \underset{\text{Ethyl 2-methylpropanoate}}{\overset{\overset{\displaystyle O}{\|}}{(CH_3)_2CHCOCH_2CH_3}} \longrightarrow \underset{\text{\textit{N}-Methyl-2-methylpropanamide}}{\overset{\overset{\displaystyle O}{\|}}{(CH_3)_2CHCNHCH_3}} + \underset{\text{Ethanol}}{CH_3CH_2OH}
$$

13.14 (b) Hydrolysis under basic conditions yields the amine and a carboxylate salt.

$$
\underset{\text{\textit{N}-Methylacetamide}}{\overset{\overset{\displaystyle O}{\|}}{CH_3CNHCH_3}} \xrightarrow[\text{heat}]{H_2O,\ NaOH} \underset{\text{Sodium acetate}}{\overset{\overset{\displaystyle O}{\|}}{CH_3CO^-\ Na^+}} + \underset{\text{Methylamine}}{CH_3NH_2}
$$

(c) The products of acidic hydrolysis of *N*-phenylbenzamide are benzoic acid and anilinium chloride.

$$
\underset{\text{\textit{N}-Phenylbenzamide}}{\overset{\overset{\displaystyle O}{\|}}{C_6H_5CNHC_6H_5}} \xrightarrow[\text{heat}]{HCl,\ H_2O} \underset{\text{Benzoic acid}}{\overset{\overset{\displaystyle O}{\|}}{C_6H_5COH}} + \underset{\text{Anilinium chloride}}{C_6H_5NH_3^+\ Cl^-}
$$

13.15 (a) The halogen that is attached to the carbonyl group is identified in the name as a separate word following the name of an acyl group.

$$
\underset{\text{Cl}}{\overset{\text{O}}{\parallel}}\text{—CCl}
$$

m-Chlorobenzoyl chloride

(b) This acyl chloride is derived from 4-methylpentanoic acid.

$$
(CH_3)_2CHCH_2CH_2\overset{\overset{\text{O}}{\parallel}}{C}Cl
$$

4-Methylpentanoyl chloride

(c) Trifluoroacetic anhydride is the anhydride of trifluoroacetic acid. Notice that it contains six fluorines.

$$
CF_3\overset{\overset{\text{O}}{\parallel}}{C}O\overset{\overset{\text{O}}{\parallel}}{C}CF_3
$$

Trifluoroacetic anhydride

(d) 1-Phenylethyl acetate is the ester of 1-phenylethanol and acetic acid.

$$
CH_3\overset{\overset{\text{O}}{\parallel}}{C}O\underset{\underset{CH_3}{\mid}}{CH}—
$$

1-Phenylethyl acetate

(e) Butyl 2-methylbutanoate is the ester of 1-butanol and 2-methylbutanoic acid.

$$
CH_3CH_2\underset{\underset{CH_3}{\mid}}{CH}\overset{\overset{\text{O}}{\parallel}}{C}OCH_2CH_2CH_2CH_3
$$

Butyl 2-methylbutanoate

(f) The parent compound is benzamide. In *N*-ethylbenzamide the ethyl substituent is bonded to nitrogen.

$$
\overset{\overset{\text{O}}{\parallel}}{—C}NHCH_2CH_3
$$

N-Ethylbenzamide

(g) Two phenyl groups are bonded to nitrogen in *N,N*-diphenylacetamide.

$$
CH_3\overset{\overset{\text{O}}{\parallel}}{C}N(C_6H_5)_2
$$

N,N-Diphenylacetamide

13.16 (a) This compound is an acyl chloride, and is named **3-chlorobutanoyl chloride**.

$$
CH_3\underset{\underset{Cl}{\mid}}{CH}CH_2\overset{\overset{\text{O}}{\parallel}}{C}Cl
$$

(b) The group attached to oxygen, in this case benzyl, is identified first in the name of the ester. As the benzyl ester of acetic acid its name is **benzyl acetate**.

$$
CH_3\overset{\overset{\text{O}}{\parallel}}{C}OCH_2—
$$

(c) The group attached to oxygen is methyl. This compound is the methyl ester of phenylacetic acid. It is **methyl phenylacetate**.

$$
CH_3O\overset{\overset{\text{O}}{\parallel}}{C}CH_2—
$$

(d) This compound is an amide. We name the corresponding acid, then replace the suffic-*oic acid* by -*amide*. The correct name is **4-methylpentanamide**.

$$CH_3CHCH_2CH_2\overset{\overset{\displaystyle O}{\|}}{C}NH_2$$
$$\underset{CH_3}{|}$$

(e) This compound is the *N*-methyl derivative of the amide in part (d). It is ***N*-methyl-4-methylpentanamide**.

$$CH_3CHCH_2CH_2\overset{\overset{\displaystyle O}{\|}}{C}NHCH_3$$
$$\underset{CH_3}{|}$$

(f) The amide nitrogen bears two methyl groups. We designate this as an *N,N*-dimethylamide. It is ***N,N*-dimethyl4-methylpentanamide**.

$$CH_3CHCH_2CH_2\overset{\overset{\displaystyle O}{\|}}{C}N(CH_3)_2$$
$$\underset{CH_3}{|}$$

13.17 (a) Thionyl chloride reacts with with carboxylic acids to form acyl chlorides.

$$C_6H_5CH_2\overset{\overset{\displaystyle O}{\|}}{C}OH \xrightarrow{SOCl_2} C_6H_5CH_2\overset{\overset{\displaystyle O}{\|}}{C}Cl$$

Phenylacetic acid   Phenylacetyl chloride

(b) Acyl chlorides undergo a rapid reaction with water to form the corresponding carboxylic acid.

$$C_6H_5CH_2\overset{\overset{\displaystyle O}{\|}}{C}Cl \quad + \quad H_2O \longrightarrow C_6H_5CH_2CO_2H \quad + \quad HCl$$

Phenylacetyl chloride   Water   Phenylacetic acid   Hydrogen chloride

(c) Esters are formed by reaction of acyl chlorides with alcohols.

$$C_6H_5CH_2\overset{\overset{\displaystyle O}{\|}}{C}Cl \quad + \quad C_6H_5CH_2OH \longrightarrow C_6H_5CH_2\overset{\overset{\displaystyle O}{\|}}{C}OCH_2C_6H_5$$

Phenylacetyl chloride   Benzyl alcohol   Benzyl phenylacetate

(d) Amines react with acyl chlorides to form amides.

$$C_6H_5CH_2\overset{\overset{\displaystyle O}{\|}}{C}Cl \quad + \quad C_6H_5CH_2NH_2 \longrightarrow C_6H_5CH_2\overset{\overset{\displaystyle O}{\|}}{C}NHCH_2C_6H_5$$

Phenylacetyl chloride   Benzylamine   *N*-Benzylphenylacetamide

(e) Esters are hydrolyzed on heating in aqueous acid solution.

$$C_6H_5CH_2\overset{\overset{\displaystyle O}{\|}}{C}OCH_2C_6H_5 \xrightarrow[\text{heat}]{HCl, H_2O} C_6H_5CH_2\overset{\overset{\displaystyle O}{\|}}{C}OH \quad + \quad C_6H_5CH_2OH$$

Benzyl phenylacetate   Phenylacetic acid   Benzyl alcohol

(f) Hydrolysis of an ester in basic solution produces an alcohol and a carboxylate salt.

$$C_6H_5CH_2\overset{\overset{O}{\|}}{C}OCH_2C_6H_5 \xrightarrow[\text{heat}]{H_2O,\ NaOH} C_6H_5CH_2\overset{\overset{O}{\|}}{C}O^-\ Na^+ \quad + \quad C_6H_5CH_2OH$$

Benzyl phenylacetate   Sodium phenylacetate   Benzyl alcohol

(g) Acidic hydrolysis of an amide gives a carboxylic acid and the salt of the corresponding amine.

$$C_6H_5CH_2\overset{\overset{O}{\|}}{C}NHCH_2C_6H_5 \xrightarrow[\text{heat}]{HCl,\ H_2O} C_6H_5CH_2\overset{\overset{O}{\|}}{C}OH \ + \ C_6H_5CH_2NH_3^+\ Cl^-$$

N-Benzylphenylacetamide   Phenylacetic acid   Benzylammonium chloride

(h) Basic hydrolysis of an amide produces an amine and a carboxylate salt.

$$C_6H_5CH_2\overset{\overset{O}{\|}}{C}NHCH_2C_6H_5 \xrightarrow[\text{heat}]{H_2O,\ NaOH} C_6H_5CH_2\overset{\overset{O}{\|}}{C}O^-\ Na^+ \ + \ C_6H_5CH_2NH_2$$

N-Benzylphenylacetamide   Sodium phenylacetate   Benzylamine

(i) Acetic anhydride reacts with alcohols to form esters.

$$CH_3\overset{\overset{O}{\|}}{C}O\overset{\overset{O}{\|}}{C}CH_3 \ + \ HO-\bigcirc \longrightarrow CH_3\overset{\overset{O}{\|}}{C}O-\bigcirc \ + \ CH_3\overset{\overset{O}{\|}}{C}OH$$

Acetic anhydride   Cyclohexanol   Cyclohexyl acetate   Acetic acid

(j) Reaction of an amine with acetic anhydride yields an amide.

$$CH_3\overset{\overset{O}{\|}}{C}O\overset{\overset{O}{\|}}{C}CH_3 \ + \ (CH_3)_2NH \longrightarrow CH_3\overset{\overset{O}{\|}}{C}N(CH_3)_2 \ + \ CH_3\overset{\overset{O}{\|}}{C}O^-\ H_2\overset{+}{N}(CH_3)_2$$

Acetic anhydride   Dimethylamine   N,N-Dimethylacetamide Dimethylammonium acetate

(k) Amines and esters react to form amides.

$$H\overset{\overset{O}{\|}}{C}OCH_2CH_3 \ + \ CH_3CH_2NH_2 \longrightarrow H\overset{\overset{O}{\|}}{C}NHCH_2CH_3 \ + \ CH_3CH_2OH$$

Ethyl formate   Ethylamine   N-Ethylformamide   Ethanol

13.18 (a) Heating an amine or ammonia and a carboxylic acid yields an amide and water.

$$NH_3 \ + \ C_6H_5CO_2H \xrightarrow{\text{heat}} C_6H_5\overset{\overset{O}{\|}}{C}NH_2 \ + \ H_2O$$

Ammonia   Benzoic acid   Benzamide   Water

(b) The reaction of acetic anhydride and 1-butanol will yield butyl acetate and acetic acid.

$$CH_3\overset{\displaystyle O}{\overset{\|}{C}}O\overset{\displaystyle O}{\overset{\|}{C}}CH_3 \quad + \quad CH_3CH_2CH_2CH_2OH \quad \longrightarrow \quad CH_3\overset{\displaystyle O}{\overset{\|}{C}}OCH_2CH_2CH_2CH_3 \quad + \quad CH_3\overset{\displaystyle O}{\overset{\|}{C}}OH$$

Acetic anhydride   1-Butanol   Butyl acetate   Acetic acid

(c) The acid-catalyzed hydrolysis of formamide will yield formic acid and an ammonium ion.

$$HC\overset{\displaystyle O}{\overset{\|}{}}NH_2 \quad + \quad H_3O^+ \quad \xrightarrow{\text{heat}} \quad HC\overset{\displaystyle O}{\overset{\|}{}}OH \quad + \quad NH_4^+$$

Formamide   Hydronium ion   Formic acid   Ammonium ion

(d) An ester is formed by reaction of a carboxylic acid and an alcohol. This reaction is a Fischer esterification.

$$(CH_3)_2CHOH \quad + \quad CH_3CH_2CO_2H \quad \xrightarrow{H^+} \quad CH_3CH_2CO_2CH(CH_3)_2$$

Isopropyl alcohol   Propanoic acid   Isopropyl propanoate

13.19 (a) Ethanol is the nucleophile that adds to the carbonyl group of benzoyl chloride to form the tetrahedral intermediate.

$$C_6H_5\overset{\displaystyle O}{\overset{\|}{C}}Cl \quad + \quad CH_3CH_2OH \quad \longrightarrow \quad C_6H_5\overset{\displaystyle OH}{\underset{\displaystyle Cl}{\overset{|}{\underset{|}{C}}}}OCH_2CH_3$$

Benzoyl chloride   Ethanol   Tetrahedral intermediate

(b) The tetrahedral intermediate formed from acetic anhydride and ethylamine has a carbon-nitrogen bond.

$$CH_3\overset{\displaystyle O}{\overset{\|}{C}}O\overset{\displaystyle O}{\overset{\|}{C}}CH_3 \quad + \quad CH_3CH_2NH_2 \quad \longrightarrow \quad CH_3\overset{\displaystyle O}{\overset{\|}{C}}\underset{\displaystyle NHCH_2CH_3}{\overset{\displaystyle OH}{\overset{|}{\underset{|}{C}}}}CH_3$$

Acetic anhydride   Ethylamine   Tetrahedral intermediate

(c) Hydroxide ion is the nucleophile in the saponification (base-promoted hydrolysis) of an ester. The neutral form of the tetrahedral intermediate is as shown in the equation.

$$C_6H_5\overset{\displaystyle O}{\overset{\|}{C}}OCH_3 \quad + \quad H_2O \quad \xrightarrow{^-OH} \quad C_6H_5\underset{\displaystyle OH}{\overset{\displaystyle OH}{\overset{|}{\underset{|}{C}}}}OCH_3$$

Methyl benzoate   Water   Tetrahedral intermediate

(d)   Ammonia adds to the carbonyl group in the formation of the tetrahedral intermediate with methyl benzoate.

$$\underset{\text{Methyl benzoate}}{\overset{\overset{\textstyle O}{\|}}{C_6H_5COCH_3}} \quad + \quad \underset{\text{Ammonia}}{NH_3} \quad \longrightarrow \quad \underset{\text{Tetrahedral intermediate}}{\overset{\overset{\textstyle OH}{|}}{\underset{\underset{\textstyle NH_2}{|}}{C_6H_5COCH_3}}}$$

13.20 Fischer esterification is the acid-catalyzed reaction of an alcohol with a carboxylic acid. Ethyl acetate is prepared by the reaction of acetic acid with ethanol.

13.21 **Step 1:** Nucleophilic addition of hydroxide ion to the carbonyl group

Step 2: Proton transfer to give neutral form of the tetrahedral intermediate

Step 3: Proton transfer from water to nitrogen of tetrahedral intermediate

Step 4: Dissociation of *N*-protonated form of tetrahedral intermediate

Step 5: Irreversible formation of formate ion by proton abstraction from formic acid

13.22 Since ester hydrolysis in base proceeds by acyl-oxygen cleavage, the ^{18}O label (shown as bold-face **O** in the equation) becomes incorporated into the carboxylate ion, acetate.

13.23 (a) Acetyl chloride is prepared by reaction of acetic acid with thionyl chloride. The first task then is to prepare acetic acid by oxidation of ethanol.

(b) Ethanol can be converted to ethyl acetate by reaction with acetic acid or acetyl chloride.

or

(c)   Reaction of acetyl chloride prepared in (a) with ammonia gives acetamide.

$$CH_3\overset{\overset{\displaystyle O}{\|}}{C}Cl \xrightarrow{\;NH_3\;} CH_3\overset{\overset{\displaystyle O}{\|}}{C}NH_2$$

Acetyl chloride   Acetamide

13.24 (a)   The target molecule is a tertiary alcohol in which two of the groups (phenyl) are the same. Thus treatment of an ester with 2 moles of Grignard reagent will give the desired tertiary alcohol.

$$C_6H_5Br \xrightarrow[\text{diethyl ether}]{Mg} C_6H_5MgBr$$

Bromobenzene   Phenylmagnesium bromide

$$2C_6H_5MgBr \;\; + \;\; CH_3CH_2CO_2CH_3 \xrightarrow[\text{2. } H_3O^+]{\text{1. diethyl ether}} CH_3CH_2\overset{\overset{\displaystyle OH}{|}}{C}(C_6H_5)_2$$

Phenylmagnesium bromide   Methyl propanoate   1,1-Diphenyl-1-propanol

(b)   Since the desired product is a secondary alcohol, phenylmagnesium bromide should be treated with an aldehyde. The appropriate aldehyde is benzaldehyde.

Phenylmagnesium bromide   Benzaldehyde   Diphenylmethanol

(c)   Benzamide is prepared by heating ammonia with benzoic acid. The required benzoic acid is obtained by carboxylation of the Grignard reagent from (a).

$$C_6H_5MgBr \xrightarrow[\text{2. } H_3O^+]{\text{1. } CO_2} C_6H_5CO_2H \xrightarrow[\text{heat}]{NH_3} C_6H_5\overset{\overset{\displaystyle O}{\|}}{C}NH_2$$

Phenylmagnesium bromide   Benzoic acid   Benzamide

The synthesis outlined above is modeled after that shown for *N*-methylbenzamide on p 390 in the text. Alternatively, benzoic acid may be converted to benzoyl chloride with thionyl chloride, which is then treated with ammonia to give benzamide.

13.25 (a)   The starting material is a cyclic anhydride. Acid anhydrides react with water to yield two carboxylic acid functions; when the anhydride is cyclic, a dicarboxylic acid results.

$$+ \quad H_2O \longrightarrow HO\overset{\overset{\displaystyle O}{\|}}{C}CH_2CH_2\overset{\overset{\displaystyle O}{\|}}{C}OH$$

Succinic anhydride   Water   Succinic acid

(b)   The starting material is a cyclic amide. Hydrolysis of amides in acid solution amides yields a carboxylic acids and an ammonium salt.

N-Methylpyrrolidone   4-(Methylammonio)butanoic acid

(c)   The starting material is a cyclic ester, a lactone. Esters undergo saponification in aqueous base to give an alcohol and a carboxylate salt.

γ-Butyrolactone   Sodium hydroxide   Sodium 4-hydroxybutanoate

(d)   Grignard reagents react with esters to give tertiary alcohols.

γ-Butyrolactone   4-Methyl-1,4-pentanediol

13.26 The starting material contains three acetate ester functions. All these ester groups undergo hydrolysis in aqueous sulfuric acid.

$(C_5H_{12}O_3)$

The name of the product is 1,2,5-pentanetriol. Also formed in the hydrolysis of the starting triacetate are three molecules of acetic acid.

13.27 The first two steps in the synthesis of novocain are formation of an acyl chloride and an ester, respectively. The reagents necessary for these conversions are:

Step 1:

$$RCOH(=O) \xrightarrow{SOCl_2} RCCl(=O)$$

Step 2:

$$RCCl(=O) \xrightarrow{HOCH_2CH_2Cl} RC(=O)OCH_2CH_2Cl$$

In these equations "R" stands for the *p*-nitrobenzoyl group.

13.28 (a) Compound A has the molecular formula C_7H_8O, which corresponds to that of *cis*-1,3-cyclopentanedicarboxylic acid ($C_7H_{14}O_4$) minus one molecule of water. It is the cyclic anhydride of the diacid.

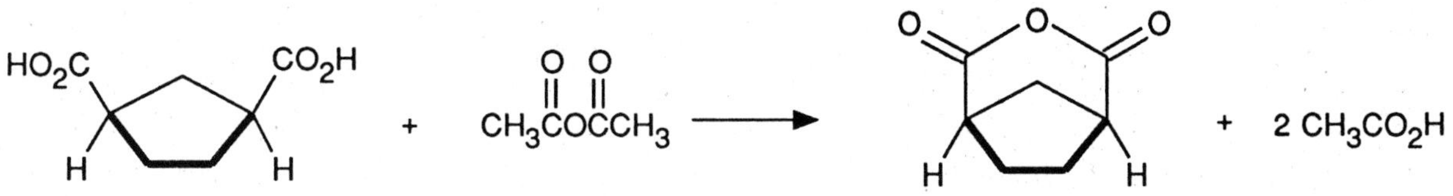

cis-1,3-Cyclopentane-dicarboxylic acid   Acetic anhydride   Compound A   Acetic acid

(b) Of the two carbonyl groups in the starting material, the ketone carbonyl is more reactive than the ester. (The ester carbonyl is stabilized by electron release from oxygen). The Grignard reagent adds to the ketone carbonyl.

$$CH_3C(=O)CH_2CH_2C(=O)OCH_2CH_3 \xrightarrow{CH_3MgI} CH_3C(CH_3)(OMgI)CH_2CH_2C(=O)OCH_2CH_3$$

Compound B has the molecular formula $C_6H_{12}O_2$. The alkoxymagnesium iodide shown above forms a cyclic ester with elimination of ethoxide ion.

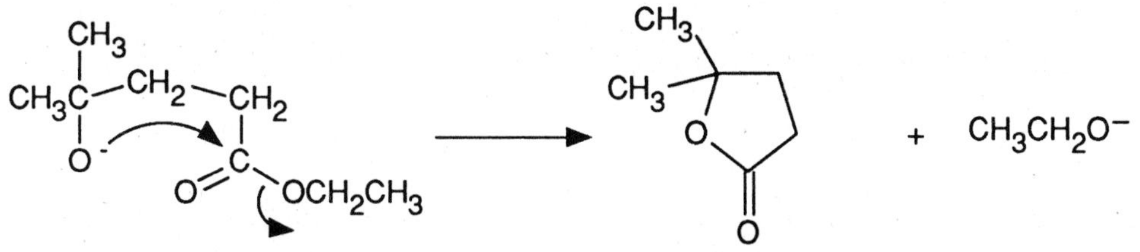

Compound B

13.29 Compound C is an ester but has within it an amine function. Acyl transfer from oxygen to nitrogen converts the ester to a more stable amide.

Compound C   Tetrahedral intermediate   Compound D
(Ar = *p*-Nitrophenyl)   (Ar = *p*-Nitrophenyl)

The tetrahedral intermediate is the key intermediate in this reaction

13.30 Nomex is a polyamide formed by the reaction of the arylamine groups of 1,3-benzenediamine with the aromatic diacyl dichloride

13.31 Compound E has the characteristic triplet-quartet pattern of an ethyl group in its ^{1}H nmr spectrum. Since these signals correspond to 10 protons, there must be two equivalent ethyl groups in the molecule. The methylene quartet of these ethyl groups appears at relatively low field (δ = 4.1 ppm), which is consistent with ethyl groups bonded to oxygen, as in —OCH$_2$CH$_3$. There is a peak at 1730 cm^{-1}, suggesting that these ethoxy groups reside in ester functions. The molecular formula $C_8H_{14}O_4$ reveals that if two ester groups are present, there can be no rings or double bonds. The remaining four hydrogens are equivalent in the ^{1}H nmr spectrum, so two equivalent CH$_2$ groups are present. Compound E is the diethyl ester of succinic acid.

Diethyl succinate (compound E)

CHAPTER 14

AMINES

GLOSSARY OF TERMS

Alkaloid An amine that occurs naturally in plants. The name "alkaloid" derives from the fact that such compounds are weak bases.

Alkylamine An amine in which the organic groups attached to nitrogen are alkyl groups.

Amine An organic derivative of ammonia. Nitrogen has three substituents which may be any combination of alkyl or aryl groups and 0, 1, or 2 hydrogens.

Arylamine An amine which has an aryl group attached to the amine nitrogen.

Azo coupling The formation of a compound of the type ArN=NAr' by reaction of an aryl diazonium salt with an arene. The arene must be strongly activated toward electrophilic aromatic substitution; that is, it must bear a powerful electron-releasing substituent such as -OH or -NR$_2$.

Basicity constant Expressed by the symbol K_b, it is a measure of base strength, especially of amines.

$$K_b = \frac{[R_3NH^+][HO^-]}{[R_3N]}$$

Diazonium ion An ion of the type shown below. Aryl diazonium ions are formed by treatment of primary aromatic amines with nitrous acid. They are extremely useful in the preparation of aryl halides, phenols, and aryl cyanides.

$$R-\overset{+}{N}\equiv N:$$

Diazotization The reaction by which a primary arylamine is converted to the corresponding diazonium ion.

Hofmann elimination The reaction that takes place when a quaternary ammonium hydroxide, especially an alkyltrimethylammonium hydroxide, is converted to an alkene on heating. The elimination occurs in the direction that gives the less-substituted double bond.

N-Nitrosoamine A compound of the type:

$$R_2N-N=O$$

The groups designated "R" may be alkyl or aryl and may be the same or different. N-Nitrosoamines are formed by nitrosation of secondary amines.

Nitrosamine See N-Nitrosoamine.

Nitrosation The reaction of a substance, usually an amine, with the species ^+NO, usually derived from nitrous acid. Primary amines yield diazonium ions, secondary amines yield N-nitrosoamines.

Primary amine An amine with a single alkyl or aryl substituent and two hydrogens. An amine of the type RNH$_2$ (primary alkylamine) or ArNH$_2$ (primary arylamine).

259

Quaternary ammonium salt A salt of the type:

$$R_4N^+ \ X^-$$

in which the positively charged ion contains a nitrogen with a total of four organic substituents (any combination of alkyl and aryl groups).

Sandmeyer reaction The reaction of an aryl diazonium ion with CuCl, CuBr, or CuCN to give respectively an aryl chloride, aryl bromide, or aryl cyanide (nitrile).

Secondary amine An amine with any combination of two alkyl or aryl substituents and one hydrogen on nitrogen. An amine of the type RNHR', RNHAr or ArNHAr'.

Tertiary amine An amine of the type R₃N with any combination of three alkyl or aryl substituents on nitrogen.

SOLUTIONS TO TEXT PROBLEMS

14.1 (b) The amino and phenyl groups are both attached to C-1 of an ethyl group.

1-Phenylethanamine, or

1-phenylethylamine

(c) This tertiary amine has two methyl groups and an ethyl group attached to nitrogen. The groups are named in alphabetical order, ignoring the *di-* prefix in dimethyl.

$$CH_3NCH_2CH_3$$
$$|$$
$$CH_3$$

N,N-Dimethylethylamine, or

N,N-dimethylethanamine

(d) This compound is a quaternary ammonium salt. There are three methyl groups attached to nitrogen, and the bromide ion is named as a separate word (as in sodium bromide, for example).

$$(CH_3)_3\overset{+}{N}H \ \overset{-}{Br}$$ Trimethylammonium bromide

14.2 (b) In alphabetical order , the substituents present on the aniline nucleus are ethyl, isopropyl, and methyl. The groups on nitrogen are preceded by *N-*.

N-Ethyl-4-isopropyl-*N*-methylaniline

(c)   This compound is named as a difluoro-substituted aniline. The ring is numbered beginning with the carbon that bears the -NH$_2$ group, since aniline is the basis of the name.

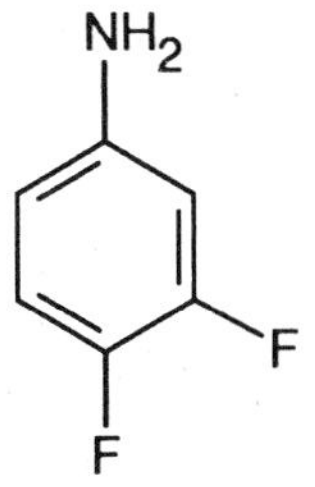

NH$_2$

3,4-Difluoroaniline

F

F

14.3   The nitrogen atom of an amine can participate in a hydrogen bond, as can each one the protons bonded to nitrogen. A primary amine has two hydrogens available for hydrogen bonding and can form more hydrogen bonds (is more associated) than a secondary amine which has only a single proton on nitrogen. A tertiary amine has no protons directly bonded to its nitrogen so two tertiary amine molecules cannot hydrogen bond to one another.

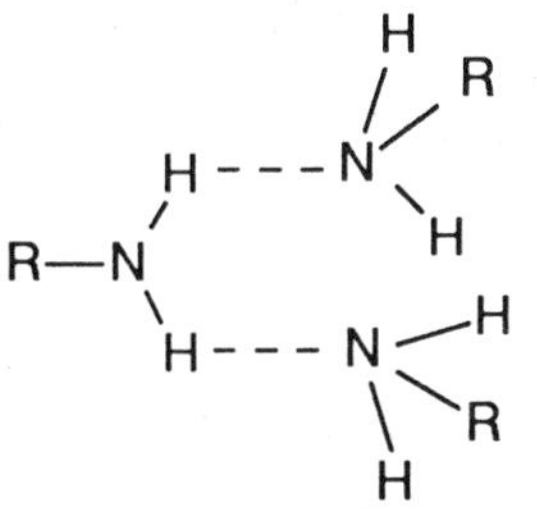

Hydrogen bonds in a primary amine (RNH$_2$) 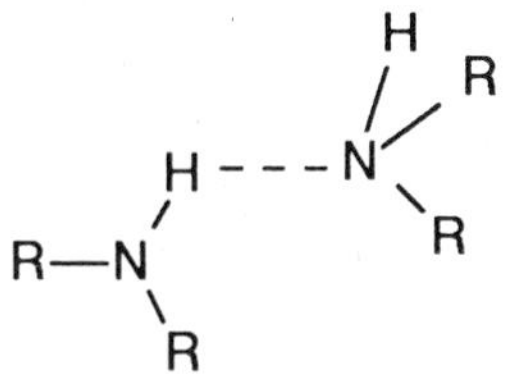   Hydrogen bonds in a secondary amine (R$_2$NH)

The greater the number of hydrogen bonds which an amine can participate in, the higher its boiling point. We predict that, among isomers, a primary amine will have the highest boiling point and a tertiary amine the lowest. The boiling points given for the C$_4$H$_9$N isomers shown below confirm this prediction.

CH$_3$CH$_2$CH$_2$NH$_2$	CH$_3$CH$_2$NHCH$_3$	(CH$_3$)$_3$N
Propylamine	*N*-Methylethylamine	Trimethylamine
(a primary amine)	(a secondary amine)	(a tertiary amine)
bp 50°C	bp 34°C	bp 3°C

14.4   Nitrogen is attached directly to the aromatic ring in tetrahydroquinoline, which is therefore an arylamine, and the nitrogen lone pair is delocalized into the π system of the aromatic ring. It is less basic than tetrahydroisoquinoline, in which the nitrogen is insulated from the ring by an *sp^3* hybridized carbon.

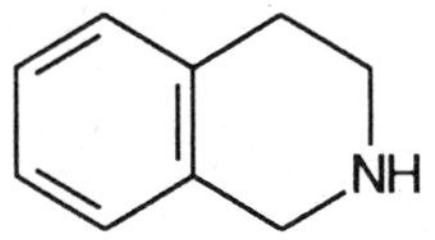

Tetrahydroisoquinoline   Tetrahydroquinoline

(an alkylamine; more basic)   (an arylamine; less basic)

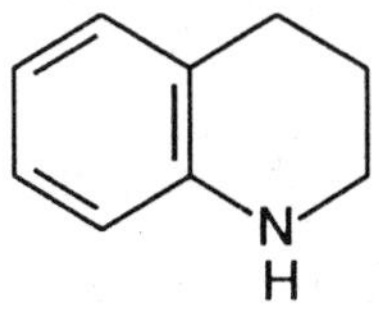

Tetrahydroisoquinoline has a K_b = 2.5 x 10^{-5} (characteristic of an alkylamine), while tetrahydroquinoline has K_b = 2.5 x 10^{-5} (characteristic of an arylamine).

14.5   *N, N*-Dioctylamine reacts with 1-bromooctane to form *N, N, N*-trioctylamine. This tertiary amine reacts with an additional mole of 1-bromooctane to form tetraoctylammonium bromide.

$[CH_3(CH_2)_6CH_2]_2NH$ + $CH_3(CH_2)_6CH_2Br$ $\longrightarrow$ $[CH_3(CH_2)_6CH_2]_3N$

N,N-Dioctylamine   1-Bromooctane   *N,N,N*-Trioctylamine

$[CH_3(CH_2)_6CH_2]_3N$ + $CH_3(CH_2)_6CH_2Br$ $\longrightarrow$ $[CH_3(CH_2)_6CH_2]_4N^+\ Br^-$

N,N,N-Trioctylamine   1-Bromooctane   Tetraoctylammonium bromide

14.6   The equation in the text describes formation of *N*-butyl-1-pentanamine by reduction of *N*-butylpentanamide. The necessary amide may be prepared from pentanoic acid by reaction of the corresponding acyl chloride with butylamine (1-butanamine).

$CH_3CH_2CH_2CH_2COH$ $\xrightarrow{SOCl_2}$ $CH_3CH_2CH_2CH_2CCl$ $\xrightarrow{CH_3CH_2CH_2CH_2NH_2}$ $CH_3CH_2CH_2CH_2CNHCH_2CH_2CH_2CH_3$

Pentanoic acid   Pentanoyl chloride   *N*-Butylpentanamide

The amide is then reduced with lithium aluminum hydride as outlined in the text equation.

$CH_3CH_2CH_2CH_2CNHCH_2CH_2CH_2CH_3$ $\xrightarrow[\text{2. } H_2O]{\text{1. } LiAlH_4}$ $CH_3CH_2CH_2CH_2CH_2NHCH_2CH_2CH_2CH_3$

N-Butylpentanamide   *N*-Butyl-1-pentanamine

14.7   The first reaction is a type of electrophilic aromatic substitution, nitration. The combination HNO_3/H_2SO_4 is the usual one for nitration of aromatic rings. The last step is reduction of the nitro group to an arylamine. This may be accomplished using iron (Fe) or tin (Sn) in hydrochloric acid.

Toluene   *p*-Nitrotoluene   Ethyl *p*-nitrophenylbenzoate   Benzocaine

14.8   The epoxide ring is readily opened by nucleophiles.

$(CH_3)_3N$: $(CH_3)_3N\!-\!CH_2CH_2O^-$ $\xrightarrow{H_2O}$ $(CH_3)_3NCH_2CH_2OH$ HO^-

Trimethylamine   Ethylene oxide   Choline

Choline has the structure shown at the bottom of the preceding page; it is a quaternary ammonium hydroxide. It cannot have a nitrogen-oxygen covalent bond as in the structure below because that structure violates the octet rule in that five bonds to nitrogen require it to 10 electrons.

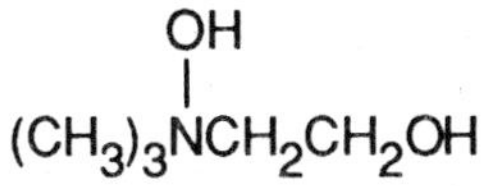

OH
(CH3)3NCH2CH2OH

Incorrect structure!

14.9 As outlined in the sample solution, the Hofmann elimination sequence involves three separate steps: (1) preparation of the quaternary ammonium iodide; (2) conversion to the quaternary ammonium hydroxide; and (3) heating the quaternary ammonium hydroxide to give the alkene.

(b)

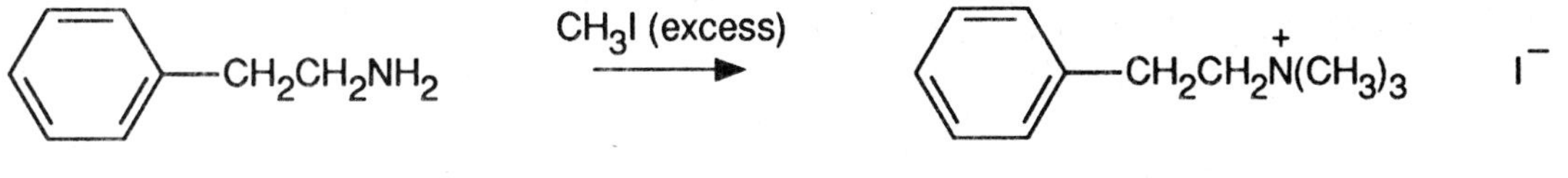

CH2CH2NH2
CH3I (excess)
CH2CH2N(CH3)3 I

2-Phenylethylamine 2-Phenylethylammonium iodide

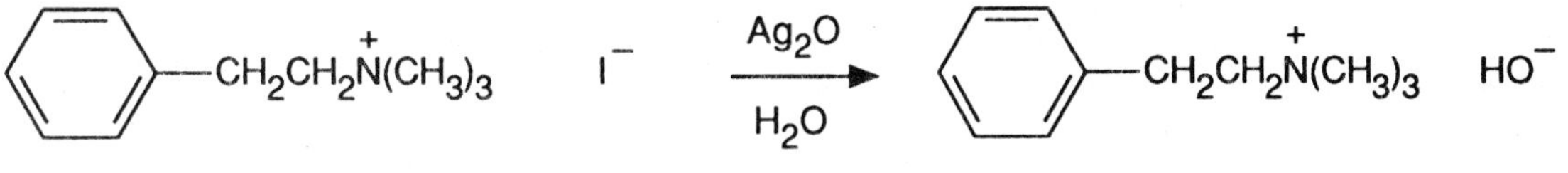

CH2CH2N(CH3)3 I
Ag2O
H2O
CH2CH2N(CH3)3 HO

2-Phenylethylammonium iodide 2-Phenylethylammonium hydroxide

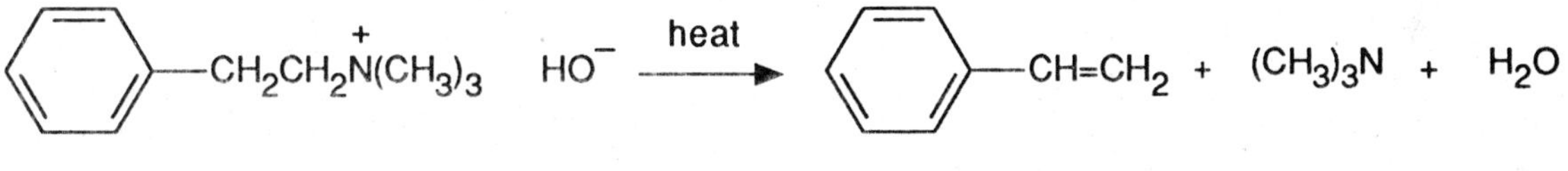

CH2CH2N(CH3)3 HO
heat
CH=CH2 + (CH3)3N + H2O

2-Phenylethylammonium hydroxide Styrene Trimethylamine Water

(c) This problem is somewhat more challenging because it involves a secondary amine. The nature of the process, however, is the same as that observed with primary amines. Here we will present the Hofmann procedure as a flow chart rather than writing each reaction individually. The alkene which is formed is **cyclohexene**.

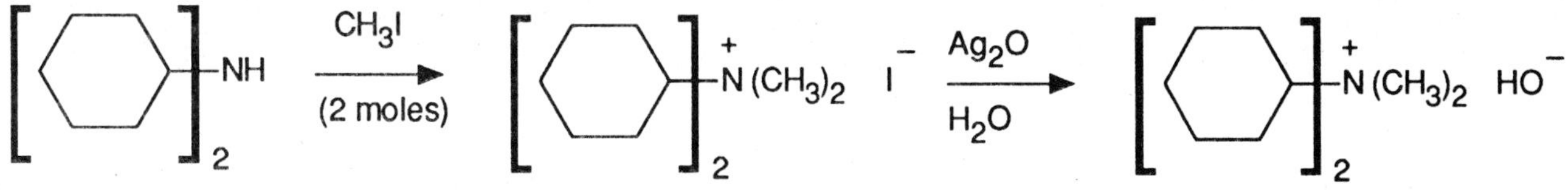

NH
CH3I
(2 moles)
N (CH3)2 I
Ag2O
H2O
N (CH3)2 HO

Dicyclohexylamine

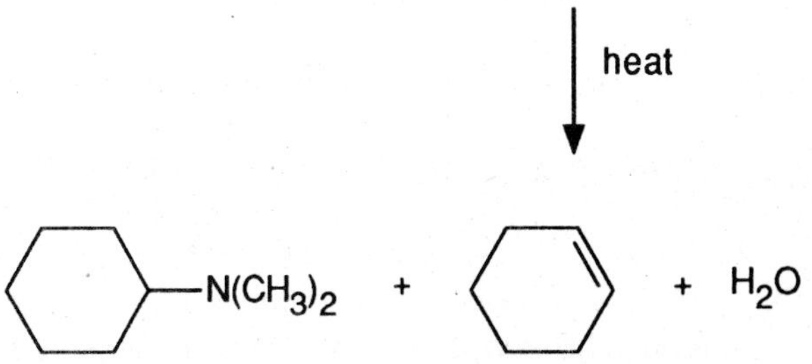

heat
N(CH3)2 + + H2O

Cyclohexene

14.10 First identify the hydrogens capable of being lost in an elimination reaction. These are hydrogens on carbon atoms which are *adjacent* to the carbon that bears the positively charged nitrogen. These hydrogens are marked with ✓ in the structural formula shown.

$$(CH_3)_3CCH_2—\overset{\overset{\displaystyle CH_3}{|}}{\underset{\underset{\displaystyle +N(CH_3)_3}{}}{C}}—CH_3 \quad \text{Two equivalent methyl groups}$$

Methylene group

Hofmann elimination reactions proceed in the direction that gives the less-substituted alkene. It is a proton from one of the two equivalent methyl groups, rather than one from the more sterically hindered methylene, which is lost on elimination.

$$(CH_3)_3CCH_2—\overset{\overset{\displaystyle CH_3}{|}}{\underset{\underset{\displaystyle N(CH_3)_3}{|}}{C}}—CH_2—H \quad \overset{-}{O}H \longrightarrow (CH_3)_3CCH_2C=CH_2 \;+\; (CH_3)_3N \;+\; H_2O$$

$$\overset{\overset{\displaystyle CH_3}{|}}{}$$

2,4,4-Trimethyl-1-pentene

(c)   There are two possible elimination products. Loss of the indicated proton in the structural formula shown gives ethylene.

✓

$$H—CH_2—CH_2—\overset{\overset{\displaystyle +CH_3}{|}}{\underset{\underset{\displaystyle CH_3}{|}}{N}}CH_2CH_2CH_2CH_3 \longrightarrow CH_2=CH_2 \;+\; (CH_3)_2NCH_2CH_2CH_2CH_3$$

Ethylene

Elimination in the opposite direction yields 1-butene.

$$CH_3CH_2\overset{\overset{\displaystyle CH_3}{|}}{\underset{\underset{\displaystyle CH_3}{|}}{\overset{+}{N}}}—CH_2—\underset{\underset{\displaystyle H}{|}}{C}HCH_2CH_3 \longrightarrow CH_3CH_2N(CH_3)_2 \;+\; CH_2=CHCH_2CH_3$$

1-Butene

The preferred order of proton removal in Hofmann elimination reactions is $CH_3 > CH_2 > CH$. **Ethylene** is the major alkene formed, the experimentally observed ratio of ethylene to 1-butene was 98:2.

(d)   In this case the competition is between attack at a CH_2 group versus attack at CH.

$$CH_3CH_2CH_2—\overset{\overset{\displaystyle CH_3}{\overset{+}{|}}}{\underset{\underset{\displaystyle CH_3}{|}}{N}}—CH_2CH(CH_3)_2$$

The CH_2 group is attacked faster than the CH group and we predict that **propene** should be the major product. This was observed to be the case when the elimination reaction was performed in the laboratory; an alkene mixture composed of $CH_3CH=CH_2$ (72 percent) and $(CH_3)_2C=CH_2$ (28 percent) was isolated in 92 percent yield.

14.11 As the example in the text illustrates, *m*-bromophenol may be prepared by diazotization of *m*-bromoaniline. The problem simplifies itself, therefore, to the preparation of *m*-bromoaniline. Recognizing that arylamines are ultimately derived from nitroarenes, we derive the retrosynthetic sequence of intermediates.

m-Bromophenol   *m*-Bromoaniline   *m*-Bromonitrobenzene   Nitrobenzene   Benzene

The desired reaction sequence is straightforward, using reactions that have been discussed previously in the text.

Benzene   Nitrobenzene   *m*-Bromonitrobenzene   *m*- Bromoaniline   *m*-Bromophenol

14.12 A reaction sequence analogous to the one described in the text for the preparation of *m*-dibromobenzene may be used. The only difference is in the last step where the diazonium salt is treated with CuCl instead of CuBr.

Nitrobenzene   *m*-Bromonitrobenzene   *m*-Bromoaniline   *m*-Bromochlorobenzene

14.13 Amines may be primary, secondary, or tertiary. The $C_4H_{11}N$ primary amines, compounds of the type $C_4H_9NH_2$ and their systematic names are:

$$CH_3CH_2CH_2CH_2NH_2 \qquad (CH_3)_2CHCH_2NH_2 \qquad \overset{NH_2}{\underset{|}{CH_3CHCH_2CH_3}} \qquad (CH_3)_3CNH_2$$

Butylamine   Isobutylamine   sec-Butylamine   tert-Butylamine

(1-butanamine)   (2-methyl-1-propanamine)   (2-butanamine)   (2-methyl-2-propanamine)

Secondary amines have the general formula R₂NH. Those of molecular formula C₄H₁₁N are:

$(CH_3CH_2)_2NH$ $CH_3NHCH_2CH_2CH_3$ $CH_3NHCH(CH_3)_2$

Diethylamine   *N*-Methylpropylamine   *N*-Methylisopropylamine

(*N*-ethylethanamine)   (*N*-methyl-1-propanamine)   (*N*-methyl-2-propanamine)

There is only one tertiary amine (R₃N) of molecular formula C₄H₁₁N.

$(CH_3)_2NCH_2CH_3$ *N,N*-dimethylethanamine   (*N,N*-Dimethylethylamine)

14.14 (a) Heptylamine is the IUPAC name for a primary amine that bears a seven-carbon chain on nitrogen.

$CH_3CH_2CH_2CH_2CH_2CH_2CH_2NH_2$ Heptylamine

(b) The name 2-ethyl-1-butanamine designates a four-carbon chain terminating in an amino group and bearing an ethyl group at C-2.

$CH_3CH_2CHCH_2NH_2$ 2-Ethyl-1-butanamine
      |
      CH_2CH_3

(c) *N*-Ethylpentylamine is a secondary amine bearing an ethyl group and a pentyl group on nitrogen.

$CH_3CH_2NHCH_2CH_2CH_2CH_2CH_3$ *N*-Ethylpentylamine

(d) Dibenzylamine is a secondary amine bearing two benzyl groups on nitrogen.

Dibenzylamine

(e) Tetraethylammonium hydroxide is a quaternary ammonium salt. There are four ethyl groups on a positively charged nitrogen. Hydroxide is the anion.

Tetraethylammonium hydroxide

(f) Aniline is the parent compound in *N*-ethyl-4-methylaniline. Thus the molecule has a benzene ring to which an amino nitrogen is attached and this nitrogen bears an ethyl substituent. A methyl group is attached to the aromatic ring para to the nitrogen.

N-Ethyl-4-methylaniline

(g) Both chlorine atoms are attached to the aromatic ring of aniline in 2,4-dichloroaniline.

2,4-Dichloroaniline

(h) Both methyl groups are attached to the nitrogen atom in *N,N*-dimethylaniline.

N(CH₃)₂

N,N-Dimethylaniline

14.16 (a) The nitrogen in *N,N*-dipropylamine is directly bonded to two carbons so the compound is a **secondary amine**. Both groups bonded to nitrogen are *propyl* (CH₃CH₂CH₂–) groups.

CH₃CH₂CH₂–N–CH₂CH₂CH₃   *N,N*–Dipropylamine
               |
               H

Alternatively, this compound could be named *N*-propylpropanamine.

(b) The nitrogen atom in this amine bears three alkyl groups; the compound is a **tertiary amine**. The group with the longest continuous chain is *butyl* (CH₃CH₂CH₂CH₂–), so the compound is named as a derivative of *butylamine*. The methyl and propyl groups are named as substituents on the nitrogen of butylamine and are cited in alphabetical order.

CH₃CH₂CH₂–N–CH₂CH₂CH₂CH₃   *N*–Methyl–*N*–propylbutylamine
               |
               CH₃

An alternative name is *N*-methyl-*N*-propylbutanamine.

(c) This compound has only a single alkyl group on nitrogen so is a **primary amine**. It is best named according to the *alkanamine* method.   The amine function takes precedence over methyl in numbering.

CH₃CHCH₂CH₂NH₂   3–Methyl-1–butanamine
    |
    CH₃

(d) Since nitrogen in this compound is attached to a benzene ring, it is best named as a derivative of *aniline*. There are two additional organic groups bonded to nitrogen so it is a **tertiary amine**. Both additional groups are *propyl* groups and their location as substituents on nitrogen is indicated by the prefixes *N,N* in the name.

—N(CH₂CH₂CH₃)₂   *N,N*–Dipropylaniline

267

(e)   Here, as in the preceding compound, the amine is best named as a derivative of *aniline*. In addition to the benzene ring, nitrogen bears a *benzyl* ($C_6H_5CH_2-$) substituent.   It is a **secondary amine**.

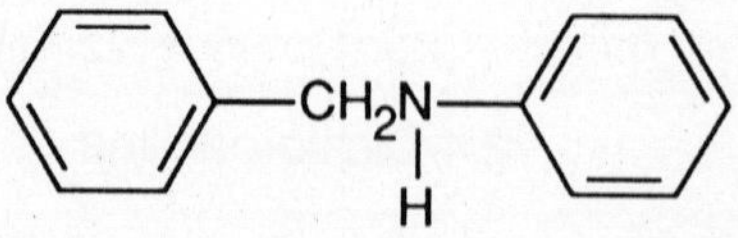

$$C_6H_5-CH_2-\underset{\underset{H}{|}}{N}-C_6H_5$$

N–Benzylaniline

14.17 (a)   An amino group is a substituent at C-4 of butanoic acid in 4-aminobutanoic acid (γ–aminobutyric acid).

$$H_2NCH_2CH_2CH_2\overset{\overset{O}{\|}}{C}OH$$

4-Aminobutanoic acid

You will learn in the following chapter that compounds which contain both an amino group and a carboxylic acid group exist as *dipolar species*. In the case of 4-aminobutanoic acid, this dipolar species has the structure

$$\overset{+}{H_3N}CH_2CH_2CH_2\overset{\overset{O}{\|}}{C}O^-$$

(b)   Mescaline has a 3,4,5-trimethoxy-substituted benzene ring at C-2 of an ethylamine unit.

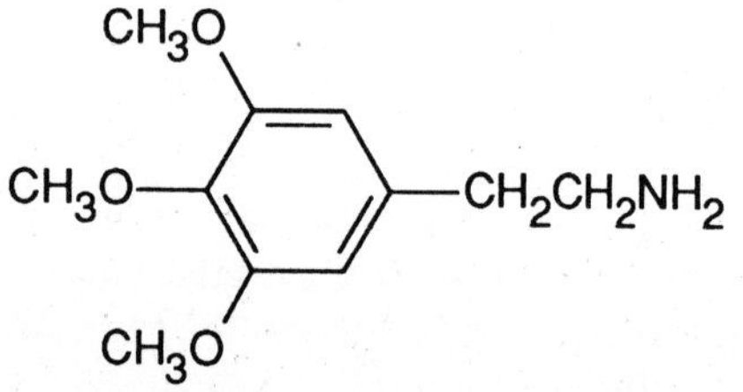

2-(3,4,5-Trimethoxyphenyl)ethylamine (mescaline)

(c)   A phenyl group and an amino group are trans to each other on a three-membered ring in this compound.

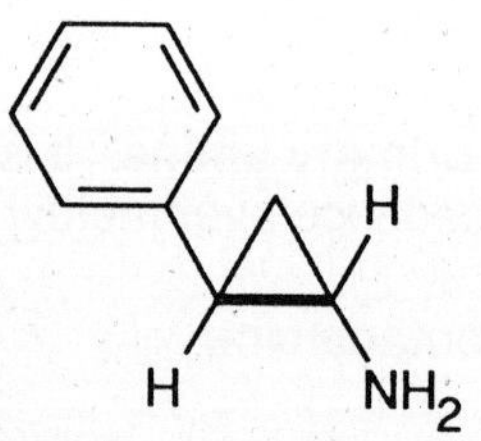

trans–2-Phenylcyclopropylamine (tranylcypromine)

(d)   This compound is a tertiary amine. It bears a benzyl a group, a methyl group, and a 2-propynyl group on nitrogen.

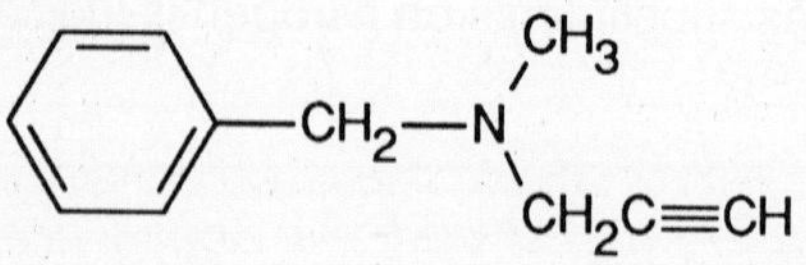

N-Benzyl-*N*-methyl–2–propynylamine

(pargyline)

(e)   The amino group is at C-2 of a three-carbon chain that bears a phenyl substituent at its terminus.

1-Phenyl-2-propanamine (amphetamine)

14.18 This problem requires you to deduce which one of the two nitrogen atoms in lidocaine is more basic. It will be the more basic of the two nitrogens which is protonated by hydrogen chloride. One of the nitrogens is part of an amide function and is a much weaker base than the amine nitrogen. Therefore, it is the amine nitrogen which is protonated and the structure of lidocaine hydrochloride is:

Lidocaine hydrochloride

this hydrogen is transferred from H-Cl

14.19 Of the group, two are amines (aniline and benzylamine) and one is an amide (acetanilide). Both of the amines are more basic than acetanilide because the electron pair on nitrogen in an amide is delocalized into the carbonyl group. Alkylamines are stronger bases than arylamines, so benzylamine is a stronger base than aniline.   The order of decreasing basicity is

Benzylamine	Aniline	Acetanilide
Strongest base, pK_b 4.65	pK_b 9.4	Weakest base, pK_b 13.6

14.20 The nitrogen of physostigmine that reacts with methyl iodide is the one marked *a* in the structure because it is the most nucleophilic.

Physostigmine   Quaternary ammonium salt

Nitrogen (*a*) is an alkylamine-type nitrogen. Of the other two nitrogens, *b* is attached to an aromatic ring and is much less basic and less nucleophilic. The third nitrogen *c* is an amide nitrogen; amides are less nucleophilic than amines.

14.21 (a) Amines are basic and are protonated by hydrogen halides.

$$C_6H_5CH_2NH_2 \quad + \quad HBr \quad \longrightarrow \quad C_6H_5CH_2\overset{+}{N}H_3 \;\; \overset{-}{Br}$$

Benzylamine   Hydrogen bromide   Benzylammonium bromide

(b) Acetic acid transfers a proton to benzylamine.

$$C_6H_5CH_2NH_2 \quad + \quad CH_3\overset{O}{\overset{\|}{C}}OH \quad \longrightarrow \quad C_6H_5CH_2\overset{+}{N}H_3 \;\; \overset{-}{O}\overset{O}{\overset{\|}{C}}CH_3$$

Benzylamine   Acetic acid   Benzylammonium acetate

(c) Acetyl chloride reacts with benzylamine to form an amide. Two moles of the amine are required. One mole acts as a nucleophile and is converted to the amide. The second mole acts as a base and is converted to the corresponding ammonium ion.

$$2\,C_6H_5CH_2NH_2 \quad + \quad CH_3\overset{O}{\overset{\|}{C}}Cl \quad \longrightarrow \quad C_6H_5CH_2NH\overset{O}{\overset{\|}{C}}CH_3 \quad + \quad C_6H_5CH_2\overset{+}{N}H_3 \;\; \overset{-}{Cl}$$

Benzylamine   Acetyl chloride   *N*-Benzylacetamide   Benzylammonium chloride

(d) Primary amines react with ketones to give imines.

$$C_6H_5CH_2NH_2 \quad + \quad CH_3\overset{O}{\overset{\|}{C}}CH_3 \quad \longrightarrow \quad C_6H_5CH_2N{=}C(CH_3)_2$$

Benzylamine   Acetone   *N*-Isopropylidenebenzylamine

(e) With excess methyl iodide amines are converted to quaternary ammonium halides.

$$C_6H_5CH_2NH_2 \quad + \quad 3\,CH_3I \quad \longrightarrow \quad C_6H_5CH_2\overset{+}{N}(CH_3)_3 \;\; \overset{-}{I}$$

Benzylamine   Methyl iodide   Benzyltrimethylammonium iodide

14.22 An arylamine such as aniline will react with each of the reagents in the previous problem in a manner comparable to the reaction observed with benzylamine.

(a)

$$C_6H_5NH_2 \quad + \quad HBr \quad \longrightarrow \quad C_6H_5\overset{+}{N}H_3 \;\; \overset{-}{Br}$$

Aniline   Hydrogen bromide   Anilinium bromide

(b)

$$C_6H_5NH_2 \quad + \quad CH_3\overset{O}{\overset{\|}{C}}OH \quad \longrightarrow \quad C_6H_5\overset{+}{N}H_3 \;\; \overset{-}{O}\overset{O}{\overset{\|}{C}}CH_3$$

Aniline   Acetic acid   Anilinium acetate

270

(c)

$$2 \; C_6H_5NH_2 \quad + \quad CH_3\overset{\overset{\displaystyle O}{\|}}{C}Cl \quad \longrightarrow \quad C_6H_5NH\overset{\overset{\displaystyle O}{\|}}{C}CH_3 \quad + \quad C_6H_5\overset{+}{N}H_3 \quad Cl^-$$

Aniline   Acetyl chloride   Acetanilide   Anilinium chloride

(d)

$$C_6H_5NH_2 \quad + \quad CH_3\overset{\overset{\displaystyle O}{\|}}{C}CH_3 \quad \longrightarrow \quad C_6H_5N=C(CH_3)_2$$

Aniline   Acetone   *N*-Isopropylideneaniline

(e)

$$C_6H_5NH_2 \quad + \quad 3 \; CH_3I \quad \longrightarrow \quad C_6H_5\overset{+}{N}H_3 \quad I^-$$

Aniline   Methyl iodide   Phenyltrimethylammonium iodide

14.23 (a)   Nitrosation of primary arylamines yields aryl diazonium salts.

$$CH_3-\!\!\left\langle\!\!\bigcirc\!\!\right\rangle\!\!-NH_2 \quad \xrightarrow[\text{H}_2\text{O, 0-5°C}]{\text{NaNO}_2,\ \text{H}_2\text{SO}_4} \quad CH_3-\!\!\left\langle\!\!\bigcirc\!\!\right\rangle\!\!-\overset{+}{N}\!\!\equiv\!\!N{:} \quad HSO_4^-$$

p-Toluidine   *p*-Methylbenzenediazonium hydrogen sulfate

The replacement reactions that can be achieved by using diazonium salts are illustrated in (b) through (f). In all cases molecular nitrogen is lost from the ring carbon to which it was attached and is replaced by another substituent. In each of the reactions, the symbol Ar stands for *p*-methylphenyl.

$$CH_3-\!\!\left\langle\!\!\bigcirc\!\!\right\rangle\!\!-\overset{+}{N}\!\!\equiv\!\!N{:} \quad HSO_4^-$$

p-Methylbenzenediazonium hydrogen sulfate

(a) $\xrightarrow[\text{heat}]{\text{H}_3\text{O}^+,\ \text{H}_2\text{O}}$ ArOH

(c) $\xrightarrow{\text{CuCl}}$ ArCl

(d) $\xrightarrow{\text{KI}}$ ArI

(e) $\xrightarrow{\text{CuBr}}$ ArBr

(f) $\xrightarrow{\text{CuCN}}$ ArCN

(g) The nitrogens of an aryl diazonium salt are retained on reaction with the electron-rich ring of a phenol. Azo coupling occurs:

$$CH_3-C_6H_4-\overset{+}{N}\equiv N: \quad HSO_4^- \quad + \quad C_6H_5-OH \quad \longrightarrow \quad CH_3-C_6H_4-N=N-C_6H_4-OH$$

14.24 (a)   The necessary conversion as analyzed retrosynthetically is shown. A key step is the formation of a secondary amine by reduction of an amide.

$$C_6H_5CH_2NHCH_3 \quad \Rightarrow \quad C_6H_5\overset{O}{\overset{\|}{C}}NHCH_3 \quad \Rightarrow \quad C_6H_5\overset{O}{\overset{\|}{C}}Cl \quad \Rightarrow \quad C_6H_5\overset{O}{\overset{\|}{C}}OH$$

Desired product   Designated starting material

The reaction sequence which will carry out the desired conversion is:

$$C_6H_5\overset{O}{\overset{\|}{C}}OH \xrightarrow{SOCl_2} C_6H_5\overset{O}{\overset{\|}{C}}Cl \xrightarrow{CH_3NH_2} C_6H_5\overset{O}{\overset{\|}{C}}NHCH_3 \xrightarrow[\text{2. } H_2O]{\text{1. } LiAlH_4} C_6H_5CH_2NHCH_3$$

Benzoic acid   Benzoyl chloride   *N*–Methylbenzamide   *N*–Methylbenzylamine

(b)   The conversion is:

$$C_6H_5CH_2Br \xrightarrow{?} C_6H_5CH_2CH_2NH_2$$

A chain extension is required since the product has one more carbon atom than the starting material. The most direct approach involves nucleophilic substitution by cyanide ion. The derived nitrile is then reduced to the desired amine.

$$C_6H_5CH_2Br \xrightarrow{NaCN} C_6H_5CH_2C\equiv N \xrightarrow{H_2,\ Ni} C_6H_5CH_2CH_2NH_2$$

Benzyl bromide   2-Phenylethanenitrile   2-Phenylethylamine

The equation shows the reduction step as a catalytic hydrogenation of the nitrile. Reduction with lithium aluminum hydride is also an effective way to achieve the reduction of a nitrile to an amine.

An alternative synthesis uses carboxylation of a Grignard reagent in the chain extension step.

$$C_6H_5CH_2Br \xrightarrow{\text{Mg, diethyl ether}} C_6H_5CH_2MgBr \xrightarrow[\text{2. } H_3O^+]{\text{1. } CO_2} C_6H_5CH_2\overset{O}{\overset{\|}{C}}OH$$

Benzyl bromide   Benzylmagnesium bromide   Phenylacetic acid

Following chain extension, the carboxylic acid function is converted to the amine by way of an amide.

The reaction sequence at the top of the page:

$$C_6H_5CH_2\overset{\overset{\displaystyle O}{\|}}{C}OH \xrightarrow{SOCl_2} C_6H_5CH_2\overset{\overset{\displaystyle O}{\|}}{C}Cl \xrightarrow{NH_3} C_6H_5CH_2\overset{\overset{\displaystyle O}{\|}}{C}NH_2 \xrightarrow[2.\ H_2O]{1.\ LiAlH_4} C_6H_5CH_2CH_2NH_2$$

Phenylacetic acid Phenylacetyl chloride Phenylacetamide 2-Phenylethylamine

14.25 (a) Analyze the conversion retrosynthetically.

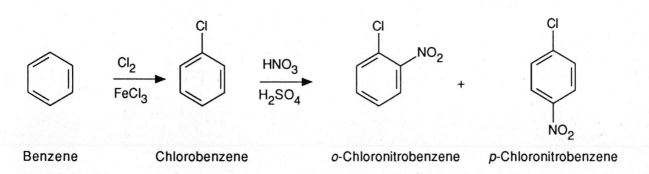

o–Isopropylaniline   *o*–Isopropylnitrobenzene   Isopropylbenzene   Benzene

The last step in the synthesis is reduction of the corresponding nitro compound. The necessary nitro arene is obtained by separating the ortho-para mixture formed during nitration of isopropylbenzene. Isopropylbenzene is prepared by the Friedel-Crafts alkylation of benzene. Shown in the forward (synthetic) mode, the sequence of reactions is:

(The synthetic scheme showing:)

Benzene $\xrightarrow[AlCl_3]{(CH_3)_2CHCl}$ Isopropylbenzene $\xrightarrow[H_2SO_4]{HNO_3}$ *o*–Isopropylnitrobenzene + para isomer

o–Isopropylnitrobenzene $\xrightarrow[\text{or, 1. Fe, HCl; 2. HO}^-]{\underset{\text{or, 1. Sn, HCl; 2. HO}^-}{H_2,\ Ni}}$ *o*–Isopropylaniline

(b) The final step in the preparation of *p*-chloroaniline is once again reduction of the corresponding nitro compound. *p*-Chloronitrobenzene is prepared by nitration of chlorobenzene.

Benzene $\xrightarrow[FeCl_3]{Cl_2}$ Chlorobenzene $\xrightarrow[H_2SO_4]{HNO_3}$ *o*-Chloronitrobenzene + *p*-Chloronitrobenzene

The para isomer is observed to comprise 69 percent of the product in this reaction (30 percent ortho; 1 percent meta). Separation of *p*-chloronitrobenzene and its reduction complete the synthesis.

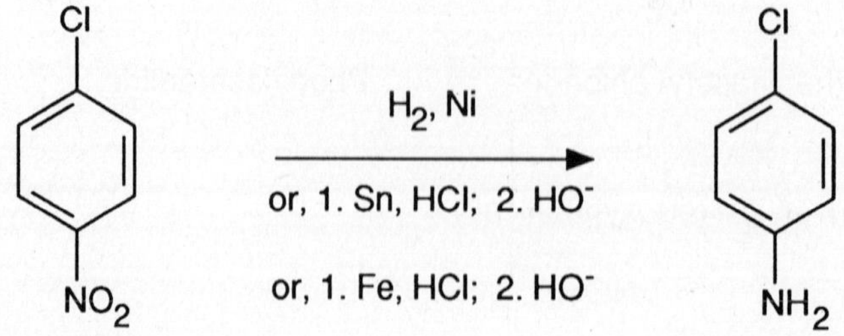

Cl
H2, Ni
or, 1. Sn, HCl; 2. HO-
or, 1. Fe, HCl; 2. HO-
NO2
Cl
NH2

p-Chloronitrobenzene *p*-Chloroaniline

Chlorination of nitrobenzene would not be a suitable route to the required intermediate because it would produce mainly *m*-chloronitrobenzene.

(c) The key to this problem is to recognize that the iodine substituent is derived from an arylamine by diazotization.

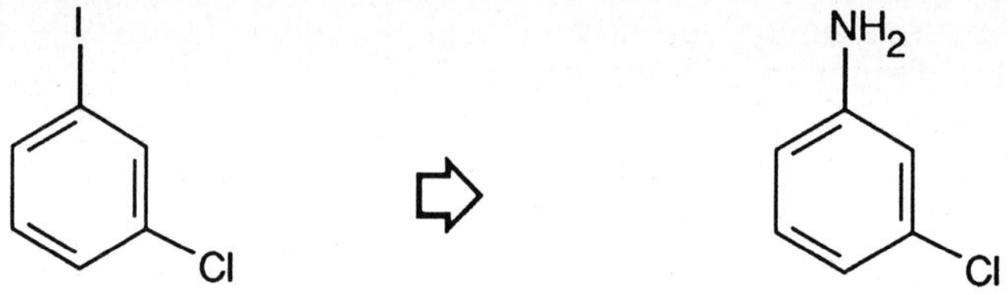

I
Cl
NH2
Cl

m-Chloroiodobenzene *m*-Chloroaniline

m-Chloroaniline is prepared by chlorination of nitrobenzene, followed by reduction.

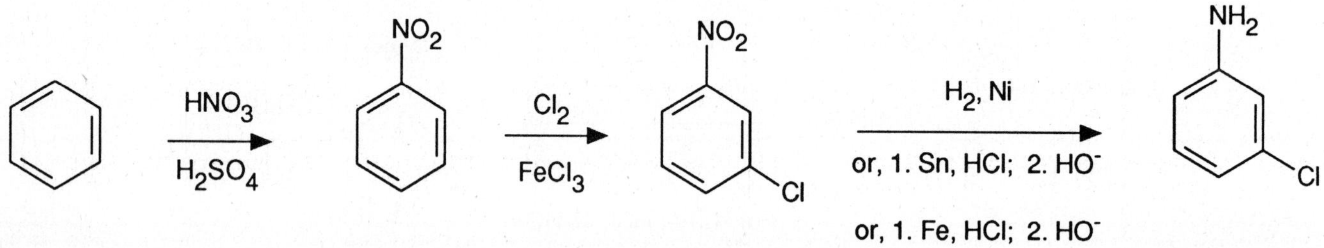

HNO3
H2SO4
NO2
Cl2
FeCl3
NO2
Cl
H2, Ni
or, 1. Sn, HCl; 2. HO-
or, 1. Fe, HCl; 2. HO-
NH2
Cl

Benzene Nitrobenzene *m*-Chloronitrobenzene *m*-Chloroaniline

Formation of the diazonium salt followed by its reaction with potassium iodide completes the synthesis.

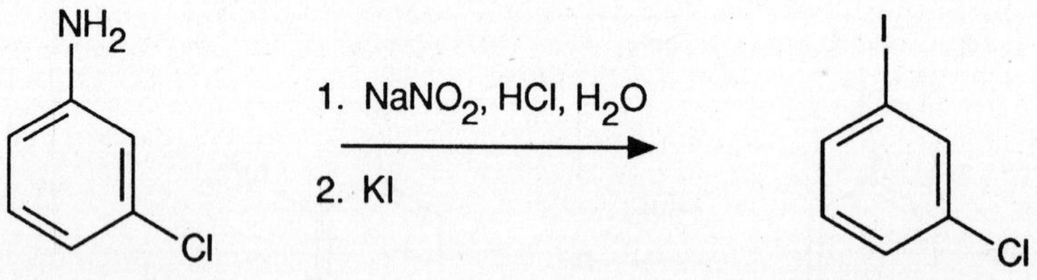

NH2
Cl
1. NaNO2, HCl, H2O
2. KI
I
Cl

m-Chloroaniline *m*-Chloroiodobenzene

(d) Phenols are obtained by hydrolysis of the corresponding diazonium salt. In this synthesis the para orientation of the substituents requires that halogenation of benzene *precede* nitration.

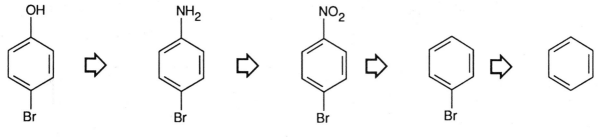

p-Bromophenol   p-Bromoaniline   p-Bromonitrobenzene   Bromobenzene   Benzene

The necessary reaction scheme to carry out the conversion is:

Benzene $\xrightarrow[\text{FeBr}_3]{\text{Br}_2}$ Bromobenzene $\xrightarrow[\text{H}_2\text{SO}_4]{\text{HNO}_3}$ p-Bromonitrobenzene + ortho isomer

p-Bromonitrobenzene $\xrightarrow{\text{H}_2, \text{Ni}}$ p-Bromoaniline $\xrightarrow[\text{2. H}_2\text{O, heat}]{\text{1. NaNO}_2, \text{HCl, H}_2\text{O}}$ p-Bromophenol

Catalytic hydrogenation is shown in the step in which the nitro group is reduced. Reduction with iron or tin in hydrochloric acid could also be used. (See, for example, problems 25a, b, and c.)

14.26 (a) *p*-Methylaniline is related to *p*-nitrotoluene by reduction of the amino group.

p-Nitrotoluene $\xrightarrow[\substack{\text{or, 1. Fe, HCl; 2. HO}^- \\ \text{or, 1. Sn HCl; 2. HO}^-}]{\text{H}_2, \text{Ni}}$ p-Nitroaniline

The preparation of *p*-nitroaniline described in this problem sets the stage for parts (b)-(e). In each of parts (b)-(e) the desired compound is prepared by way of the diazonium salt formed from *p*-nitroaniline. This diazonium salt is prepared as shown.

275

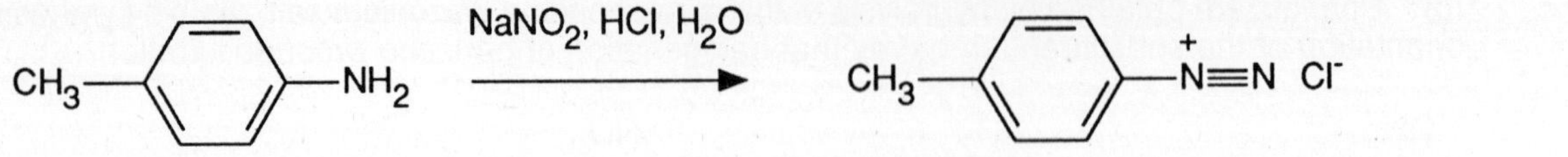

p-Methylaniline → (NaNO₂, HCl, H₂O) → p-Methylbenzenediazonium chloride

(b-e)   Each of the following conversions may be accomplished by reacting a diazonium salt with the appropriate reagent.

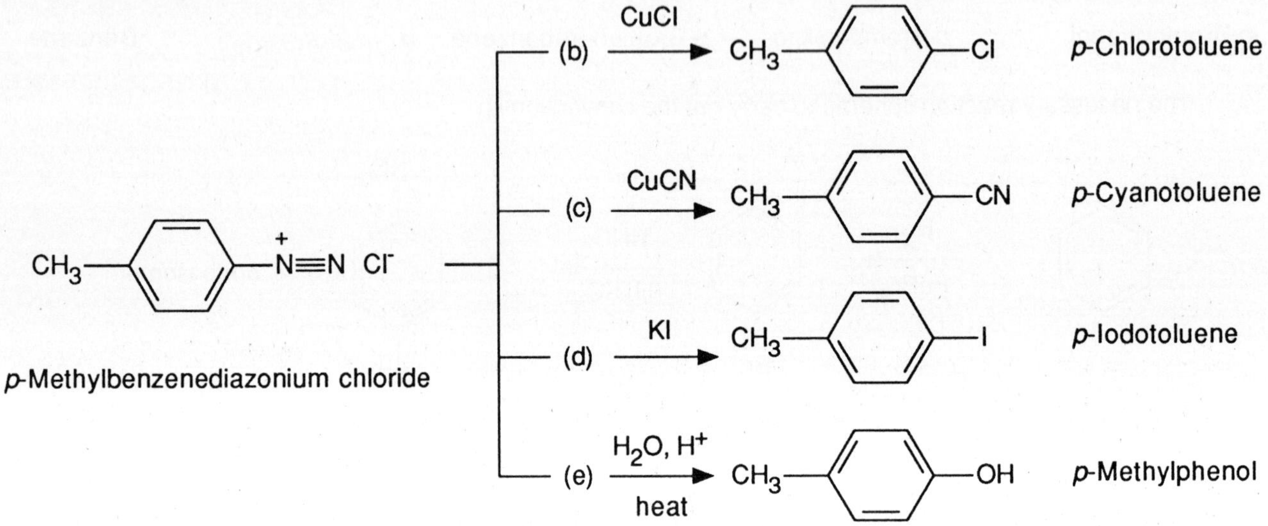

p-Methylbenzenediazonium chloride

(b) CuCl → p-Chlorotoluene

(c) CuCN → p-Cyanotoluene

(d) KI → p-Iodotoluene

(e) H₂O, H⁺, heat → p-Methylphenol

14.27 The reactions in this problem describe a Hofmann elimination sequence. Primary amines react with an excess of methyl iodide to form a quaternary ammonium iodide salt. Reaction with moist silver oxide gives the corresponding hydroxide salt. Heating the hydroxide salt leads to elimination, forming an alkene.

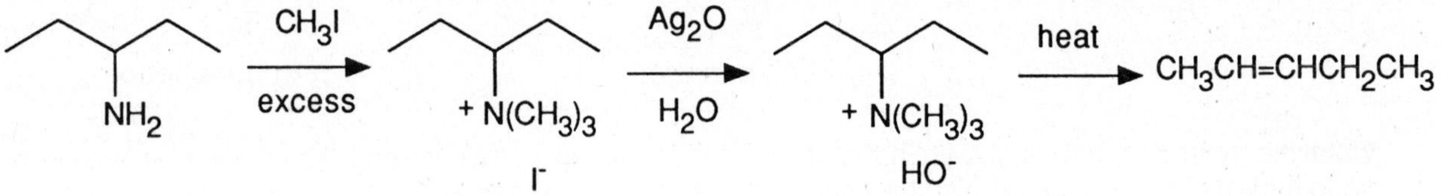

Compound A   Compound B   Compound C (cis + trans)

14.28 The desired alkene may be obtained by a Hofmann elimination sequence starting with 1-octanamine and the amine is available by reduction of 1-octanamide with lithium aluminum hydride.

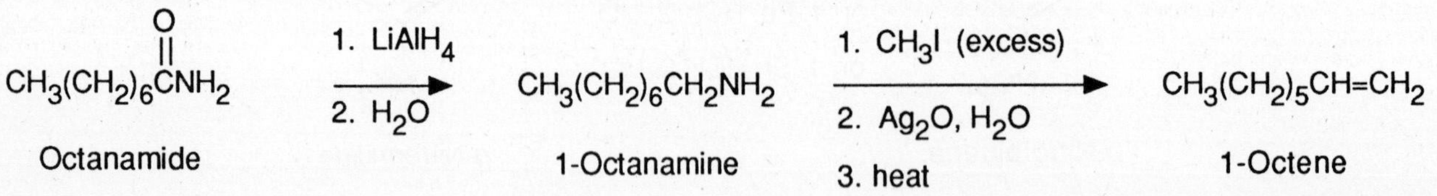

$CH_3(CH_2)_6CNH_2$ (Octanamide) — 1. LiAlH₄ 2. H₂O → $CH_3(CH_2)_6CH_2NH_2$ (1-Octanamine) — 1. CH₃I (excess) 2. Ag₂O, H₂O 3. heat → $CH_3(CH_2)_5CH=CH_2$ (1-Octene)

276

14.29 (a) 4-Methylpiperidine can participate in intermolecular hydrogen bonding in the liquid phase.

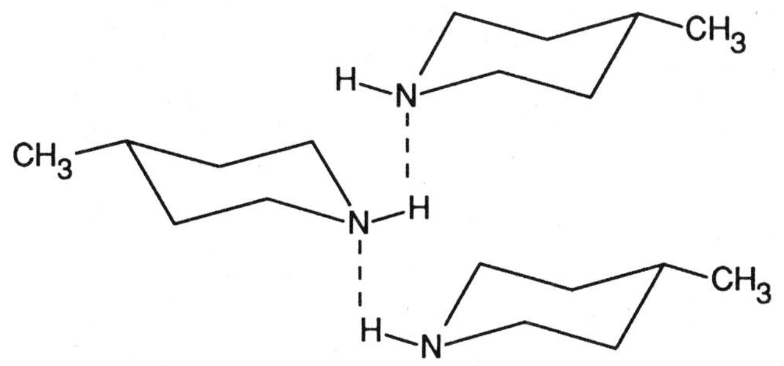

These hydrogen bonds must be broken in order for individual 4-methylpiperidine molecules to escape into the gas phase. *N*-Methylpiperidine lacks a proton bonded to nitrogen and so cannot engage in intermolecular hydrogen bonding. Less energy is required to transfer a molecule of *N*-methylpiperidine to the gaseous state, and therefore it has a lower boiling point than 4-methylpiperidine.

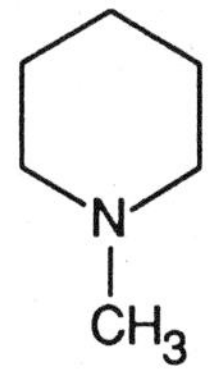

N-Methylpiperidine
lacks hydrogens bonded to nitrogen;
cannot form hydrogen bonds to
other *N*-methylpiperidine molecules

(b)   The two products are diastereomeric quaternary ammonium chlorides that differ in the configuration at nitrogen.

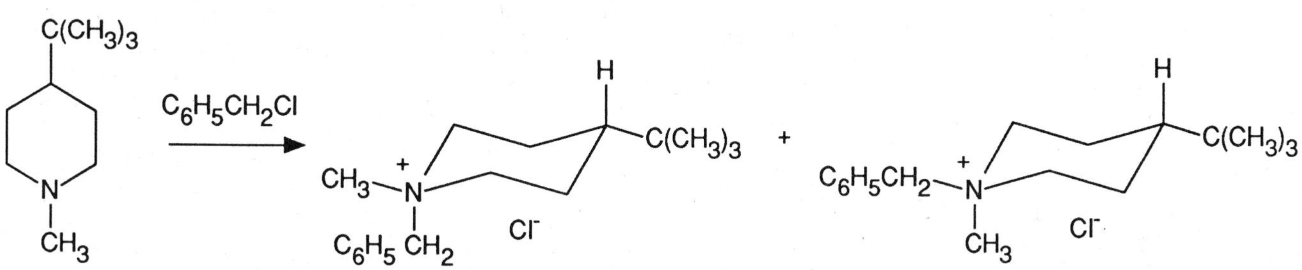

14.30 The reaction sequence in the problem describes two successive Hofmann eliminations. We are told that compound D is a heterocycle, that is, the nitrogen atom is part of a ring. Each Hofmann elimination produces one of the double bonds of the final butadiene product. The reaction sequence is

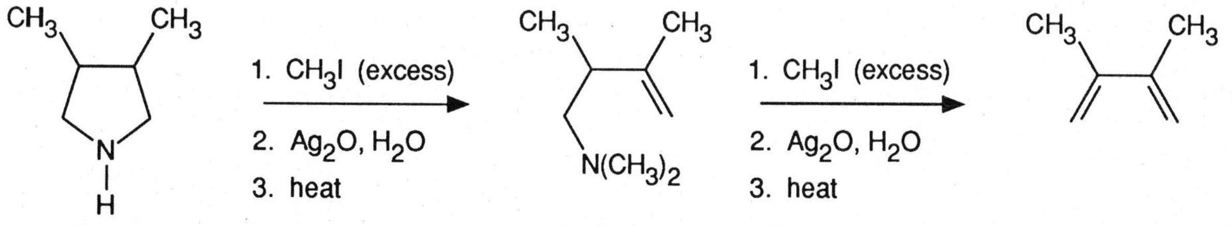

Compound D

1. CH₃I (excess)
2. Ag₂O, H₂O
3. heat

Compound E

1. CH₃I (excess)
2. Ag₂O, H₂O
3. heat

2,3-Dimethyl-1,3-butadiene

277

14.31 Tetramethylammonium hydroxide cannot undergo Hofmann elimination. The only reaction that can take place is nucleophilic substitution.

Tetramethylammonium hydroxide　　　Trimethylamine　Methanol

14.32 Primary and secondary amines exhibit absorptions due to N-H stretching in the 3200-3400 cm^{-1} region of the infrared spectrum. Primary amines have two bands in this region, secondary amines have one. Tertiary amines show no absorption in this region.

Two peaks　　　　　　One peak　　　　　　No peak

14.33 The infrared spectrum of compound F exhibits only a single peak at 3330 cm^{-1} in the N-H stretching region. This suggests that it is a secondary amine, since primary amines show two peaks in this region and tertiary amines none. The ^{1}H nmr spectrum reveals the presence of five aromatic protons at δ = 7.3 ppm, so compound F must be a monosubstituted benzene. The benzene ring accounts for all four sites of unsaturation suggested by the molecular formula $C_8H_{11}N$. The remaining peaks in the nmr spectrum are singlets, and the integral ratio 5:2:3:2 is consistent with the formulation of compound F as *N*-methylbenzylamine.

CHAPTER *15*

CARBOHYDRATES

GLOSSARY OF TERMS

Aldose   A carbohydrate which contains an aldehyde carbonyl group in its open-chain form.

Amino sugar   A carbohydrate in which one of the hydroxyl groups has been replaced by an amino group.

Amylopectin   Present in starch. Amylopectin is a polymer of $\alpha(1,4)$ linked glucose units, as is amylose (*see below*). Unlike amylose, amylopectin contains branches of 24-30 glucose units connected to the main chain by an $\alpha(1,6)$ linkage.

Amylose   The water-dispersible component of starch. It is a polymer of $\alpha(1,4)$ linked glucose units.

Anomeric carbon   The carbon atom in a furanose or pyranose form that is derived from the carbonyl carbon of the open-chain form. It is the ring carbon that is bonded to two oxygens.

Benedict's reagent   A solution containing the citrate complex of $CuSO_4$. It is used to test for the presence of reducing sugars.

Branched-chain carbohydrate   A carbohydrate in which the main carbon chain bears a carbon substituent in place of a hydrogen or hydroxyl group.

Cellobiose   A disaccharide in which two glucose units are joined by a $\beta(1,4)$ linkage. Cellobiose is obtained by the hydrolysis of cellulose.

Cellulose   A polymer in which thousands of glucose units are joined by a $\beta(1,4)$ linkage.

Deoxy sugar   A carbohydrate in which one of the hydroxyl groups has been replaced by a hydrogen.

Disaccharide   A carbohydrate which yields two monosaccharide units (which may be the same or different) on hydrolysis.

Furanose form   A five-membered ring arising via cyclic hemiacetal formation between the carbonyl group and a hydroxyl group of a carbohydrate.

Glycogen   A polymer of glucose present in animals. Similar in structure to amylopectin.

Glycoside   A carbohydrate derivative in which the hydroxyl group at the anomeric position has been replaced by some other group. An *O*-glycoside is an ether of a carbohydrate in which the anomeric position bears an alkoxy group.

Haworth formula   Planar representations of furanose and pyranose forms.

Hexose   An aldose with six carbon atoms.

Ketose   A carbohydrate which contains a ketone carbonyl group in its open-chain form.

Lactose   Also known as milk sugar. Lactose is a disaccharide formed by a β glycosidic linkage between C-4 of glucose and C-1 of galactose.

Maltose A disaccharide obtained from starch in which two glucose units are joined by an α(1,4) glycosidic link.

Monosaccharide A carbohydrate that cannot be hydrolyzed further to yield a more simple carbohydrate.

Mutarotation The change in optical rotation that occurs when a single form of a carbohydrate is allowed to equilibrate to a mixture of isomeric hemiacetals.

Oligosaccharide A carbohydrate that gives 3-10 monosaccharides on hydrolysis.

Osazone A compound of the type shown, formed by reaction of a carbohydrate with a hydrazine derivative.

$$CH=N-NHR$$
$$C=N-NHR$$

Pentose An aldose with five carbon atoms.

Polysaccharide A carbohydrate which yields "many" monosaccharide units on hydrolysis.

Pyranose form A six-membered ring arising via cyclic hemiacetal formation between the carbonyl group and a hydroxyl group of a carbohydrate.

Reducing sugar A carbohydrate that can be oxidized with substances such as Tollens' reagent and Benedict's reagent. In general, a carbohydrate with a hydroxyl group at the anomeric position will give a positive test with these reagents.

Sucrose A disaccharide of glucose and fructose in which the two monosaccharides are joined at their anomeric positions.

Tetrose An aldose with four carbon atoms.

SOLUTIONS TO TEXT PROBLEMS

15.1 (b) Redraw the Fischer projection so as to show the orientation of the groups in three dimensions.

$$HOCH_2 \underset{OH}{\overset{H}{|}} CHO \quad \text{is equivalent to} \quad HOCH_2 \blacktriangleright \underset{OH}{\overset{H}{C}} \blacktriangleleft CHO$$

Reorient the three-dimensional representation, putting the aldehyde group at the top and the CH₂OH group at the bottom.

$$HOCH_2 \blacktriangleright \underset{OH}{\overset{H}{C}} \blacktriangleleft CHO \quad \xrightarrow{\text{turn 90°}} \quad H \cdots C \cdots OH$$

What results is not equivalent to a proper Fischer projection because the horizontal bonds are directed "back" when they should be "forward." The opposite is true for the vertical bonds. In order to make the drawing correspond to a proper Fischer projection, we need to rotate it 180° around the vertical axis.

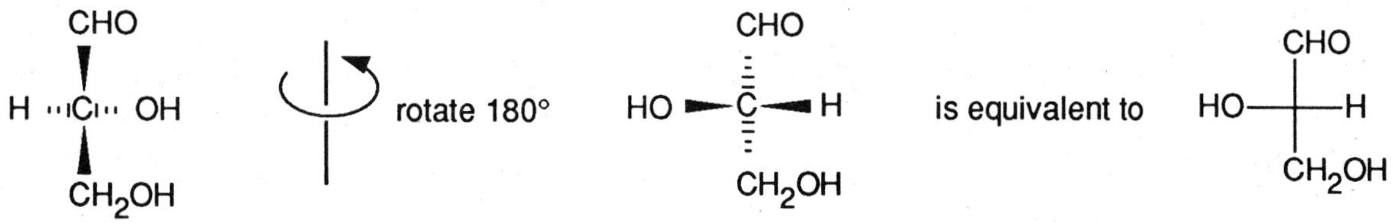

CHO

H ···C··· OH

CH$_2$OH

rotate 180°

CHO

HO ►C◄ H

CH$_2$OH

is equivalent to

CHO

HO———H

CH$_2$OH

Now, having the molecule arranged properly, we see that it is **L-glyceraldehyde**.

(c) Again proceed by converting the Fischer projection into a three-dimensional representation.

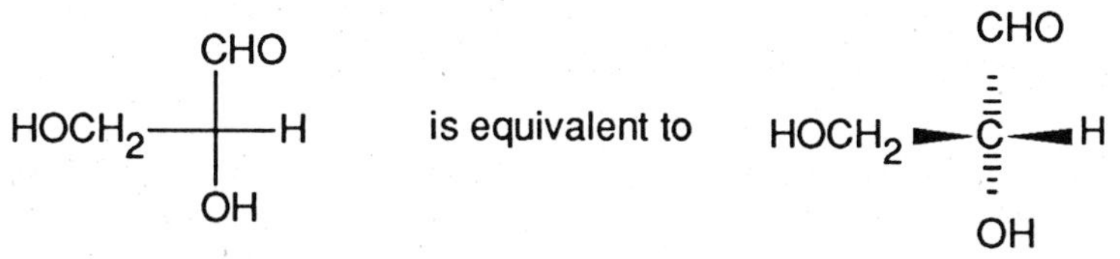

CHO

HOCH$_2$———H

OH

is equivalent to

CHO

HOCH$_2$ ►C◄ H

OH

Look at the drawing from a perspective that permits you to see the carbon chain oriented vertically with the aldehyde at the top and the CH$_2$OH at the bottom. Both groups should point away from you. When examined from this perspective, the hydrogen is to the left and they hydroxyl to the right with both pointing toward you.

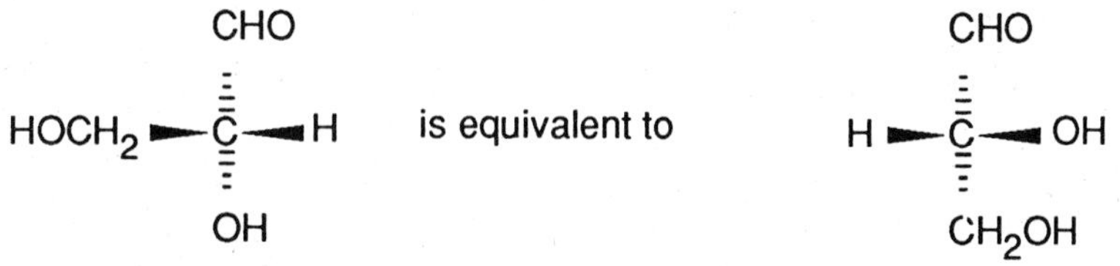

CHO

HOCH$_2$ ►C◄ H

OH

is equivalent to

CHO

H ►C◄ OH

CH$_2$OH

The molecule is **D-glyceraldehyde**.

15.2 In order to match the compound with the Fischer projection formulas of the aldotetroses, first redraw it in an eclipsed conformation.

H   OH   O
          ‖
          CH

HOCH$_2$ HO   H

⇄

H   OH
       H

HOCH$_2$ OH

O=CH

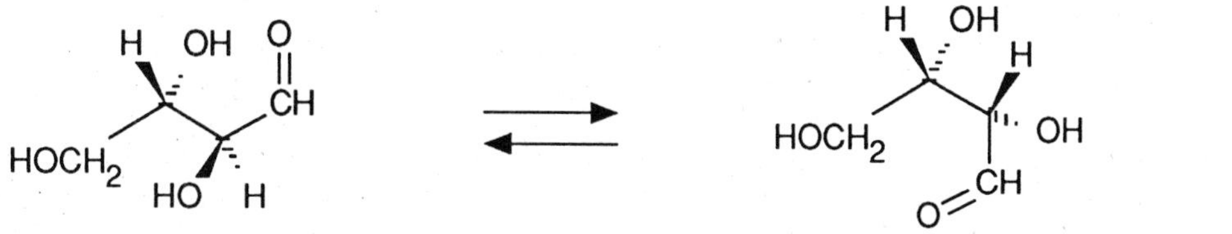

Staggered conformation

Same molecule in eclipsed conformation

The eclipsed conformation shown, when oriented so that the aldehyde carbon is at the top, vertical bonds back, and horizontal bonds pointing outward from the chiral centers, is readily transformed into the Fischer projection of **L-erythrose**.

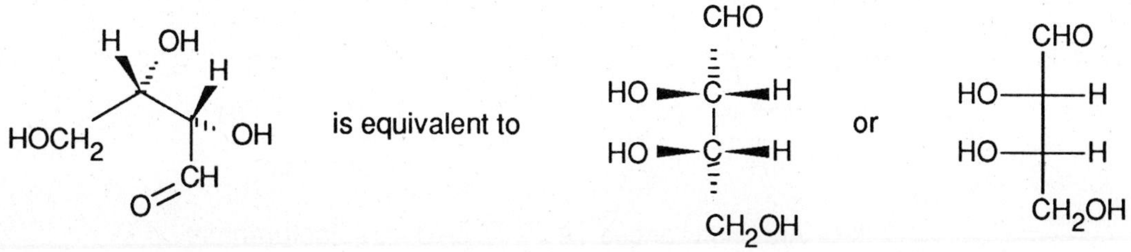

H   OH
       H

HOCH$_2$ OH

O=CH

is equivalent to

CHO

HO ►C◄ H

HO ►C◄ H

CH$_2$OH

or

CHO

HO———H

HO———H

CH$_2$OH

281

15.3 L-Arabinose is the mirror image of D-arabinose, the structure of which is given in Figure 15.3. The configuration at *each* chiral center of D-arabinose must be reversed to transform it into L-arabinose.

```
        CHO                        CHO
HO ──┼── H              H ──┼── OH
  H ──┼── OH           HO ──┼── H
  H ──┼── OH           HO ──┼── H
       CH₂OH                     CH₂OH
```

D–(–)–Arabinose   L–(+)–Arabinose

15.4 The configuration at C-5 is opposite to that of D-glyceraldehyde. Therefore, this particular carbohydrate belongs to the L series. Comparing it with the Fischer projection formulas of the eight D-aldohexoses in Figure 15.3 reveals it to be the mirror image of D-(+)-talose; it is **L-(-)-talose**.

```
        CHO                        CHO
HO ──┼── H              H ──┼── OH
HO ──┼── H              H ──┼── OH
HO ──┼── H              H ──┼── OH
  H ──┼── OH           HO ──┼── H
       CH₂OH                     CH₂OH
```

D–Talose   L–Talose

15.5 (b) The Fischer projection formula of D-arabinose may be found in text Figure 15.3 (p 436). The Fischer projection and the eclipsed representation ("coiled" form) corresponding to it are:

```
        CHO
HO ──┼── H
  H ──┼── OH
  H ──┼── OH
       CH₂OH
```

D-Arabinose   Coiled form of D–arabinose   Conformation of coiled form suitable for furanose ring formation

Cyclic hemiacetal formation between the carbonyl group and the C-4 hydroxyl yields the α- and β-furanose forms of D-arabinose.

β–D–Arabinofuranose   α–D–Arabinofuranose

(c) The mirror image of D-arabinose [from (b)] is L-arabinose.

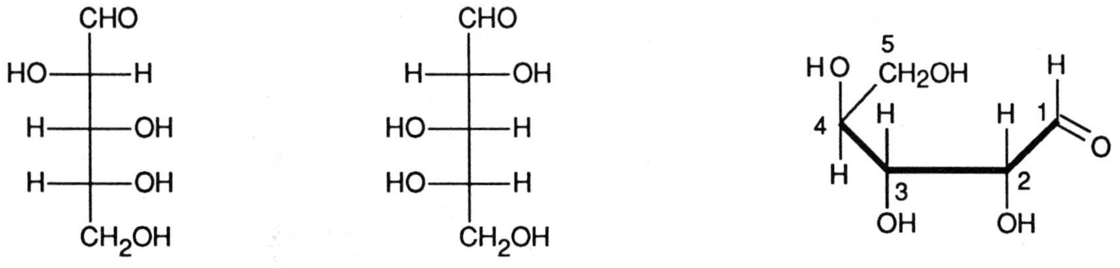

CHO
HO——H
H——OH
H——OH
CH₂OH

D–Arabinose

CHO
H——OH
HO——H
HO——H
CH₂OH

L–Arabinose

Coiled form of L–arabinose

The C-4 atom of the coiled form of L-arabinose must be rotated 120° in a clockwise sense so as to bring its hydroxyl group into the proper orientation for furanose ring formation.

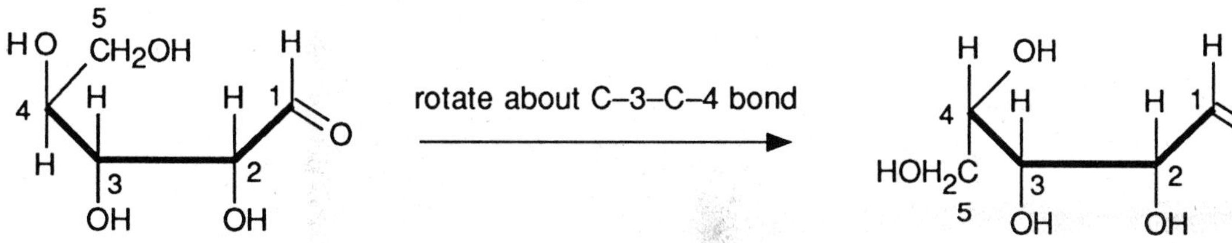

rotate about C–3–C–4 bond

Cyclization gives the α– and β-furanose forms of L-arabinose.

α–D–Arabinofuranose

β–D–Arabinofuranose

In the L-series the anomeric hydroxyl is up in the α isomer and down in the β isomer.

15.6 The pyranose forms of D-ribose are formed by addition of the C-5 hydroxyl to the carbonyl group. As with the furanose forms, two pyranose forms, α and β, are possible. The pyranose forms are constructed by first writing the eclipsed conformation of D-ribose in a form suitable for attack of the C-5 hydroxyl on the carbonyl carbon.

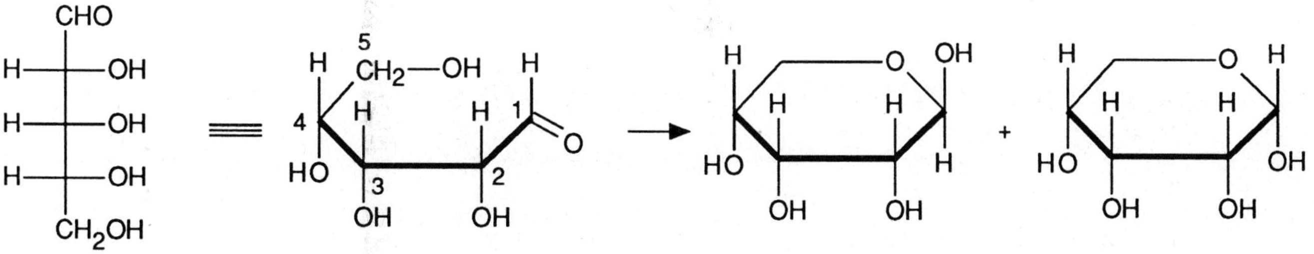

CHO
H——OH
H——OH
H——OH
CH₂OH

D–Ribose

β–D–Ribopyranose

α–D–Ribopyranose

15.7 To reveal the structure of a carbohydrate, replace the group attached to the anomeric carbon with –OH. Adenosine is an *N*-glycoside formed from the β-furanose form of D-ribose.

Adenosine   β–D–Ribofuranose

The carbohydrate in sinigrin is the same as that in linamarin, the β-pyranose form of D-glucose.

Sinigrin   β–D–Glucopyranose

15.8 Those disaccharides which contain a hydroxyl (OH) group at the anomeric position are reducing sugars. The anomeric carbon is the one which is derived from the carbonyl group of the open-chain form; it is the only carbon in the molecule that is bonded to two oxygens.

(b)   The structure of lactose is shown on page 447. It contains a glycosidic linkage between C-1 of galactose and C-4 of glucose. A hydroxyl group is present at the anomeric carbon of the glucose portion of lactose. Lactose is a reducing sugar.

(c)   The structure of sucrose is shown on page 446. Its glycosidic bond connects the anomeric carbon of glucose to the anomeric carbon of fructose. There is no OH substituent on either of the anomeric carbons, and sucrose is not a reducing sugar.

(d)   As with maltose in (a), the glycosidic linkage of cellobiose is between C-1 (the anomeric carbon) of one glucose unit and C-4 of another. The anomeric carbon of one glucose unit (the one on the right in Figure 15.6 on page 447) bears an OH group. Cellobiose is a reducing sugar.

15.9   (b)   The phenylosazone derived from L-erythrose is the same as the one from L-threose. The two carbohydrates differ in configuration only at C-2.

L–Erythrose   L–Threose

(c)   The carbohydrate that differs in configuration from D-allose only at C-2 is D-altrose. Both yield the same osazone.

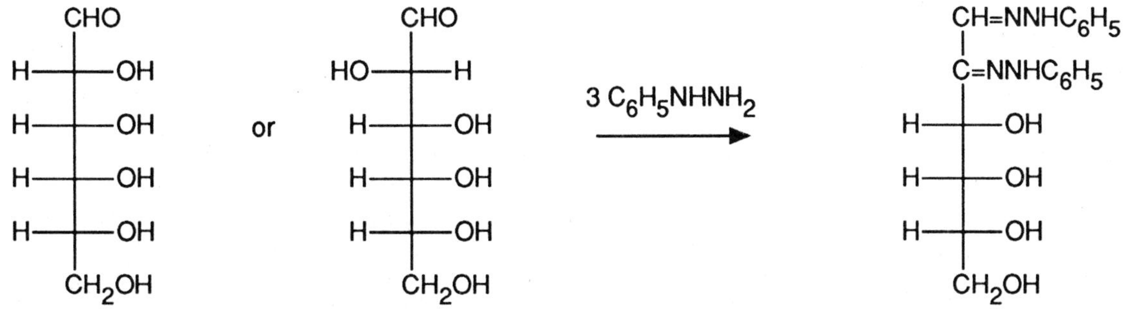

```
     CHO                    CHO                                    CH=NNHC6H5
H────OH              HO────H                                     C=NNHC6H5
H────OH      or       H────OH         3 C6H5NHNH2          H────OH
H────OH               H────OH         ────────────→       H────OH
H────OH               H────OH                              H────OH
   CH2OH                 CH2OH                                  CH2OH

  D–Allose             D–Altrose
```

15.10 (a)   Excluding compounds with rings or carbon-carbon double bonds, the only constitutional isomer of glyceraldehyde is the ketone 1,3-dihydroxy-2-propanone, also called dihydroxyacetone.

$$HOCH_2\overset{\overset{\displaystyle O}{\|}}{C}CH_2OH \qquad \text{1,3-Dihydroxy-2-propanone}$$

(b)   The configuration of the highest-numbered chiral center is the same as L-glyceraldehyde in any L carbohydrate. The Fischer projection of the only L-ketose, called L-erythrulose, is:

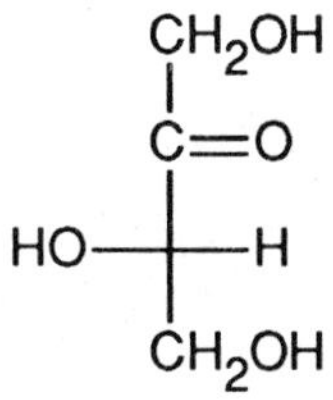

```
     CH2OH
      C=O
HO────H
     CH2OH
```

(c)   The enantiomer of L-erythrulose is D-erythrulose, having the opposite configuration at the chiral center.

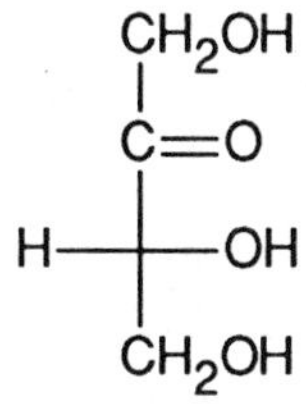

```
     CH2OH
      C=O
 H────OH
     CH2OH
```

15.11 (a)   The structure shown in Figure 15.3 is D-(+)-xylose; therefore (-)-xylose must be its mirror image and has the L-configuration at C-4.

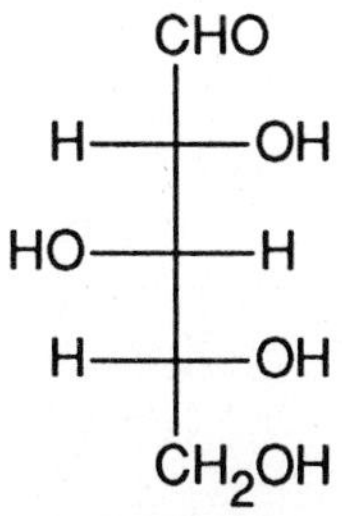

```
       CHO                        CHO
  H────OH                  HO────H
 HO────H                    H────OH
  H────OH                  HO────H
    CH2OH                     CH2OH

 D–(+)–Xylose              L–(–)–Xylose
```

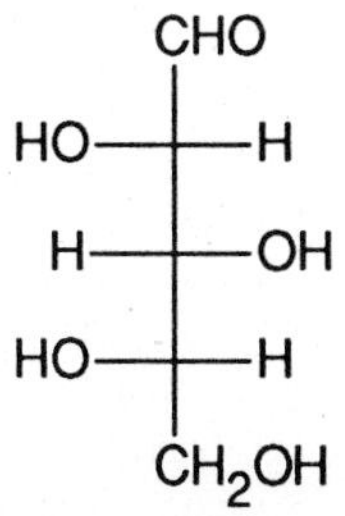

(b) Redraw the Fischer projection of D-xylose in its coiled form.

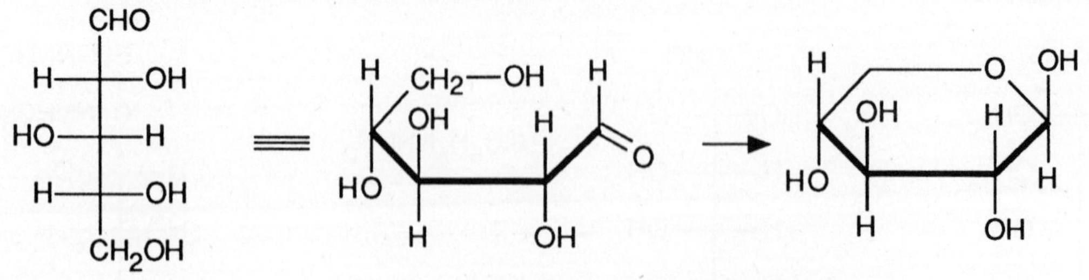

CHO
H——OH
HO——H
H——OH
CH₂OH

D–Xylose

Haworth formula of β–D–Xylopyranose

The pyranose form arises by closure to a six-membered cyclic hemiacetal, with the C-5 hydroxyl group undergoing nucleophilic addition to the carbonyl. In the β-pyranose form of D-xylose the anomeric hydroxyl group is up.

The preferred conformation of β-D-xylopyranose is a chair with all the hydroxyl groups equatorial.

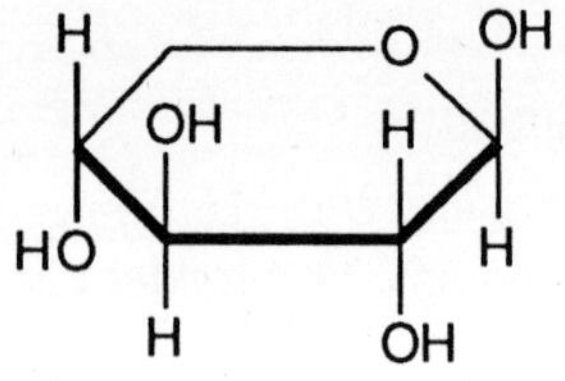

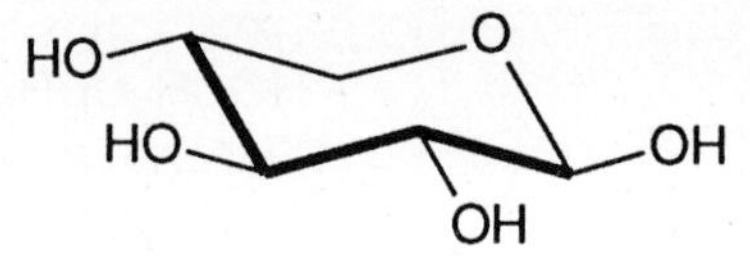

Haworth formula of β–D–Xylopyranose   is better represented as   Chair conformation of β–D–Xylopyranose

(c) Methyl β-D-xylopyranoside is the methyl glycoside corresponding to the structure just drawn in (b). The glycosidic group –OCH₃ has replaced the –OH.

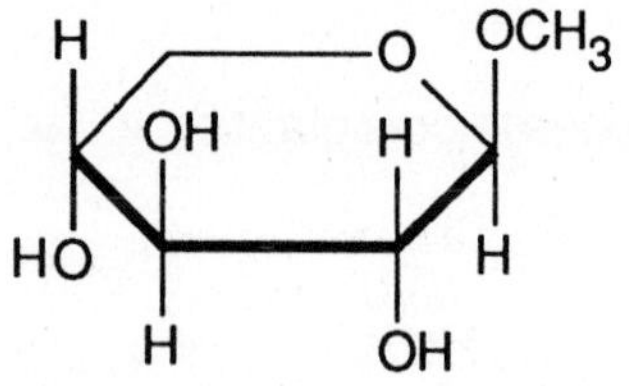

also represented as

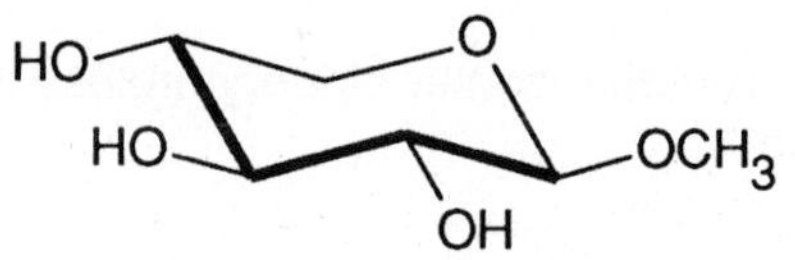

(d) With excess phenylhydrazine, D-xylose is converted to its osazone.

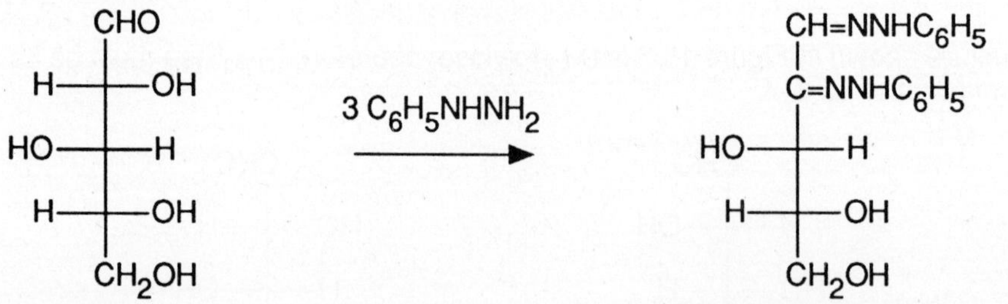

CHO
H——OH
HO——H
H——OH
CH₂OH

D–Xylose

$3 \ C_6H_5NHNH_2$ →

CH=NNHC₆H₅
C=NNHC₆H₅
HO——H
H——OH
CH₂OH

D–Xylose phenylosazone

286

15.12 (a) Begin by drawing a Haworth formula for the β-pyranose form of D-galactose. Follow the procedure outlined in Figure 15.5 for D-glucose.

D–Galactose   β–D–Galactopyranose

Next, redraw the planar Haworth formula more realistically as a chair conformation, choosing the one that has the CH₂OH group equatorial.

Galactose differs from glucose in configuration at C-4. The C-4 hydroxyl is axial in β-D-galactopyranose while it is equatorial in β-D-glucopyranose.

The Fischer projection and the Haworth formula for D-mannose are:

D–Mannose   β–D–Mannopyranose

The Haworth formula is more realistically drawn as the following chair conformation.

β–D–Mannopyranose

Mannose differs from glucose in configuration at C-2. All hydroxyl groups are equatorial in β-D-glucopyranose; the hydroxyl at C-2 is axial in β-D-mannopyranose.

(b)   The α-pyranose forms of D-galactose and D-mannose differ from the β-pyranose forms by the configuration of the anomeric hydroxyl group. Their Haworth formulas are as shown.

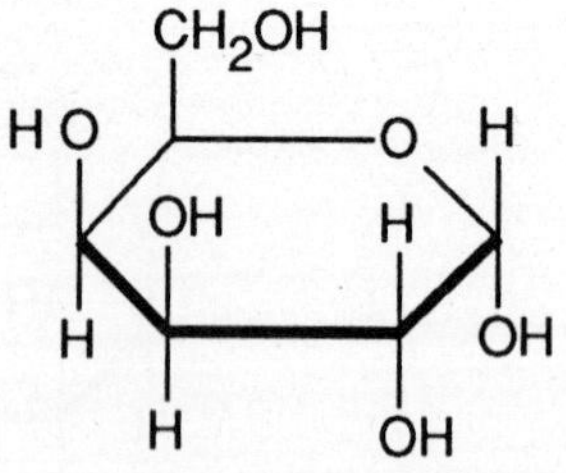

α–D–Galactopyranose

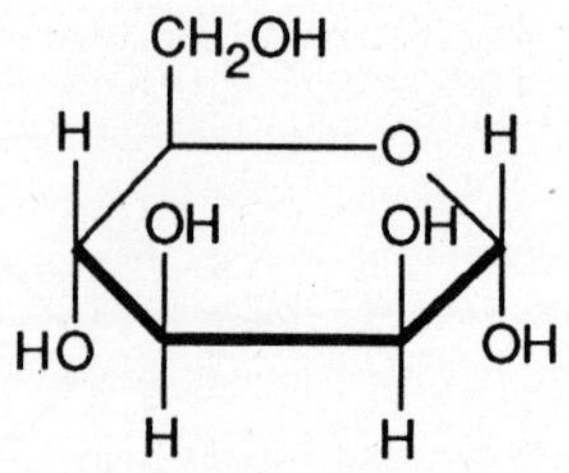

α–D–Mannopyranose

15.13 A proper Fischer projection of an aldose has a vertical carbon chain with the aldehyde function at the top. The Fischer projections given need to be revised to suit these requirements.   Exchanging the positions of two substituents reverses the configuration at a chiral center. However, exchanging the positions of three substituents retains the original configuration.   All of the Fischer projections given can be converted to proper Fischer projections by exchanging the positions of three substituents on one of the chiral centers.

(a)

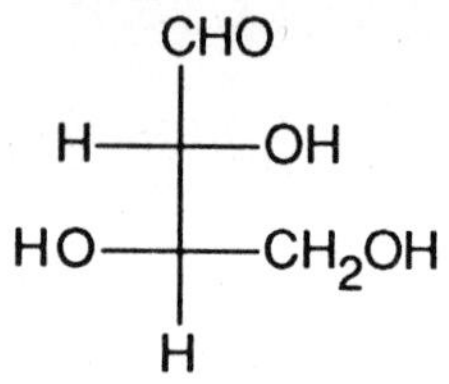

becomes

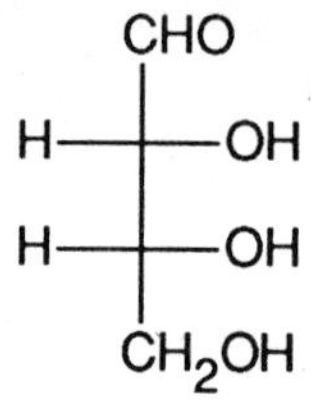

D–Erythrose

(b)

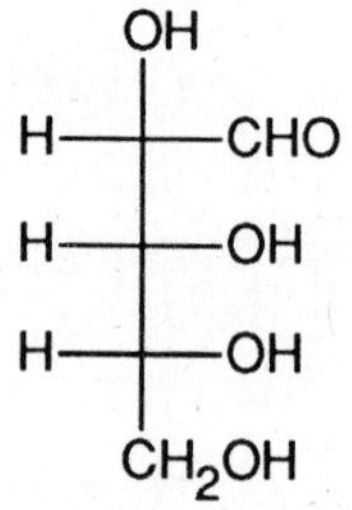

becomes

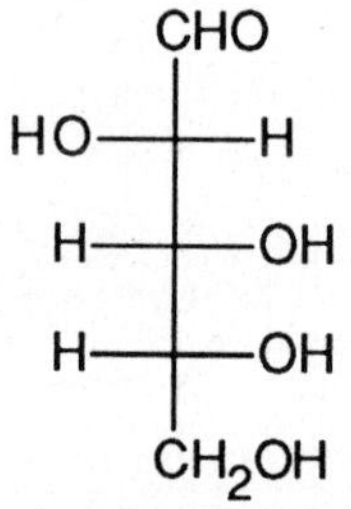

D–Arabinose

(c)

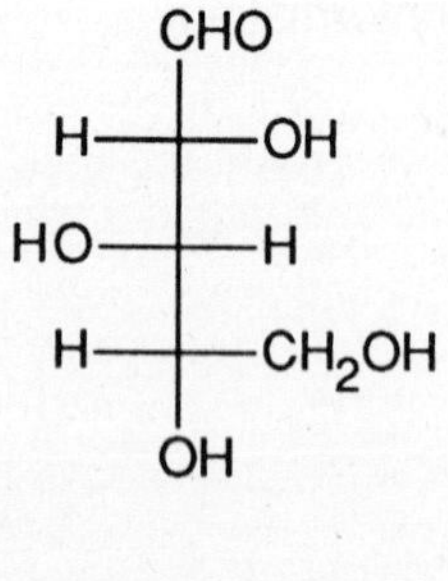

becomes

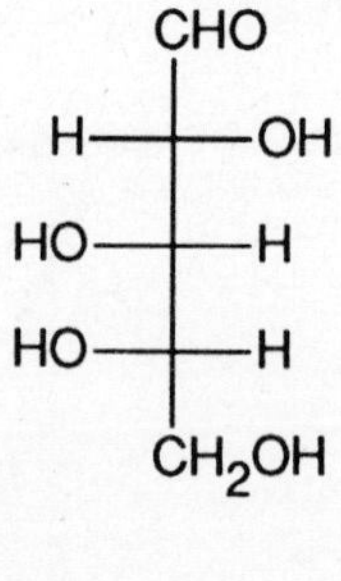

L –Arabinose

288

15.14 The hemiacetal function opens to give an intermediate containing a free aldehyde function. Cyclization of the intermediate can produce either the α or the β configuration at this center.

Key intermediate formed by cleavage of hemiacetal

Maltose with β configuration at C-1

Maltose with α configuration at C-1

Only the configuration of the free hemiacetal function is affected in this process. The α configuration of the glycosidic linkage remains unchanged.

15.15 (a) To unravel a pyranose form, locate the anomeric carbon and mentally convert the hemiacetal linkage to a carbonyl group and a hydroxyl function.

Convert the open-chain form to a proper Fischer projection. This sugar is D-ribose.

equivalent to

(b) Proceeding as in (a) reveals this sugar to be L-arabinose.

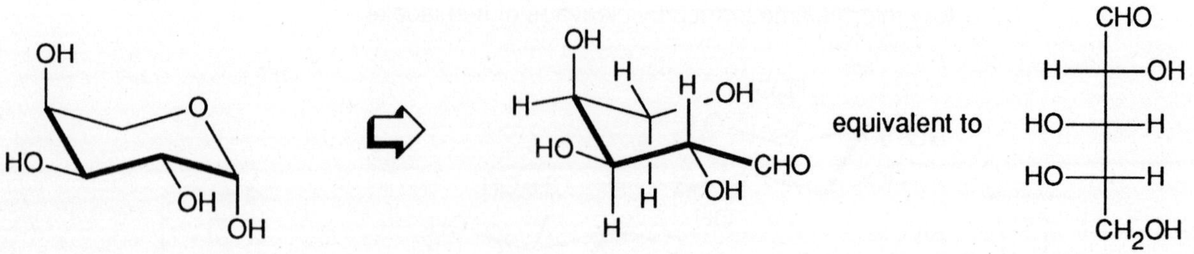

equivalent to

(c) The sugar in this part is an aldohexose. First convert the hemiacetal linkage to the open-chain form.

The open-chain form is converted to a Fischer projection as above, however it is necessary to rotate the groups on C-5 to obtain the proper orientation.

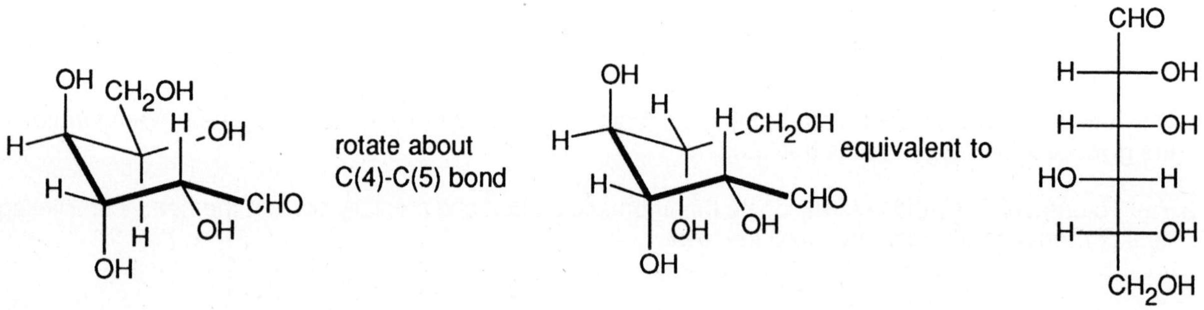

rotate about C(4)-C(5) bond

equivalent to

The sugar is D-gulose.

(d) Proceeding as before reveals this sugar to be D-arabinose.

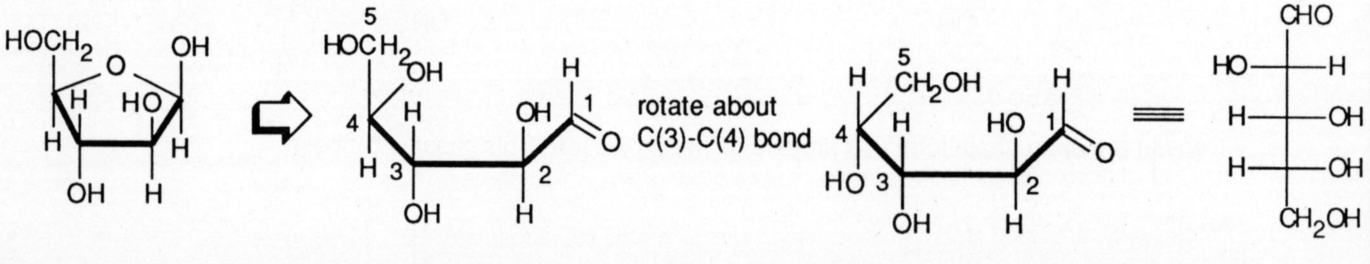

rotate about C(3)-C(4) bond

15.16 Sugars having a free anomeric hydroxyl group are reducing sugars, and react with Benedict's solution. Since trehalose does not react with Benedict's solution, the glycosidic linkage of trehalose must involve both anomeric carbons of the disaccharide.

15.17 (a) The hydroxyl group at C-3 in D-ribose is replaced by hydrogen in 3-deoxy-D-ribose.

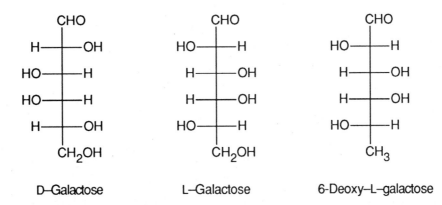

D–Ribose
(from Figure 15.3)

3-Deoxy–D–ribose
(cordycepose)

(b)   Since L-fucose is 6-deoxy-L-galactose, first write the Fischer projection of D-galactose, then transform it to its mirror image, L-galactose. Transform the C-6 CH_2OH group to CH_3 to produce 6-deoxy-L-galactose.

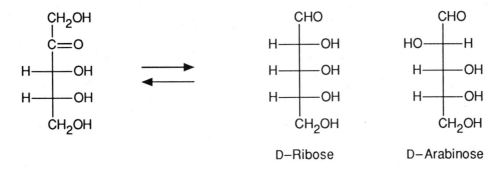

D–Galactose   L–Galactose   6-Deoxy–L–galactose

15.18 The two aldoses differ in their configuration at C-2. They are D-ribose and D-arabinose.

CH_2OH
|
C=O
H——OH
H——OH
CH_2OH

⇌

CHO
H——OH
H——OH
H——OH
CH_2OH

D–Ribose

CHO
HO——H
H——OH
H——OH
CH_2OH

D–Arabinose

15.19   In order to give the same phenylosazone as D-xylose, the carbohydrate must differ from D-xylose only in configuration at C-2. There is only one possibility and that is **D-lyxose**.

CHO
H——OH
HO——H
H——OH
CH_2OH

D–Xylose

or

CHO
HO——H
HO——H
H——OH
CH_2OH

D–Lyxose

$\xrightarrow{3\ C_6H_5NHNH_2}$

CH=NNHC_6H_5
|
C=NNHC_6H_5
HO——H
H——OH
CH_2OH

291

15.20 Sodium borohydride reduction does not change the configuration of any chiral groups; the only change is –CHO → CH₂OH.

(a)

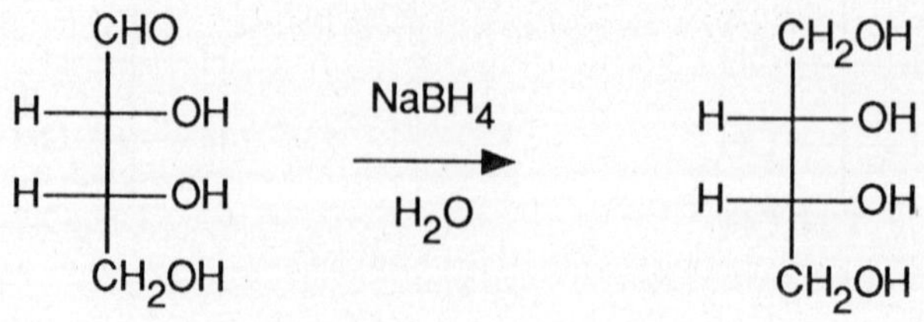

```
      CHO                              CH2OH
  H ──┼── OH          NaBH4        H ──┼── OH
                    ─────────►
  H ──┼── OH           H2O         H ──┼── OH
      CH2OH                            CH2OH
```

D–Erythrose

(b)

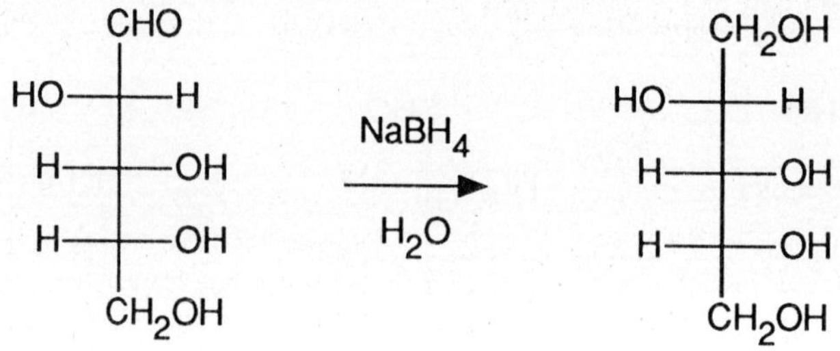

```
      CHO                              CH2OH
 HO ──┼── H          NaBH4       HO ──┼── H
                    ─────────►
  H ──┼── OH           H2O         H ──┼── OH
  H ──┼── OH                        H ──┼── OH
      CH2OH                            CH2OH
```

D–Arabinose

(c)

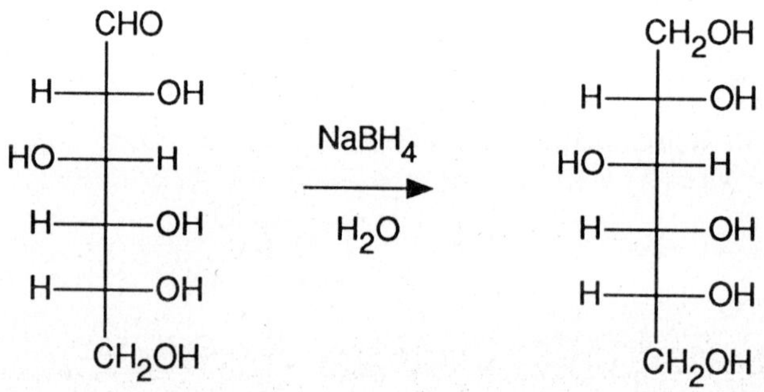

```
      CHO                              CH2OH
  H ──┼── OH                        H ──┼── OH
 HO ──┼── H          NaBH4       HO ──┼── H
                    ─────────►
  H ──┼── OH           H2O         H ──┼── OH
  H ──┼── OH                        H ──┼── OH
      CH2OH                            CH2OH
```

D–Glucose

(d)

```
      CHO                              CH2OH
  H ──┼── OH                        H ──┼── OH
 HO ──┼── H          NaBH4       HO ──┼── H
                    ─────────►
 HO ──┼── H           H2O        HO ──┼── H
  H ──┼── OH                        H ──┼── OH
      CH2OH                            CH2OH
```

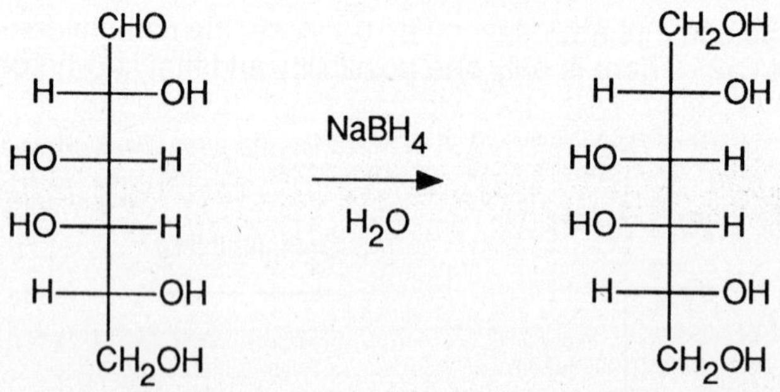

D–Galactose

15.21 A molecule that contains a plane of symmetry cannot be chiral. Erythritol and galactitol each contain a plane of symmetry (perpendicular to the page and passing through the dotted line shown in the Fischer projection). The chiral alditols formed in the preceding problem are **arabinitol** and **glucitol**.

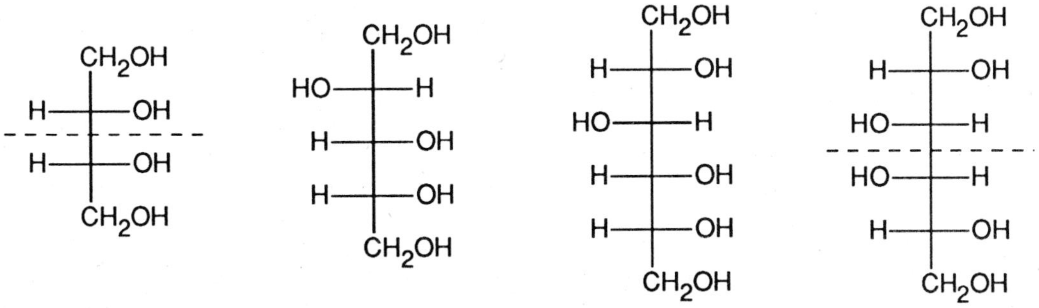

CH_2OH	CH_2OH	CH_2OH	CH_2OH
H——OH	HO——H	H——OH	H——OH
- - - - - - -	H——OH	HO——H	HO——H
H——OH	H——OH	H——OH	HO——H
CH_2OH	H——OH	H——OH	H——OH
	CH_2OH	CH_2OH	CH_2OH
Erythritol	**Arabinitol**	**Glucitol**	**Galactitol**
(achiral)	(chiral)	(chiral)	(achiral)

15.22 When the ketone function of D–fructose is reduced, a new chiral center is generated. The hydroxyl group can be to the left or to the right in the Fischer projection.   The two alditols formed are shown below. Their names are **glucitol** and **mannitol**.

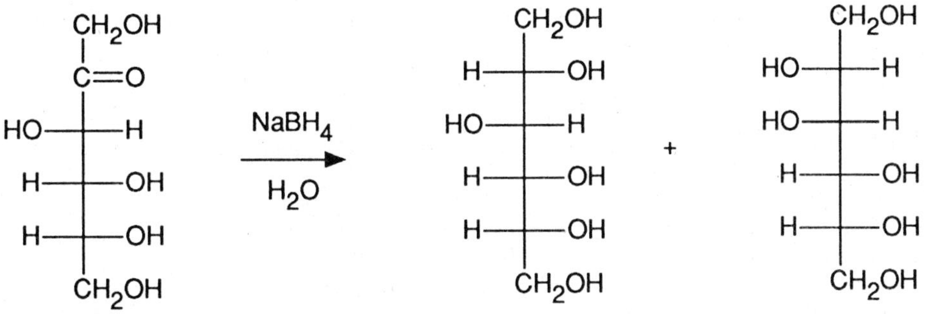

CH_2OH		CH_2OH		CH_2OH
C=O		H——OH		HO——H
HO——H	$NaBH_4$	HO——H	+	HO——H
H——OH	$\longrightarrow$	H——OH		H——OH
H——OH	H_2O	H——OH		H——OH
CH_2OH		CH_2OH		CH_2OH
Fructose		**Glucitol**		**Mannitol**

15.23 (a) Carbon-2 is the only chiral center in D-apiose.

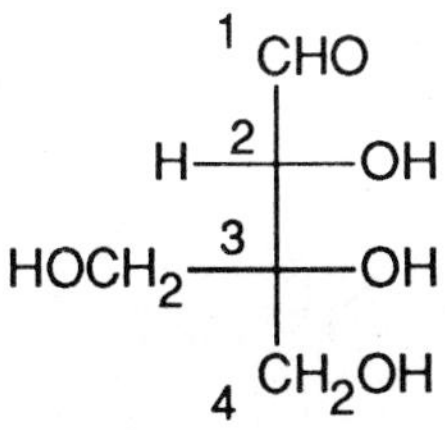

$$^1CHO$$
$$H—^2\!\!—OH$$
$$HOCH_2—^3\!\!—OH$$
$$_4\,CH_2OH$$

D–Apiose

Carbon-3 is not a chiral center; it bears two identical CH_2OH substituents.

(b) When D-apiose is converted to an osazone, the chiral center at C-2 is transformed to C=NNHC$_6$H$_5$. The osazone of D-apiose is not optically active.

D–Apiose   → $3 \, C_6H_5NHNH_2$ →   Phenylosazone of apiose
(achiral; optically inactive)

(c) The alditol obtained on reduction of D-apiose retains the chiral center. It is chiral and optically active.

D-Apiose   →   NaBH$_4$ →   D-Apiitol (optically active)

(d) Cyclic hemiacetal formation in D-apiose involves addition of a CH$_2$OH hydroxyl group to the aldehyde carbonyl.

There are three chiral centers in the furanose form. These are the anomeric carbon C-1, the original chiral center C-2, and the new chiral center at C-3.

15.24 (a) The hydrate forms by addition of water (H–OH) to the aldehyde carbonyl group.

D–Glucose   +   H$_2$O   →   D–Glucose hydrate

(b) Five-membered rings correspond to furanose forms. The problem states that furanose forms account for 0.29% of the glucose molecules in solution.

(c) The anomeric hydroxyl group and the CH_2OH group are cis the β-pyranose form of glucose.

these groups are cis to one another

15.25 The equation describing the equilibrium is:

α–D–Mannopyranose   Open-chain form of D-mannose   β–D–Mannopyranose

$[\alpha]_D$ +29.3° $[\alpha]_D$ -17.0°

Let A = percent α isomer; 100-A = percent β isomer. Then:

$$A\,(+29.3°) \; + \; (100 - A)\,(-17.0°) \; = \; 100\,(14.2°)$$

$$46.3\,A \; = \; 3120$$

$$\text{Percent } \alpha \text{ isomer} = 67\%$$

$$\text{Percent } \beta \text{ isomer} = (100 - A) = 33\%$$

CHAPTER 16

AMINO ACIDS, PEPTIDES AND PROTEINS

GLOSSARY OF TERMS

Active site   The region where chemically significant groups of an enzyme interact with a substrate to produce chemical change.

α-Amino acid   A carboxylic acid that contains an amino group at the α-carbon atom. α-Amino acids are the building blocks of peptides and proteins. An α-amino acid normally exists as a *zwitterion*.

$$RCHCO_2^-$$
$$|$$
$$+NH_3$$

L-Amino acid   A description of the stereochemistry at the α-carbon atom of a chiral amino acid. The Fischer projection of an α-amino acid has the amino group on the left when the carbon chain is vertical with the carboxyl group at the top.

$$H_3N \overset{+}{\underset{R}{\overset{CO_2^-}{—|—}}} H$$

Amino acid residues   Individual amino acids components of a peptide or protein.

Chymotrypsin   A digestive enzyme that catalyzes the hydrolysis of proteins. Chymotrypsin selectively catalyzes the cleavage of the peptide bond between the carboxyl group of phenylalanine, tyrosine, or tryptophan and some other amino acid.

Dipeptide   A compound in which two α-amino acids are linked by an amide bond between the amino group of one and the carboxyl group of the other.

$$\overset{O}{\overset{||}{RCHC}}—NHCHCO_2^-$$
$$\underset{+NH_3}{|} \qquad \underset{R'}{|}$$

Disulfide bridge   A S—S bond between the sulfur atoms of two cysteines.

Edman degradation   A method for determining the N-terminal amino acid of a peptide or protein. It involves treating the material with phenyl isothiocyanate ($C_6H_5N=C=S$), cleaving with acid, then identifying the phenylthiohydantoin (PTH-derivative) produced.

Electrophoresis   A method for separating and identifying amino acids based on their tendency to migrate to a positive or negatively charged electrode at a particular pH.

Enzyme   A protein that catalyzes a chemical reaction in a living system.

α-Helix   One type of protein secondary structure. It is a right-handed helix characterized by hydrogen bonds between NH and C=O groups. It contains approximately 3.6 amino acids per turn.

Isoelectric point   The pH at which the concentration of the zwitterionic form of an amino acid is a maximum. At a pH below the isoelectric point the dominant species is a cation. At higher pH, an anion predominates. At the isoelectric point the amino acid has no net charge.

Merrifield method   (see *Solid-phase peptide synthesis*)

Metalloenzyme   An enzyme in which a metal ion at the active site contributes in a chemically significant way to the catalytic activity.

Peptide   Structurally, a molecule composed of two or more α–amino acids joined by peptide bonds.

Peptide bond   An amide bond between the carboxyl group of one α–amino acid and the amino group of another.

$$\cancel{\;\;}-NHCHC-NHCHC-\cancel{\;\;}$$

(The darkened bond is the peptide bond.)

β-Pleated sheet   A type of protein secondary structure characterized by hydrogen bonds between NH and C=O groups of adjacent parallel peptide chains. The individual chains are in an extended "zig-zag" conformation.

Polypeptide   A polymer made up of "many" (more than 8-10) amino acid residues.

Primary structure   The sequence of amino acids in a peptide or protein.

Protein   A naturally occurring polymer which typically contains 100-300 amino acid residues.

Quaternary structure   A description of the way in which two or more protein chains are organized in a protein (such as hemoglobin) which contains more than one protein.

Sanger's reagent   The compound 1-fluoro-2,4-dinitrobenzene used in N-terminal amino acid identification.

Secondary structure   The conformation with respect to nearest-neighbor amino acids in a peptide or protein. The α–helix and the β–pleated sheet are examples of protein secondary structural types.

Solid-phase peptide synthesis   A method for peptide synthesis in which the C-terminal amino acid is covalently attached to an inert solid support and successive amino acids attached via peptide bond formation. At the completion of the synthesis the polypeptide is removed from the support.

Strecker synthesis   A method for preparing amino acids in which the first step is reaction of an aldehyde with ammonia and hydrogen cyanide to give an amino nitrile which is then hydrolyzed.

$$RCH \xrightarrow[HCN]{NH_3} RCHC\equiv N \xrightarrow{\text{hydrolysis}} RCHCO_2^-$$

C-Terminus   The amino acid at the end of a peptide or protein chain that has its carboxyl group intact; i.e., the carboxyl group is not part of a peptide bond.

N-Terminus   The amino acid at the end of a peptide or protein chain that has its α–amino group intact; i.e., the α–amino group is not part of a peptide bond.

Tertiary structure   A description of how a protein chain is folded.

Tripeptide   A compound in which three α–amino acids are linked by peptide bonds.

Trypsin   A digestive enzyme that catalyzes the hydrolysis of proteins. Trypsin selectively catalyzes the cleavage of the peptide bond between the carboxyl group of lysine or arginine and some other amino acid.

Zwitterion   The form in which neutral amino acids actually exist. The amino group is in its protonated form and the carboxyl group is present as a carboxylate ion.

$$RCHCO_2^-$$
$$+NH_3$$

16.1 GABA has an amino group attached to C-4 of butanoic acid. It is an ammonium carboxylate in the zwitterion form.

$$\overset{+}{H_3N}CH_2CH_2CH_2CO_2^-$$ 4-Aminobutanoic acid (GABA)

16.2 The amino acids in Table 16.1 that have more than one chiral center are isoleucine and threonine. The chiral centers are marked with an asterisk in the structural formulas shown.

$$CH_3CH_2\overset{*}{C}H-\overset{*}{C}HCO_2^-$$
$$\underset{CH_3}{|}\ \underset{\underset{+}{NH_3}}{|}$$

Isoleucine

$$CH_3\overset{*}{C}H-\overset{*}{C}HCO_2^-$$
$$\underset{OH}{|}\ \underset{\underset{+}{NH_3}}{|}$$

Threonine

16.3 (b) A solution of pH 5.5 corresponds to the isoelectric point of phenylalanine (p 5.48). The zwitterion is the predominant species at this pH.

$$C_6H_5CH_2CHCO_2^-$$
$$\underset{\underset{+}{NH_3}}{|}$$

(c) A pH of 9.0 represents a basic solution, therefore the predominant species is an amino carboxylate anion.

$$C_6H_5CH_2CHCO_2^-$$
$$\underset{NH_2}{|}$$

16.4 In the Strecker synthesis an aldehyde is treated with ammonia and a source of cyanide ion. The resulting amino nitrile is hydrolyzed to an amino acid.

$$(CH_3)_2CHCH\overset{O}{\overset{\|}{}} \xrightarrow[HCN]{NH_3} (CH_3)_2CHCHC\equiv N\ \underset{NH_2}{\overset{|}{}} \xrightarrow[\text{2. HO}^-]{\text{1. }H_3O^+} (CH_3)_2CHCHCO_2^- \ \underset{+\ NH_3}{|}$$

2-Methylpropanal   Valine

As actually carried out, the aldehyde was converted to the amino nitrile by treatment with an aqueous solution containing ammonium chloride and potassium cyanide. Hydrolysis was achieved in aqueous hydrochloric acid and gave valine as its hydrochloride salt in 65 percent overall yield.

16.5 (b) Alanine is the N-terminal amino acid in Ala-Phe. Its carboxyl group is joined to the nitrogen of phenylalanine by a peptide bond.

$$\overset{+}{H_3N}CHC\overset{O}{\overset{\|}{}}-NHCHCO_2^-$$
$$\underset{CH_3}{|}\qquad \underset{CH_2C_6H_5}{|}$$ Ala-Phe

Alanine ¦ Phenylalanine

(c)   The positions of the amino acids are reversed in Phe-Ala. Phenylalanine is the N terminus and alanine is the C terminus.

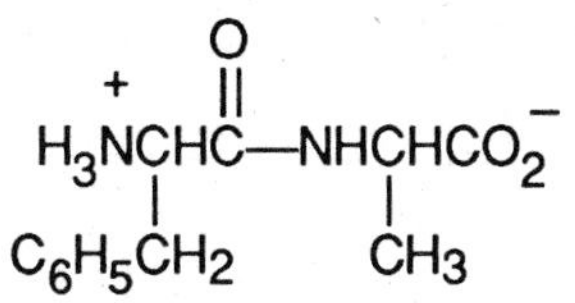

$$\overset{+}{H_3}NCHC\overset{\displaystyle O}{\overset{\|}{}}\!-\!NHCHCO_2^-\qquad \text{Phe-Ala}$$

with side chains $C_6H_5CH_2$ and CH_3

Phenylalanine ┆ Alanine

(d)   The carboxyl group of glycine is joined by a peptide bond to the amino group of glutamic acid.

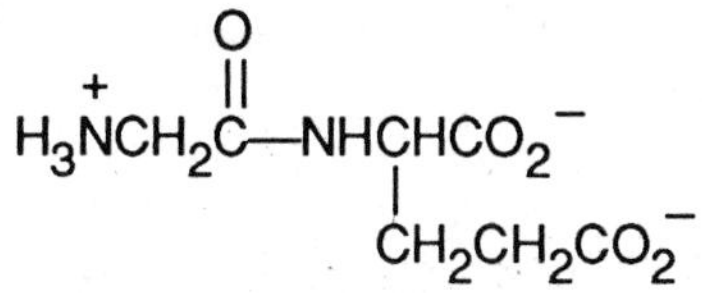

$$\overset{+}{H_3}NCH_2C\overset{\displaystyle O}{\overset{\|}{}}\!-\!NHCHCO_2^-\qquad \text{Gly–Glu}$$

side chain $CH_2CH_2CO_2^-$

Glycine ┆ Glutamic acid

The dipeptide is written in its anionic form because the carboxyl group of the side chain is ionized at pH 7. Alternatively, it could have been written as a neutral zwitterion with a $CH_2CH_2CO_2H$ side chain.

(e)   The peptide bond in Lys-Gly is between the carboxyl group of lysine and the amino group of glycine.

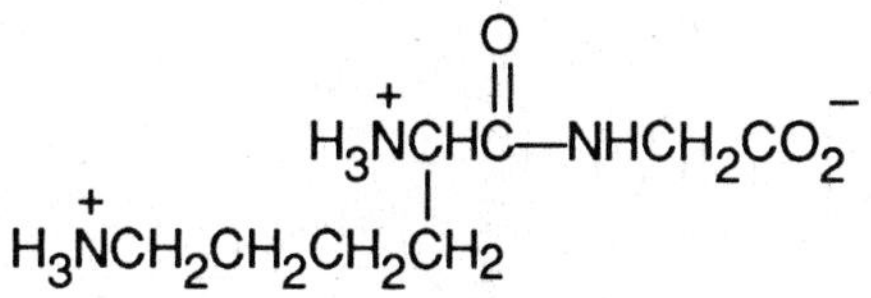

$$\overset{+}{H_3}NCHC\overset{\displaystyle O}{\overset{\|}{}}\!-\!NHCH_2CO_2^-\qquad \text{Lys–Gly}$$

side chain $\overset{+}{H_3}NCH_2CH_2CH_2CH_2$

Lysine ┆ Glycine

The amino group of the lysine side chain is protonated at pH 7 so the dipeptide is written above in its cationic form. It could have also been written as a neutral zwitterion with the side chain $H_2NCH_2CH_2CH_2CH_2$.

16.6   The dipeptide Asp-Phe has the structure shown. Aspartame is its C-terminal methyl ester.

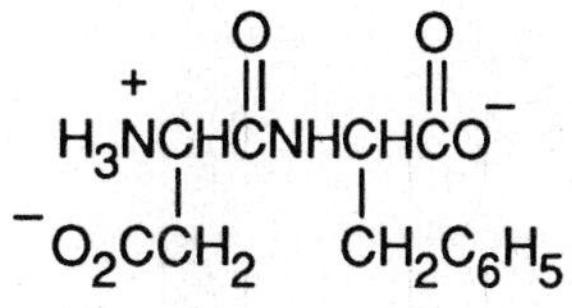

$$\overset{+}{H_3}NCHC\overset{\displaystyle O}{\overset{\|}{}}NHCHC\overset{\displaystyle O}{\overset{\|}{}}O^-$$
side chains $^-O_2CCH_2$ and $CH_2C_6H_5$

Asp–Phe

$$\overset{+}{H_3}NCHC\overset{\displaystyle O}{\overset{\|}{}}NHCHC\overset{\displaystyle O}{\overset{\|}{}}OCH_3$$
side chains $^-O_2CCH_2$ and $CH_2C_6H_5$

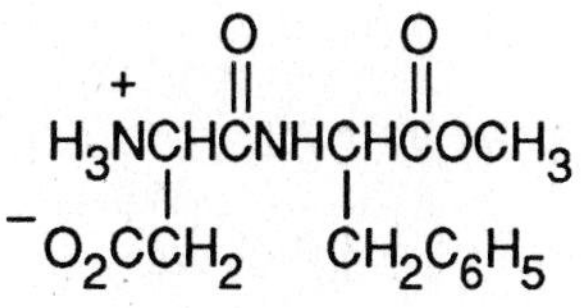

Aspartame

16.7 The structure of leucine enkephalin is given in Figure 16.3 (p 465) as Tyr–Gly–Gly–Phe–Leu. The Edman degradation cleaves the N-terminal amino acid from a peptide to give a PTH derivative. PTH derivatives have the structure shown where R is the substituent at the α carbon atom of the amino acid (the "side chain").

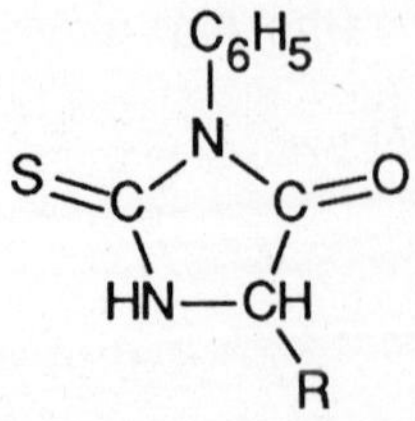

PTH derivative of $H_3\overset{+}{N}CHCO_2^-$
$\qquad\qquad\qquad\qquad R$

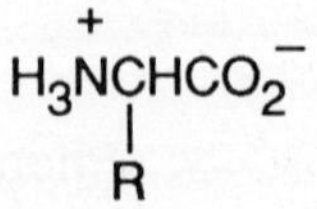

The problem asks for the structure of the PTH derivatives obtained from the first three cycles of the Edman degradation. Thus the first cycle produces the PTH derivative of tyrosine and the second and third cycles both give the PTH derivative of glycine.

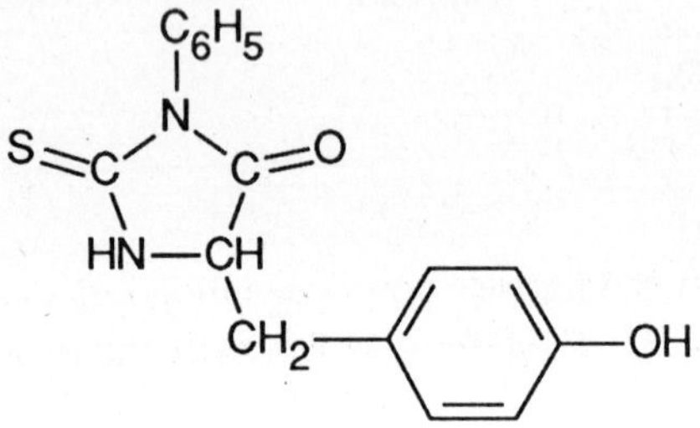

PTH derivative of tyrosine

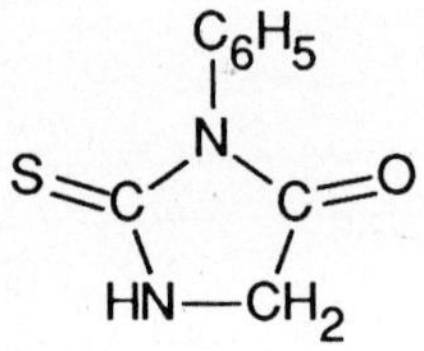

PTH derivative of glycine

16.8 Trypsin catalyzes cleavage of peptides on the carboxyl side (to the right) of lysine (Lys) and arginine (Arg). Trypsin will catalyze cleavage of the N-terminal Arg in bradykinin.

$$\text{Arg}\mid\text{Pro–Pro–Gly–Phe–Ser–Pro–Phe–Arg} \xrightarrow[\text{H}_2\text{O}]{\text{Trypsin}} \text{Arg} + \text{Pro–Pro–Gly–Phe–Ser–Pro–Phe–Arg}$$

Chymotrypsin catalyzes peptide cleavage on the carboxyl side of phenylalanine (Phe), tyrosine (Tyr), and tryptophan (Trp). Two phenylalanine residues are present in bradykinin.

$$\text{Arg–Pro–Pro–Gly–Phe}\mid\text{Ser–Pro–Phe}\mid\text{Arg} \xrightarrow[\text{H}_2\text{O}]{\text{Chymotrypsin}} \text{Arg–Pro–Pro–Gly–Phe} + \text{Ser–Pro–Phe} + \text{Arg}$$

16.9 The structure of the peptide may be revealed by aligning the fragments where there is overlap. The problem is more easily solved by starting with the larger fragments, then following with the smaller ones.

Hydrolysis fragments: Leu–Phe–Pro
                                        Val–Leu–Phe
                            Pro–Val
                                              Pro–Ala

Original peptide: Pro–Val–Leu–Phe–Pro–Ala

16.10 The peptide bond of Ala-Leu connects the carboxyl group of alanine and the amino group of leucine. Therefore we need to protect the amino group of alanine and the carboxyl group of leucine.

Protect the amino group of alanine as its benzyloxycarbonyl derivative.

$$\underset{\underset{\displaystyle CH_3}{|}}{\overset{+}{H_3N}CHCO_2^-} \;+\; C_6H_5CH_2O\overset{\overset{\displaystyle O}{||}}{C}Cl \longrightarrow C_6H_5CH_2O\overset{\overset{\displaystyle O}{||}}{C}NH\underset{\underset{\displaystyle CH_3}{|}}{CH}CO_2H$$

Alanine   Benzyloxycarbonyl chloride   Z–Protected alanine

Protect the carboxyl group of leucine as its benzyl ester:

$$\underset{\underset{\displaystyle CH_2CH(CH_3)_2}{|}}{\overset{+}{H_3N}CHCO_2^-} \;+\; C_6H_5CH_2OH \xrightarrow[\text{2. HO}^-]{\text{1. H}^+,\ \text{heat}} \underset{\underset{\displaystyle CH_2CH(CH_3)_2}{|}}{H_2NCH}\overset{\overset{\displaystyle O}{||}}{C}OCH_2C_6H_5$$

Leucine   Benzyl alcohol   Leucine benzyl ester

Coupling of two amino acids is achieved by DCCI-promoted amide bond formation between the free amino group of leucine benzyl ester and the free carboxyl group of Z-protected alanine.

$$C_6H_5CH_2O\overset{\overset{\displaystyle O}{||}}{C}NH\underset{\underset{\displaystyle CH_3}{|}}{CH}CO_2H \;+\; \underset{\underset{\displaystyle CH_2CH(CH_3)_2}{|}}{H_2NCH}\overset{\overset{\displaystyle O}{||}}{C}OCH_2C_6H_5 \xrightarrow{\text{DCCI}}$$

$$C_6H_5CH_2O\overset{\overset{\displaystyle O}{||}}{C}NH\underset{\underset{\displaystyle CH_3}{|}}{CH}\overset{\overset{\displaystyle O}{||}}{C}NH\underset{\underset{\displaystyle CH_2CH(CH_3)_2}{|}}{CH}\overset{\overset{\displaystyle O}{||}}{C}OCH_2C_6H_5$$

Z–Protected alanine   Leucine benzyl ester   Protected dipeptide

Both the benzyloxycarbonyl protecting group and the benzyl ester protecting group may be removed by treatment with hydrogen and palladium. This step completes the synthesis of Ala-Leu.

$$C_6H_5CH_2O\overset{\overset{\displaystyle O}{||}}{C}NH\underset{\underset{\displaystyle CH_3}{|}}{CH}\overset{\overset{\displaystyle O}{||}}{C}NH\underset{\underset{\displaystyle CH_2CH(CH_3)_2}{|}}{CH}\overset{\overset{\displaystyle O}{||}}{C}OCH_2C_6H_5 \xrightarrow[\text{Pd}]{\text{H}_2} \overset{+}{H_3N}\underset{\underset{\displaystyle CH_3}{|}}{CH}\overset{\overset{\displaystyle O}{||}}{C}NH\underset{\underset{\displaystyle CH_2CH(CH_3)_2}{|}}{CH}CO_2^-$$

Protected dipeptide   Ala-Leu

16.11 Essential amino acids are those we can not make from simple starting materials in our bodies. We must obtain them from our diet.

16.12 The amino group in L-threonine points to the left as it does in the Fischer projections of all L-amino acids. The problem states that the carbon that bears the hydroxyl group, however, has the D configuration, so the hydroxyl group at C-3 must point to the right.

$$
\begin{array}{c}
CO_2^{-} \\
H_3\overset{+}{N}\!\!-\!\!\!-\!\!\!-\!\!H \\
H\!-\!\!\!-\!\!\!-\!OH \\
CH_3
\end{array}
$$

L-Threonine

16.13 At low pH (acidic solution) the predominant species is an ammonium-carboxylic acid. At the isoelectric point, the zwitterion predominates. In basic solution (high pH) the amino carboxylate is the predominant species.

$$
\begin{array}{c}
CH_3 \\
| \\
CH_3CH_2CHCHCO_2H \\
| \\
\overset{+}{N}H_3
\end{array}
\qquad
\begin{array}{c}
CH_3 \\
| \\
CH_3CH_2CHCHCO_2^{-} \\
| \\
\overset{+}{N}H_3
\end{array}
\qquad
\begin{array}{c}
CH_3 \\
| \\
CH_3CH_2CHCHCO_2^{-} \\
| \\
NH_2
\end{array}
$$

Isoleucine (pH 1)　　　　　Isoleucine (pH 6)　　　　　Isoleucine (pH 13)

16.14 (a) At low pH the solution is strongly acidic and the predominant species is the protonated form of the amino acid. Both amino groups of lysine are protonated at a pH of 1.

$$
\begin{array}{c}
\overset{+}{H_3}NCH_2CH_2CH_2CH_2CHCO_2H \\
| \\
\overset{+}{N}H_3
\end{array}
$$

Lysine (pH 1)

(b) At very high pH, both nitrogens are present as NH2 groups and the carboxyl group is deprotonated.

$$
\begin{array}{c}
H_2NCH_2CH_2CH_2CH_2CHCO_2^{-} \\
| \\
NH_2
\end{array}
$$

Lysine (pH 13)

16.15 (a) At very low pH the amino group and both carboxyl groups of glutamic acid are protonated.

$$
\begin{array}{c}
HO_2CCH_2CH_2CHCO_2H \\
| \\
\overset{+}{N}H_3
\end{array}
$$

Glutamic acid (pH 1)

(b) At very high pH all three groups are deprotonated.

$$
\begin{array}{c}
{}^{-}O_2CCH_2CH_2CHCO_2^{-} \\
| \\
NH_2
\end{array}
$$

Glutamic acid (pH 13)

16.16 (a)   The Strecker synthesis begins with an aldehyde. In the synthesis of phenylalanine, the appropriate aldehyde is phenylacetaldehyde.

$$C_6H_5CH_2\overset{\overset{\displaystyle O}{\|}}{C}H \xrightarrow[\text{NaCN}]{\text{NH}_4\text{Cl}} C_6H_5CH_2\underset{\underset{\displaystyle NH_2}{|}}{C}HC\equiv N \xrightarrow[\text{2. HO}^-]{\text{1. H}_2\text{O, HCl, heat}} C_6H_5CH_2\underset{\underset{\displaystyle +NH_3}{|}}{C}HCO_2^-$$

Phenylacetaldehyde   Phenylalanine

(b)   Substitution by ammonia of 2-bromo-3-phenylpropanoic acid leads directly to formation of phenylalanine.

$$C_6H_5CH_2\underset{\underset{\displaystyle Br}{|}}{C}HCO_2H \quad + \quad 2\ NH_3 \longrightarrow C_6H_5CH_2\underset{\underset{\displaystyle +NH_3}{|}}{C}HCO_2^- \quad + \quad NH_4Br$$

2-Bromo-3-phenylpropanoic acid   Ammonia   Phenylalanine   Ammonium bromide

16.17 The structure of 6-aminopenicillanic acid is shown twice. In the representation on the left the atoms which are derived from cysteine are shown in bold, while on the right the atoms derived from valine are emphasized. The side chains in cysteine and valine are $-CH_2SH$ and $CH(CH_3)_2$, respectively.

16.18 Dipeptides which can be formed from one molecule each of valine and alanine are Val–Ala and Ala–Val.

Valylalanine (Val–Ala)   Alanylvaline (Ala–Val)

16.19 Four stereoisomeric forms of the dipeptide Val-Ala would be obtained from racemic valine and alanine. Each amino acid has one chiral center so the starting mixture would contain D-valine, L-valine, D-alanine, and L-alanine. The possible combinations are:

D-Val-D-Ala	L-Val-L-Ala
D-Val-L-Ala	L-Val-D-Ala

16.20 The structure of leucine enkephalin is given in Figure 16.3 as Tyr–Gly–Gly–Phe–Leu. Replacing the C-terminal leucine with methionine gives the amino acid sequence of methionine enkephalin as Tyr–Gly–Gly–Phe–Met.

16.21 (a) The structure of the peptide is shown below with dashed lines added to indicate the points of connection between the four amino acids. As can be seen from the structural formula, the side chains are appended to the peptide backbone. Identify the amino acids by examining the side chains and comparing them to the formulas of the α-amino acids in Table 16.1.

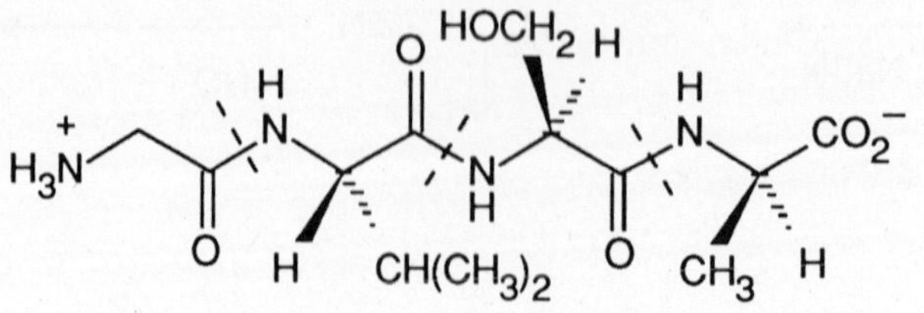

Gly   Val   Ser   Ala

The amino acid sequence is Gly–Val–Ser–Ala. The peptide could also be named glycylvalylserylalanine.

(b) The alanine residue in this peptide has the D configuration. A convenient way to check this is to determine its configuration by the Cahn-Ingold-Prelog method. In this case, the configuration is *R* which corresponds to D for all of the chiral amino acids except cysteine.

16.22 Complete hydrolysis of a peptide reveals its amino acid composition, not the sequence of amino acids. Thus alanine, leucine, and proline may be in any order. The possibilities are:

Ala–Leu–Pro	Leu–Ala–Pro	Pro–Ala–Leu
Ala–Pro–Leu	Leu–Pro–Ala	Pro–Leu–Ala

16.23 Carboxypeptidase releases the C-terminal amino acid, which in the statement of the problem is identified as proline. Formation of its PTH derivative identifies leucine as the N-terminal amino acid. The structure of the tripeptide is therefore Leu-Ala-Pro.

16.24 (a) Sanger's reagent is 1-fluoro-2,4-dinitrobenzene. Reaction of the peptide with this reagent followed by hydrolysis forms a derivative of the N-terminal amino acid. In the case of leucine enkephalin, the 2,4-dinitrophenyl derivative of tyrosine is formed.

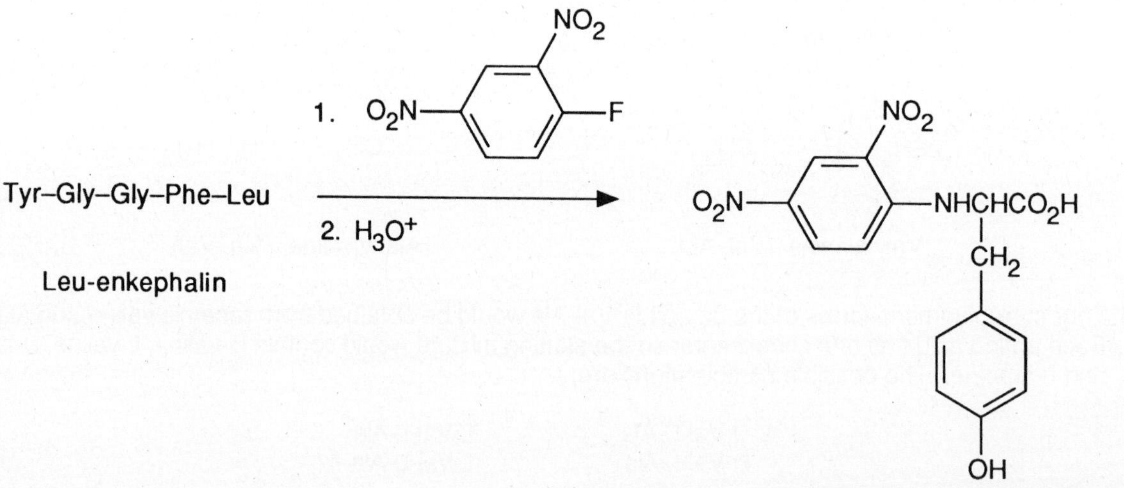

Tyr–Gly–Gly–Phe–Leu

Leu-enkephalin

1. O_2N—〈 〉—F (with NO_2)

2. H_3O^+

2,4-DNP derivative of tyrosine

(b) In problem 16.7 the products of the first three cycles of the Edman degradation were described. Recall that the Edman degradation cleaves amino acids from the N-terminus. Thus the first cycle of the Edman degradation of Tyr–Gly–Gly–Phe–Leu gives the PTH derivative of tyrosine, and the second and

304

third cycles both give the PTH derivative of glycine. The fourth cycle gives the PTH derivative of phenylalanine.

PTH derivative of phenylalanine

(c)   Carboxypeptidase releases the C-terminal amino acid from a peptide. That amino acid is leucine in leucine enkephalin.

16.25 Trypsin cleaves peptides at the carboxyl group of arginine or lysine. Therefore:

$$\text{Ser–Ala–Arg} \vert \text{Phe–Gly–Ala} \quad \xrightarrow[\text{Trypsin}]{\text{H}_2\text{O}} \quad \text{Ser–Ala–Arg} + \text{Phe–Gly–Ala}$$

16.26 The experimental data establish the following concerning the tetrapeptide:

(1)   Total hydrolysis reveals that the four amino acids present are Leu, Gly, Phe, and Val.

(2)   The Edman degradation identifies the N-terminus as Phe.

(3)   Since partial hydrolysis gives a tripeptide containing Val, Gly, and Phe and since Phe is the N-terminal amino acid, the first three residues must be either Phe–Val–Gly or Phe–Gly–Val.

(4)   Since a dipeptide containing Leu and Gly was obtained from partial hydrolysis and since Gly is present among the first three amino acids reading from the N-terminus, Leu must be the C-terminal amino acid. The peptide must end in the sequence Gly–Leu at the C-terminus.

The only sequence that accommodates all of the data is **Phe–Val–Gly–Leu**.

16.27 Lysine contains two amino groups, both of which may react with benzyloxycarbonyl chloride.

$$\overset{+}{\text{H}_3}\text{NCH}_2\text{CH}_2\text{CH}_2\text{CH}_2\overset{\underset{\displaystyle +\text{NH}_3}{|}}{\text{CH}}\text{CO}_2^- + 2\ \text{C}_6\text{H}_5\text{CH}_2\overset{\text{O}}{\overset{\|}{\text{O}}}\text{CCl} \longrightarrow \text{C}_6\text{H}_5\text{CH}_2\text{O}\overset{\text{O}}{\overset{\|}{\text{C}}}\text{NHCH}_2\text{CH}_2\text{CH}_2\text{CH}_2\overset{\underset{\displaystyle \text{NHCOCH}_2\text{C}_6\text{H}_5}{|}}{\text{CH}}\text{CO}_2\text{H}$$

Lysine   Benzyloxycarbonyl chloride

16.28 The peptide bond of Val-Ala connects the carboxyl group of valine and the amino group of alanine. Therefore we need to protect the amino group of valine and the carboxyl group of alanine.

Protect the amino group of valine as its benzyloxycarbonyl derivative.

305

$$\overset{+}{\text{H}_3}\text{NCHCO}_2^-$$ + $\text{C}_6\text{H}_5\text{CH}_2\text{OCCl}$ (with C=O) $\longrightarrow$ $\text{C}_6\text{H}_5\text{CH}_2\text{OCNHCHCO}_2\text{H}$ (with C=O)

under first: $\text{CH(CH}_3)_2$; under last: $\text{CH(CH}_3)_2$

| Valine | Benzyloxycarbonyl chloride | Z–Protected valine |

Protect the carboxyl group of alanine as its benzyl ester:

$$\overset{+}{\text{H}_3}\text{NCHCO}_2^-$$ + $\text{C}_6\text{H}_5\text{CH}_2\text{OH}$ $\xrightarrow[\text{2. HO}^-]{\text{1. H}^+, \text{ heat}}$ $\text{H}_2\text{NCHCOCH}_2\text{C}_6\text{H}_5$ (with C=O)

under first: CH_3 ; under last: CH_3

| Alanine | Benzyl alcohol | Alanine benzyl ester |

Coupling of two amino acids is achieved by DCCI-promoted amide bond formation between the free amino group of alanine benzyl ester and the free carboxyl group of Z-protected valine.

$\text{C}_6\text{H}_5\text{CH}_2\text{OCNHCHCO}_2\text{H}$ + $\text{H}_2\text{NCHCOCH}_2\text{C}_6\text{H}_5$ $\xrightarrow{\text{DCCI}}$ $\text{C}_6\text{H}_5\text{CH}_2\text{OCNHCHCNHCHCOCH}_2\text{C}_6\text{H}_5$

under: $\text{CH(CH}_3)_2$; CH_3 ; $(\text{CH}_3)_2\text{CH}$ $\quad$ CH_3

| Z–Protected valine | Alanine benzyl ester | Protected dipeptide |

Both the benzyloxycarbonyl protecting group and the benzyl ester protecting group may be removed by treatment with hydrogen and palladium. This step completes the synthesis of Val-Ala.

$\text{C}_6\text{H}_5\text{CH}_2\text{OCNHCHCNHCHCOCH}_2\text{C}_6\text{H}_5$ $\xrightarrow[\text{Pd}]{\text{H}_2}$ $\overset{+}{\text{H}_3}\text{NCHCNHCHCO}_2^-$

under: $(\text{CH}_3)_2\text{CH}$ $\quad$ CH_3 ; $(\text{CH}_3)_2\text{CH}$ $\quad$ CH_3

| Protected dipeptide | Val-Ala |

16.29 A benzyloxycarbonyl protected amino acid is an amide. Since peptides are also amides, removal of the protecting group by hydrolysis might also result in unwanted hydrolysis of the newly formed peptide.

16.30 (a) 1-Fluoro-2,4-dinitrobenzene reacts with the tripeptide to give a derivative in which the N-terminus bears a 2,4-dinitrophenyl group.

O_2N—(ring with NO$_2$)—F + $\overset{+}{\text{H}_3}\text{NCHCNHCH}_2\text{CNHCHCO}_2^-$ $\longrightarrow$ (dinitrophenyl ring with O$_2$N and NO$_2$)—NHCHCNHCH$_2$CNHCHCO$_2$H

under reactant: $\text{CH}_2\text{CH(CH}_3)_2$ $\quad$ CH_2OH ; under product: $\text{CH}_2\text{CH(CH}_3)_2$ $\quad$ CH_2OH

| 1-Fluoro-2,4-dinitrobenzene | Leu–Gly–Ser | DNP–Leu–Gly–Ser |

(b)   Hydrolysis of the product in (a) cleaves the peptide bonds. Leucine is isolated as its 2,4-dinitrophenyl (DNP) derivative, but glycine and serine are isolated as the free amino acids.

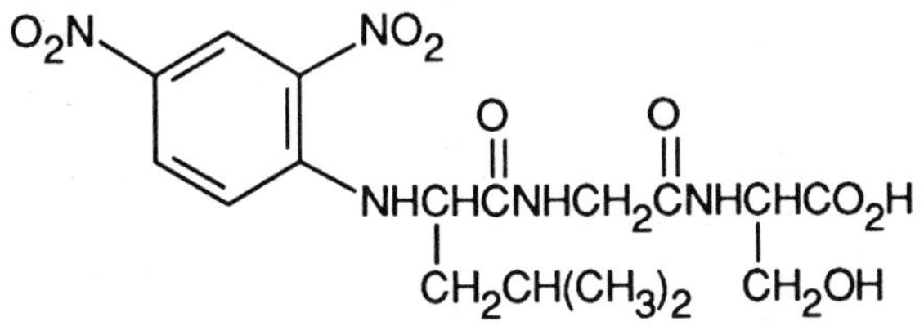

DNP–Leu–Gly–Ser

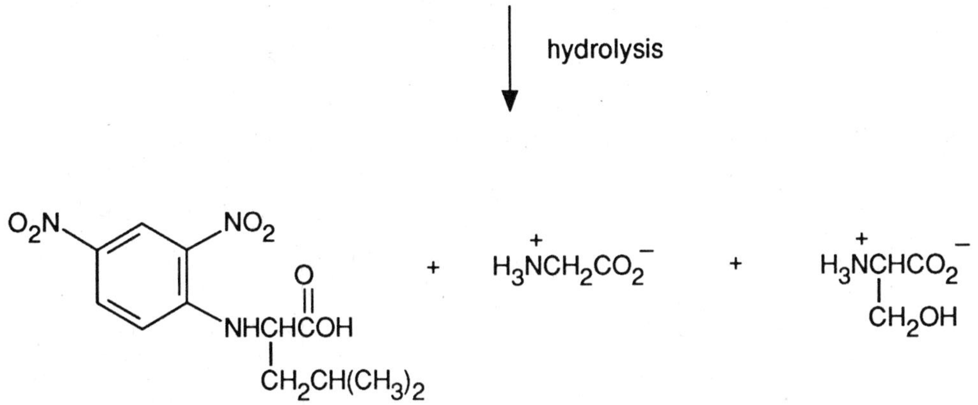

hydrolysis

DNP–Leu   Gly   Ser

(c)   Phenyl isothiocyanate is a reagent used to identify the N-terminal amino acid of a peptide by the Edman degradation. The N-terminal amino acid is cleaved as a phenylthiohydantoin (PTH) derivative, the remainder of the peptide remaining intact.

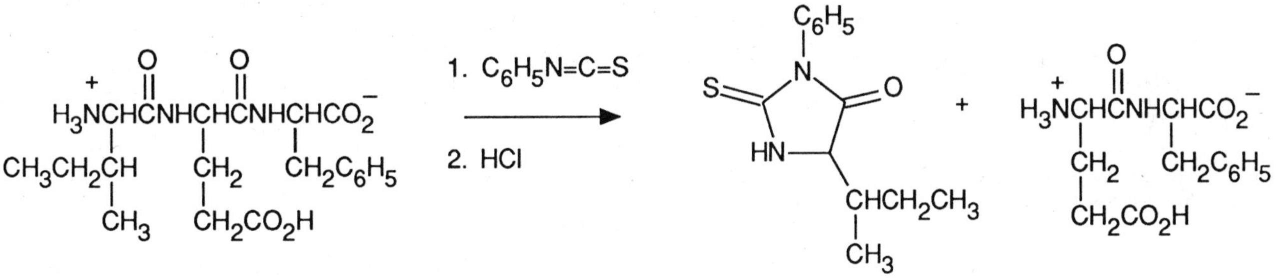

1. $C_6H_5N=C=S$

2. HCl

Ile–Glu–Phe   PTH derivative of isoleucine   Glu–Phe

(d)   Benzyloxycarbonyl chloride reacts with amino groups to convert them to amides. The only free amino group in Val-Ser-Ala is the N terminus.

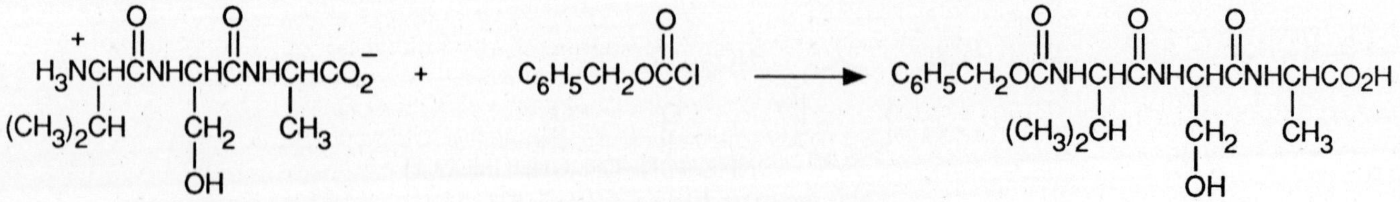

$$\overset{+}{H_3N}CHCNHCHCNHCHCO_2^- \;+\; C_6H_5CH_2OCCl \;\longrightarrow\; C_6H_5CH_2OCNHCHCNHCHCNHCHCO_2H$$

Val–Ser–Ala

Z–Val–Ser–Ala

(e)   Coupling of the Z-protected tripeptide from (d) and the benzyl ester of valine is achieved by DCCI-promoted amide bond formation between the free amino group of valine and the free carboxyl group of the protected tripeptide.

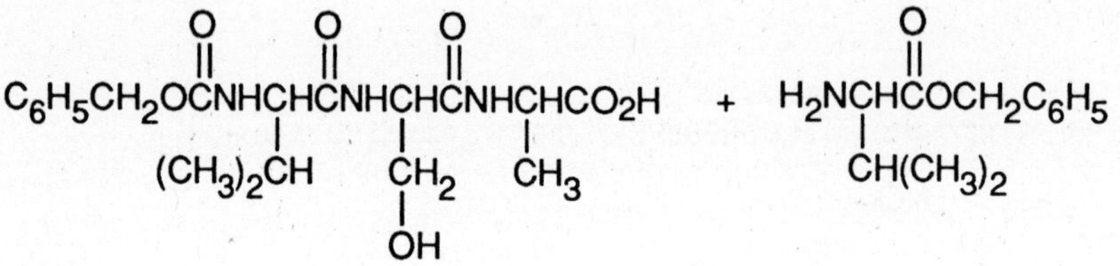

$$C_6H_5CH_2OCNHCHCNHCHCNHCHCO_2H \;+\; H_2NCHCOCH_2C_6H_5$$

Z–Val–Ser–Ala   Valine benzyl ester

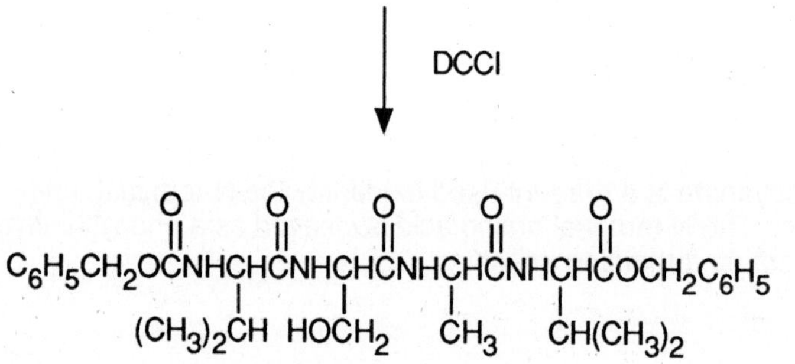

DCCI

$$C_6H_5CH_2OCNHCHCNHCHCNHCHCNHCHCOCH_2C_6H_5$$

Z–Val–Ser–Ala–Val

(f)   Both the benzyloxycarbonyl protecting group and the benzyl ester protecting group are removed by reaction of the newly formed protected tetrapeptide with hydrogen and a palladium catalyst. The product is a tetrapeptide.

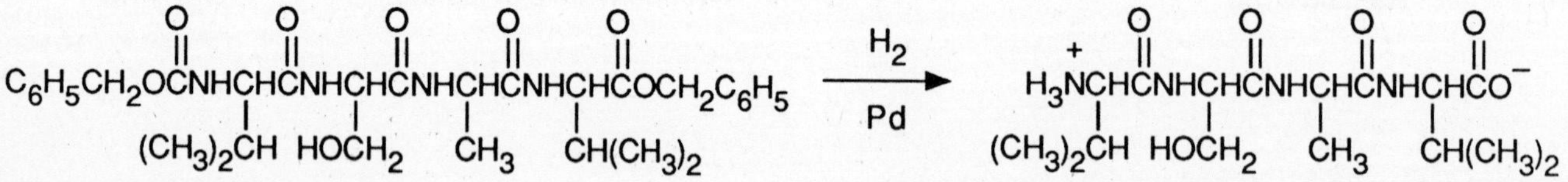

$$C_6H_5CH_2OCNHCHCNHCHCNHCHCNHCHCOCH_2C_6H_5 \;\xrightarrow[Pd]{H_2}\; \overset{+}{H_3N}CHCNHCHCNHCHCNHCHCO^-$$

Z–Val–Ser–Ala–Val   Val–Ser–Ala–Val

308

16.31 The two common protein secondary structures are the α-helix and the β-pleated sheet. The helical structure of an α-helix is stabilized by hydrogen bonds between amide groups within a protein chain. Hydrogen bonds between amide groups of adjacent chains stabilize the β-pleated sheet structure.

16.32 Bovine insulin (Figure 16.5) contains three disulfide bridges. One of these bridges connects two cysteines within a single chain, and the other two connect the two peptide chains. Reduction of the disulfide bridges would produce two peptide chains. One of these, called the "A chain" in Figure 16.5, would have four sulfhydryl groups. The "B chain" of bovine insulin would have two sulfhydryl groups.

16.33 The four types of attractive forces that contribute to the tertiary structure of a protein are:

(1)   Ionic attractions
(2)   Hydrogen bonds
(3)   Hydrophobic attractions (van der Waals attractions)
(4)   Disulfide bridges

16.34 Enzymes are proteins that act in an aqueous medium, either in solution or as colloids. They are more likely to have a globular shape. Structural proteins such as those in skin, tendons, or hair tend to adopt a fibrous structure and are insoluble in water.

16.35 Somatostatin is stated to be a tetradecapeptide and so is composed of 14 amino acids. The fact that Edman degradation gave the PTH derivative of alanine identifies this as the N-terminal amino acid. A major piece of information is the amino acid sequence of a hexapeptide obtained by partial hydrolysis.

Ala–Gly–Cys–Lys–Asn–Phe

Using this as a starting point and searching for overlaps with the other hydrolysis products gives the entire sequence.

Ala–Gly–Cys–Lys–Asn–Phe
                    Asn–Phe–Phe–Trp–Lys
                            Phe–Trp
                                Lys–Thr–Phe
                                    Thr–Phe–Thr–Ser–Cys
                                            Thr–Ser–Cys

Ala–Gly–Cys–Lys–Asn–Phe–Phe–Trp–Lys–Thr–Phe–Thr–Ser–Cys
  1   2   3   4   5   6   7   8   9   10   11   12 13 14

The disulfide bridge in somatostatin is between cysteine 3 and cysteine 14. Thus the primary structure is:

```
                          ⟋ Lys–Asn–Phe–Phe–Trp–Lys
                         ⟋                         |
        Ala–Gly–Cys                                |
                         ⟍                         |
                          ⟍ S–S–Cys–Ser–Thr–Phe–Thr
```

CHAPTER 17

LIPIDS

GLOSSARY OF TERMS

Acetyl coenzyme A The substance that acts as the source of acetyl groups in biosynthetic processes involving acetate. Acetyl coenzyme A is a thiol ester abbreviated as:

$$\underset{\displaystyle CH_3\overset{\displaystyle O}{\overset{\displaystyle \|}{C}}SCoA}{}$$

Anabolic steroid A steroid that promotes muscle growth.

Androgen A male sex hormone.

Bile acid Steroid derivatives biosynthesized in the liver that aid digestion by emulsifying fats.

Biological isoprene unit Isopentenyl pyrophosphate, biosynthesized from acetate, is the source of the branched five-carbon fragment from which is the biological precursor to terpenes and steroids.

Cholesterol The most abundant steroid in animals and the biological precursor to other naturally occurring steroids including the bile acids, sex hormones, and corticosteroids.

Essential oils Mixtures of terpenes, esters, alcohols, and other volatile organic substances that comprise pleasant-smelling oils of plants.

Estrogen A female sex hormone.

Fats and Oils Both are triesters of glycerol. Fats are solids at room temperature, oils are liquids.

Fatty acid The carboxylic acids obtained by hydrolysis of fats and oils. Fatty acids typically have unbranched chains and contain an even number of carbon atoms in the range 12-20 carbons. They may include one or more double bonds.

Iodine number A measure of the degree of unsaturation equal to the number of grams of I_2 absorbed per 100 g of fat or oil. An iodine number of zero indicates no double bonds are present. The higher the iodine number, the greater the number of double bonds.

Isoprene unit The characteristic five-carbon structural unit found in terpenes.

Lipid According to the customary classification of biologically important organic natural products (proteins, carbohydrates, nucleic acids, lipids), lipids are the compounds characterized by their high solubility in nonpolar organic solvents.

Lipid bilayer   Cell membranes are an arrangement of two layers of phospholipids in which the polar termini are located at the inner and outer membrane-water interfaces and the hydrophobic hydrocarbon tails cluster on the inside.

Mevalonic acid   An intermediate in the biosynthesis of steroids from acetyl coenzyme A.

Phospholipid   A diacylglycerol bearing a cholinephosphate "head group."   Also known as "phosphatidylcholine."

Squalene   A triterpene from which steroids are biosynthesized.

Steroid   A type of lipid present in both plants and animals characterized by a nucleus of four-fused rings (three are six-membered, one is five-membered). Cholesterol is the most abundant steroid in animals.

Terpene   Compounds that can be analyzed as cluster of isoprene units (see above). Terpenes with 10 carbons are classified as *monoterpenes*, those with 15 are *sesquiterpenes*, those with 20 are *diterpenes*, and those with 30 are *triterpenes*.

Triacylglycerol   A derivative of glycerol (1,2,3-propanetriol) in which the three oxygens bear acyl groups derived from fatty acids.

Wax   A mixture of water-repellent substances that form a protective coating on the leaves of plants, the fur of animals, the feathers of birds, *etc.* A principal component of a wax is often an ester in which both the acyl portion and alkyl portion contain long chains.

SOLUTIONS TO TEXT PROBLEMS

17.1   (b) and (c)   The triacylglycerol shown in Figure 17.1a, with an oleyl group at C-2 of the glycerol unit and two stearyl groups at C-1 and C-3, yields stearic and oleic acids (as their carboxylate ions) in a 2:1 molar ratio on hydrolysis in base. A constitutionally isomeric structure in which the oleyl group is attached to C-1 of glycerol would yield the same hydrolysis products. (The double bond is cis in oleic acid.)

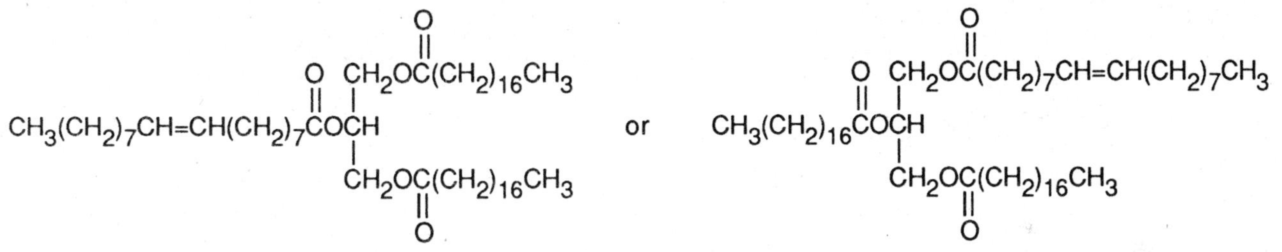

2-Oleyl-1,3-distearylglycerol   1-Oleyl-2,3-distearylglycerol

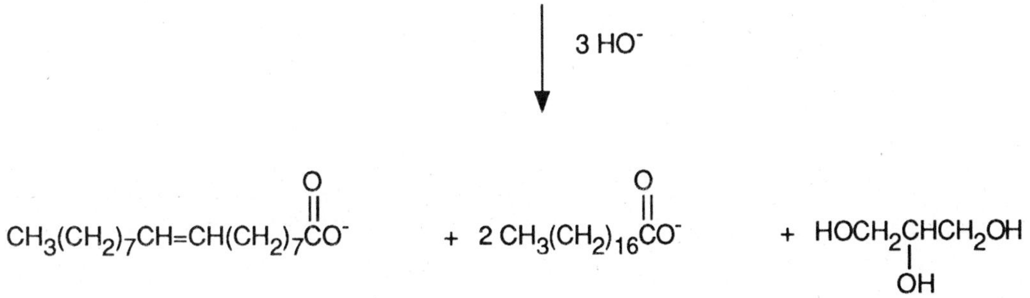

3 HO⁻

Oleate   Stearate   Glycerol

17.2   The groups designated R and R' in the general structure shown for lecithin are nonpolar alkyl and alkenyl groups and associate with nonpolar fats in the food by attractive van der Waals forces. A fat molecule is held in suspension because the lecithin to which it is bound by van der Waals ("hydrophobic") forces is in turn associated with water and dispersed through the water. The negatively charged oxygens of lecithin form hydrogen bonds to water and the positively charged nitrogen is attracted to the negatively polarized oxygen of water.

$$
\begin{array}{c}
\text{O} \\
\| \\
\text{CH}_2\text{OCR}
\end{array}
$$

$$
\begin{array}{c}
\text{O} \\
\| \\
\text{R'COCH}
\end{array}
$$

$$
\text{CH}_2\text{OPO}_2^-
$$

$$
+ \\
\text{OCH}_2\text{CH}_2\text{N(CH}_3)_3
$$

17.3   Oleyl oleate is an ester whose acyl group is derived from oleic acid (Table 17.1). The systematic name of the ester is *cis*-9-octadecenyl *cis*-9-octadecenoate. Alternatively the stereochemistry may be designated in both cases as *Z* rather than *cis*.

$$
\text{CH}_3(\text{CH}_2)_6\text{CH}_2 \quad \text{CH}_2(\text{CH}_2)_6\overset{\overset{\text{O}}{\|}}{\text{C}}\text{OCH}_2(\text{CH}_2)_6\text{CH}_2 \quad \text{CH}_2(\text{CH}_2)_6\text{CH}_3
$$

C=C   C=C

H   H   H   H

17.4   By analogy to the reaction shown on p 490 in the text in which 7-dehydrocholesterol is converted to vitamin D_3, the structure of vitamin D_2 can be deduced from that of ergosterol.

Ergosterol   Vitamin D_2

17.5   Isoprene units are structural units of the type shown.

Functional groups and multiple bonds are ignored when structures are examined for the presence of isoprene units. The isoprene units of α-selinene and farnesol are shown in the text, and are not repeated here. Similarly, the isoprene units of α-phellandrene were illustrated in the sample solution in the text.

Menthol (same carbon skeleton as α-phellandrene but different functionality). Two equally correct answers are possible.

Citral

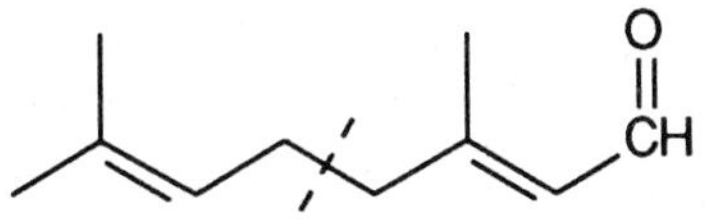

Abscicic acid

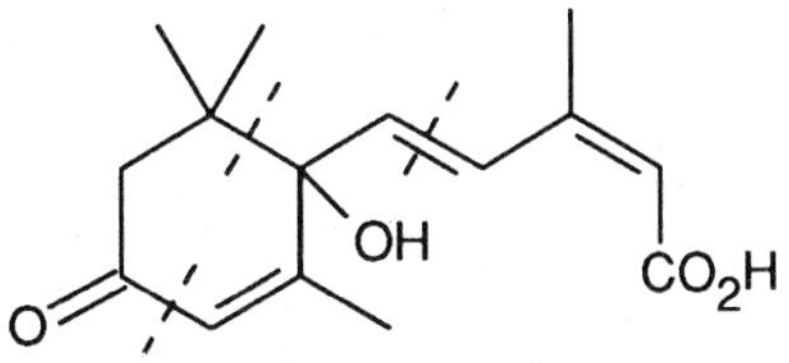

Cembrene (two equally correct answers)

or

Vitamin A

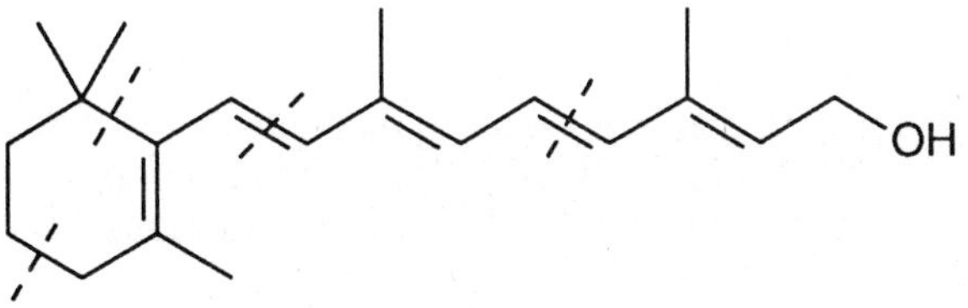

17.6   Geranyl pyrophosphate is an allylic pyrophosphate and, like dimethylallyl pyrophosphate, can act as an alkylating agent toward a molecule of isopentenyl pyrophosphate.

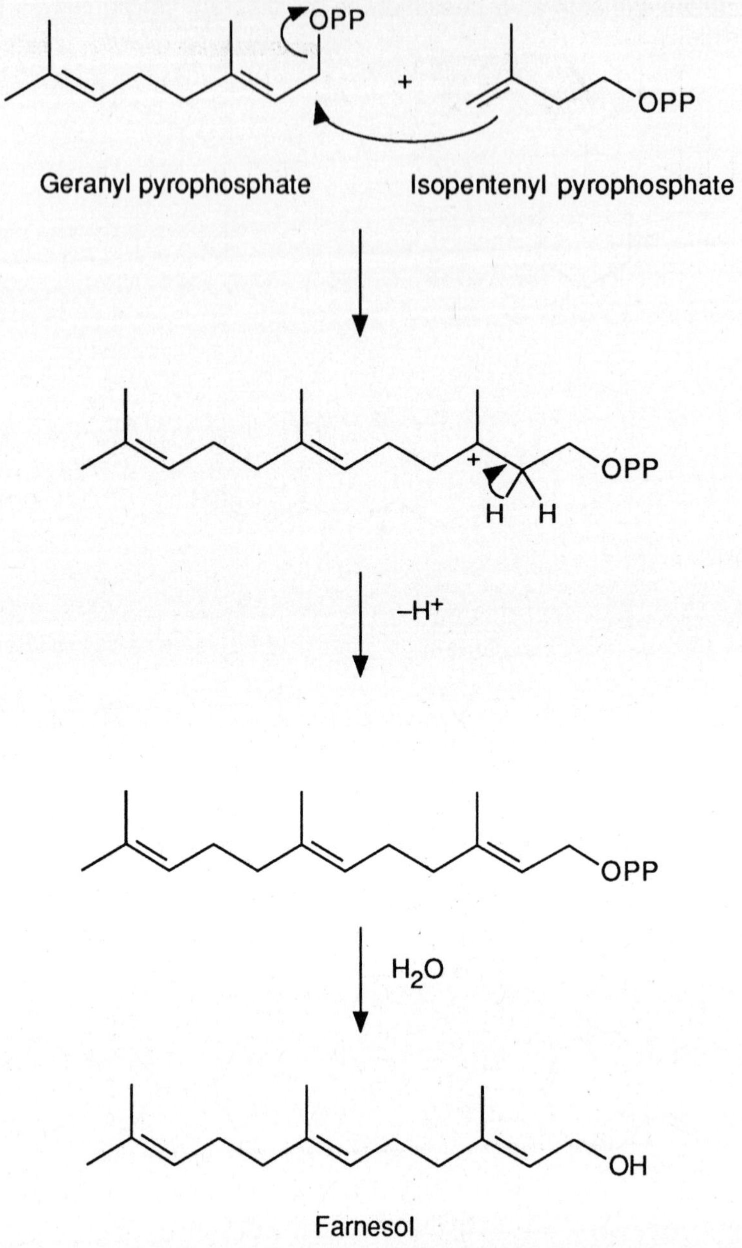
OPP
OPP
Geranyl pyrophosphate
Isopentenyl pyrophosphate
OPP
H H
$-H^+$
OPP
H_2O
OH
Farnesol

17.7 (a) Fats and oils are both triacylglycerols or, more commonly, mixtures of triacylglycerols. More of the acyl groups in oils have one or more double bonds than in fats. Fats tend to be more "saturated" than oils.

(b) A fat is a triester of glycerol. A wax is an ester of a fatty acid and a long-chain alcohol. The structure of the wax from problem 17.3 is reproduced below. While the acyl and alkyl portions of the wax shown contain equal numbers of carbons, this need not always be the case.

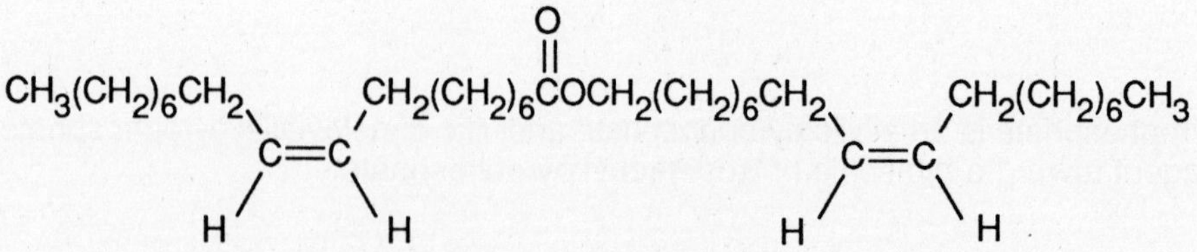
$CH_3(CH_2)_6CH_2$
$CH_2(CH_2)_6\overset{O}{\overset{\parallel}{C}}OCH_2(CH_2)_6CH_2$
$CH_2(CH_2)_6CH_3$
C=C
C=C
H
H
H
H

(c)   A triacylglycerol is a triester of glycerol in which all three acyl groups are derived from fatty acids. One of the acyl groups in a phospholipid has been replaced by a phosphate bearing a choline group.

$$
\begin{array}{c}
\text{O}\\
\|\\
\text{CH}_2\text{OCR}\\
\text{O} \quad |\\
\|\quad\\
\text{R'COCH}\\
|\\
\text{CH}_2\text{OPO}_2^-\\
| \qquad +\\
\text{OCH}_2\text{CH}_2\text{N(CH}_3)_3
\end{array}
$$

17.8   Unlike triacylglycerols which have only acyl groups bonded to the three oxygens of glycerol, a phospholipid has two acyl groups and a highly polar "head group" derived from choline attached to these oxygens. The general structure of a phospholipid is shown in the the solution to part (c) of the preceding problem. The charged head group can hydrogen bond to water through the negatively charged oxygens of its phosphate. The positively charged nitrogen of choline is also solvated with water. These interactions between the phospholipid and water increase its solubility.

17.9   (a)   Base-promoted hydrolysis of trioleylglycerol yields three moles of oleic acid and one mole of glycerol. At the pH under which the reaction is carried out, oleic acid exists as its carboxylate ion.

$$
\text{CH}_3(\text{CH}_2)_7\text{CH=CH(CH}_2)_7\overset{\displaystyle O}{\overset{\displaystyle \|}{\text{C}}}\text{OCH}\;\begin{array}{c}\overset{\displaystyle O}{\overset{\displaystyle \|}{}}\\ \text{CH}_2\text{OC(CH}_2)_7\text{CH=CH(CH}_2)_7\text{CH}_3\\ |\\ \text{CH}_2\text{OC(CH}_2)_7\text{CH=CH(CH}_2)_7\text{CH}_3\\ \|\\ O\end{array}\qquad + \;3\;\text{HO}^- \longrightarrow
$$

Trioleylglycerol

$$
3\;\text{CH}_3(\text{CH}_2)_7\text{CH=CH(CH}_2)_7\overset{\displaystyle O}{\overset{\displaystyle \|}{\text{C}}}\text{O}^- \quad + \quad \text{HOCH}_2\underset{\text{OH}}{\underset{|}{\text{CH}}}\text{CH}_2\text{OH}
$$

Oleate   Glycerol

(b)   Hydrogen adds to the double bonds of trioleoylglycerol to convert it to the saturated fat tristearoylglycerol (tristearin).

$$
\text{CH}_3(\text{CH}_2)_7\text{CH=CH(CH}_2)_7\overset{O}{\overset{\|}{\text{C}}}\text{OCH}\;\begin{array}{c}\overset{O}{\overset{\|}{}}\\ \text{CH}_2\text{OC(CH}_2)_7\text{CH=CH(CH}_2)_7\text{CH}_3\\ |\\ \text{CH}_2\text{OC(CH}_2)_7\text{CH=CH(CH}_2)_7\text{CH}_3\\ \|\\ O\end{array}\quad + \;3\;\text{H}_2 \;\xrightarrow{\text{Pt}}\; \text{CH}_3(\text{CH}_2)_{16}\overset{O}{\overset{\|}{\text{C}}}\text{OCH}\;\begin{array}{c}\overset{O}{\overset{\|}{}}\\ \text{CH}_2\text{OC(CH}_2)_{16}\text{CH}_3\\ |\\ \text{CH}_2\text{OC(CH}_2)_{16}\text{CH}_3\\ \|\\ O\end{array}
$$

Trioleoylglycerol   Tristearoylglycerol

17.10 The iodine number is a measure of the degree of unsaturation on a fat or oil and is related to the number of carbon-carbon double bonds it contains. The structures of trioleylglycerol and tristearylglycerol are shown in the preceding problem. Trioleylglycerol with three carbon-carbon double bonds is more unsaturated and has a higher iodine number than tristearylglycerol which has none. In fact, the iodine number of tristearylglycerol is zero.

17.11 The choline-containing "head group" of lecithin is hydrophilic. The nonpolar alkyl chains of the two acyl groups are hydrophobic (lipophilic).

17.12 Since hydrolysis of linolein yields glycerol and linoleic acid, linolein is a triacylglycerol in which the acyl groups are derived from linoleic acid. The structure of linoleic acid is given in Table 17.1.

17.13 (a) Oleic acid is *cis*-9-octadecenoic acid. Since elaidic acid is a stereoisomer of oleic acid, it has the same constitution but must contain a trans double bond.

(b) Ricinoleic acid has the same carbon skeleton as oleic acid, with a hydroxyl group at C-12. Remember, the carbon chain is counted beginning with the carboxyl group.

17.14 (a)   Cetyl palmitate (hexadecyl hexadecanoate) is an ester in which both the acyl group and the alkyl group contain 16 carbon atoms.

$$CH_3(CH_2)_{14}\overset{\overset{\displaystyle O}{\displaystyle \|}}{C}OCH_2(CH_2)_{14}CH_3$$

(b)   The ester hexadecyl hexadecanoate can be prepared from hexadecanoic acid and 1-hexadecanol. Hexadecanoic acid is the starting material given. 1-Hexadecanol may be prepared from hexadecanoic acid by reduction with lithium aluminum hydride. The ester is then prepared by allowing the acid and alcohol to react in the presence of an acid catalyst.

$$CH_3(CH_2)_{14}\overset{\overset{\displaystyle O}{\displaystyle \|}}{C}OH \xrightarrow[\text{2. H}_2\text{O}]{\text{1. LiAlH}_4} CH_3(CH_2)_{14}CH_2OH$$

Hexadecanoic acid   1-Hexadecanol

$$CH_3(CH_2)_{14}\overset{\overset{\displaystyle O}{\displaystyle \|}}{C}OH \;+\; CH_3(CH_2)_{14}CH_2OH \xrightarrow{\;H^+\;} CH_3(CH_2)_{14}\overset{\overset{\displaystyle O}{\displaystyle \|}}{C}OCH_2(CH_2)_{14}CH_3$$

Hexadecanoic acid   1-Hexadecanol   Hexadecyl hexadecanoate

Alternatively, the desired ester could be prepared from 1-hexadecanol and hexadecanoyl chloride (prepared from hexadecanoic acid and thionyl chloride $SOCl_2$).

17.15 The structure of vitamin D is shown on page 490 of the text. It need not be shown here since its features relevant to fat solubility are easily described. Vitamin D contains 27 carbon atoms and only one polar group (the hydroxyl at C-3). It is therefore much like a hydrocarbon in respect to its overall polarity and dissolves better in nonpolar materials than polar ones. The fat deposits in our body can be considered a nonpolar solvent, one that is suitable for dissolving vitamin D.

17.16 The number of stereoisomers is given by 2^n where n is the number of chiral centers. Cholesterol has eight chiral centers marked on the formula below with asterisks. Thus there are $2^8 = 256$ possible stereoisomers having this constitution. One of them is cholesterol and the other 255 are stereoisomers of cholesterol.

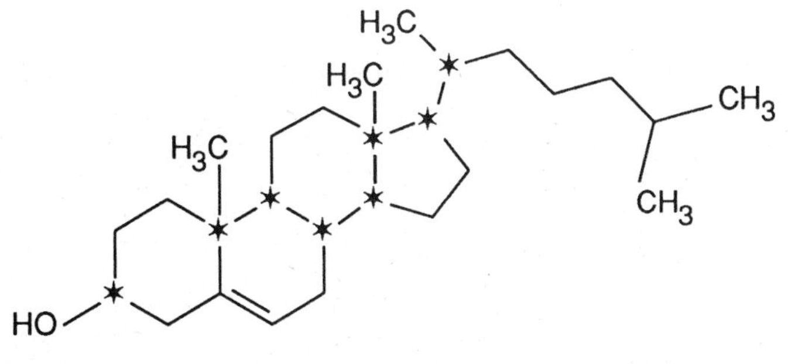

17.17 The functional groups present in cholesterol are the hydroxyl group in the A ring (see Figure 17.3 on p 490), and the double bond in the B ring.

(a) Bromine adds to the double bond of cholesterol to give a vicinal dibromide.

(b) The double bond is reduced by catalytic hydrogenation.

(c) Acetic anhydride reacts with the hydroxyl group to give an acetate.

318

(d) This reagent (Collins' reagent) oxidizes the hydroxyl group to the corresponding ketone.

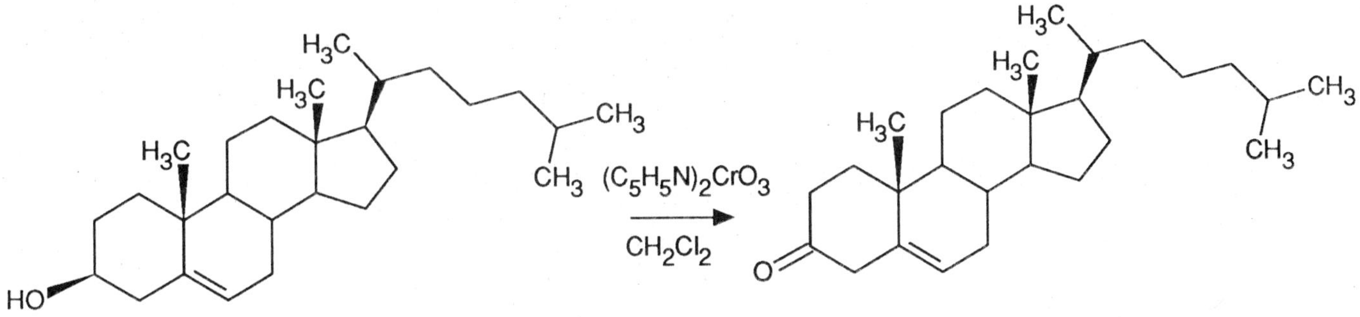

17.18 The isoprene units are revealed by finding the ⟨fragment⟩ fragments in the carbon skeleton. The carbons united by a tail-to-tail union of isoprene groups are indicated by asterisks.

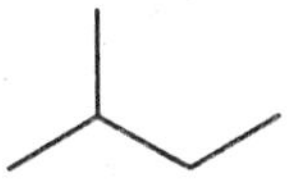

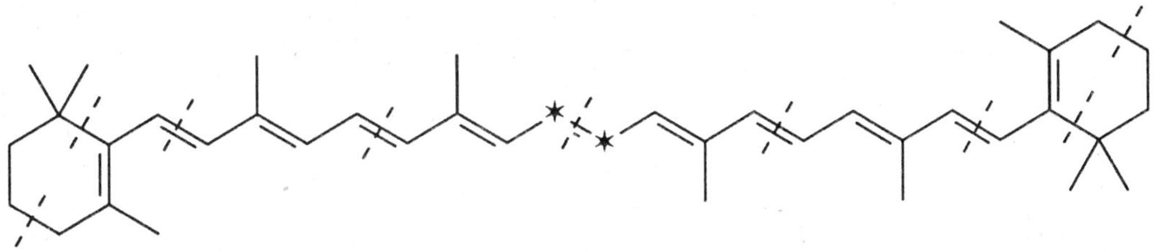

β-Carotene

17.19 The isoprene units are identified as in the previous problem. Functional groups and multiple bonds are ignored when structures are examined for the presence of isoprene units.

(a) **Dendrolasin** has three isoprene units.

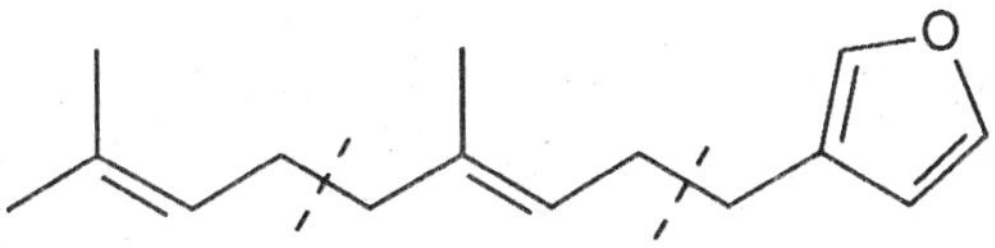

(b) γ-**Bisabolene** may be divided into three isoprene units. Two equally correct solutions are possible.

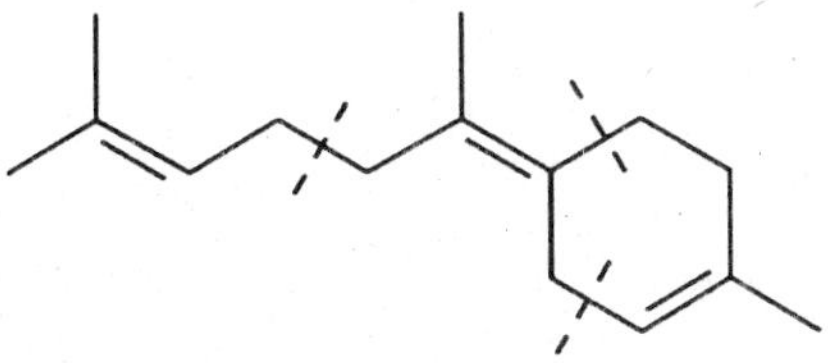

or

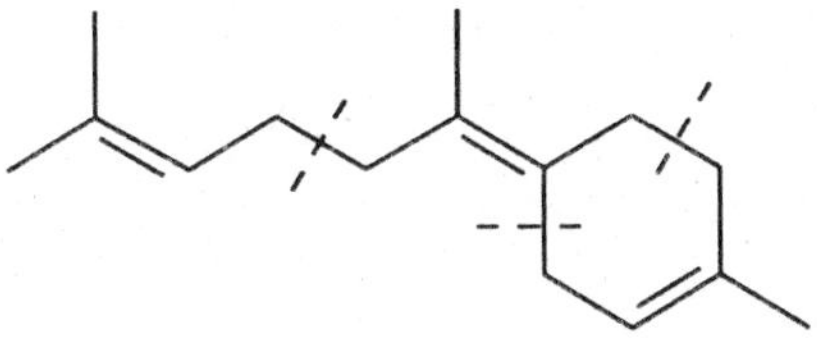

17.20 The overall transformation

to

simply requires conversion of the alcohol function to some suitable leaving group, followed by substitution by an appropriate nucleophile. A bromide leaving group is appropriate.

3-Methyl-3-buten-1-ol   4-Bromo-2-methyl-1-butene   3-Methyl-3-butenyl methyl thioether

17.21 A reasonable mechanism is protonation of the isolated carbon-carbon double bond, followed by cyclization.

17.22 Ester formation between the carboxyl group and the hydroxyl group at C-5 of mevalonic acid gives the cyclic ester mevalonolactone.

Mevalonic acid

Mevalonolactone

Mevalonolactone contains a stable six-membered ring. Had the hydroxyl group at C-3 been involved, a less stable four-membered ring would have resulted.

17.23 The two pinenes are regioisomers, and result from loss of a proton from the same tertiary carbocation:

CHAPTER *18*

NUCLEIC ACIDS

GLOSSARY OF TERMS

Adenosine triphosphate (ATP)   A nucleotide having the structure shown. It is involved in energy storage and generation in living systems. Enzyme-catalyzed hydrolysis of the phosphate ester portion releases energy.

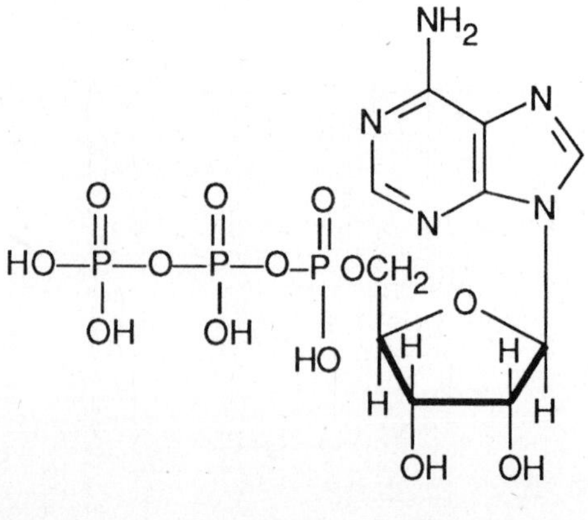

Anticodon   A sequence of three bases in a molecule of tRNA that is complementary to the codon of mRNA for a particular amino acid.

Base   The heterocyclic aromatic derived from purine or pyrimidine that is bonded to the anomeric position of ribose or 2-deoxyribose in a nucleoside.

Base pair   The term given to the purine of a nucleotide and its complementary pyrimidine. Adenine (A) is complementary to thymine (T), and guanine (G) is complementary to cytosine (C).

Codon   A set of three successive nucleotides in mRNA which is unique for a particular amino acid. The 64 codons possible from combinations of A, T, G, and C code for the 20 amino acids from which proteins are constructed.

Cyclic AMP   A cyclic nucleotide derived from adenosine in which the oxygens at C-3 and C-5 of the ribose portion of the molecule are incorporated into a cyclic phosphate. Cyclic AMP is a regulator of several biological processes.

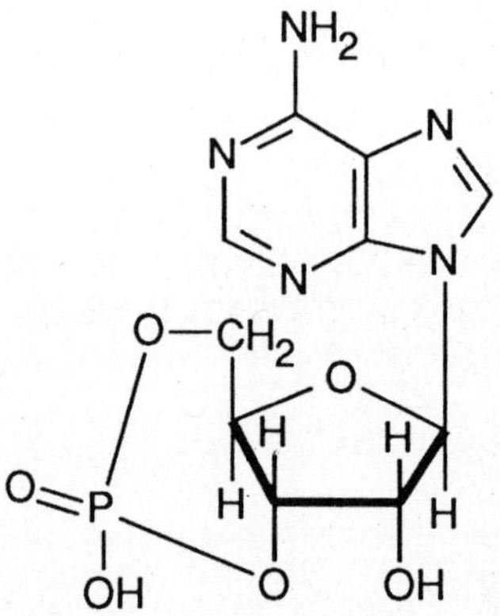

DNA (*deoxyribonucleic acid*)   A polynucleotide of 2'-deoxyribose present in the nuclei of cells which serves to store and replicate genetic information. Genes are DNA.

Double helix   The form in which DNA normally occurs in living systems. Two complementary strands of DNA are associated with one another by hydrogen bonds between their base pairs and each DNA adopts a helical shape.

Messenger RNA (mRNA)   A polynucleotide of ribose that "reads" the sequence of bases in DNA and interacts with tRNAs in the ribosomes to promote protein biosynthesis.

Nucleic acid   A polynucleotide present in the nuclei of cells.

Nucleoside   The combination of a purine or pyrimidine base and a carbohydrate, usually ribose or 2-deoxyribose.

Nucleotide   The phosphate ester of a nucleoside.

Purine The heterocyclic aromatic shown.

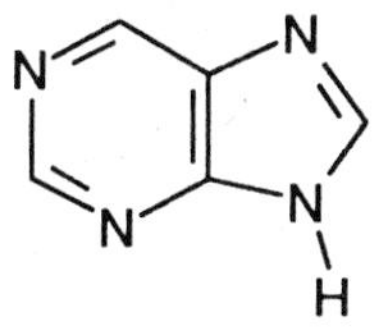

Pyrimidine   The heterocyclic aromatic shown.

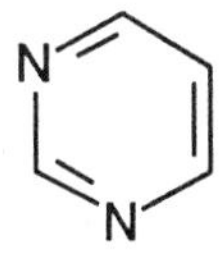

RNA (*ribonucleic acid*)   A polynucleotide of ribose.

Transcription   The construction of a strand of mRNA complementary to a DNA template.

Transfer RNA (tRNA)   A polynucleotide of ribose which is bound at one end to a particular amino acid depending on its anticodon. The amino acid is incorporated into a growing peptide chain.

Translation   The "reading" of mRNA by various tRNAs, each one of which is unique for a particular amino acid.

SOLUTIONS TO TEXT PROBLEMS

18.1   The numbering of the ring in uracil and its derivatives parallels that in pyrimidine (given in text and shown below).

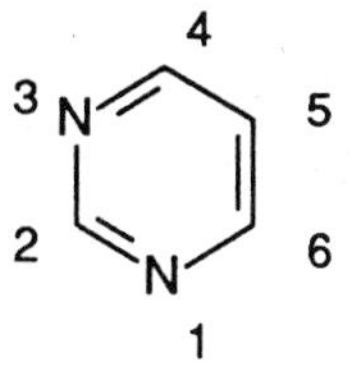

Pyrimidine 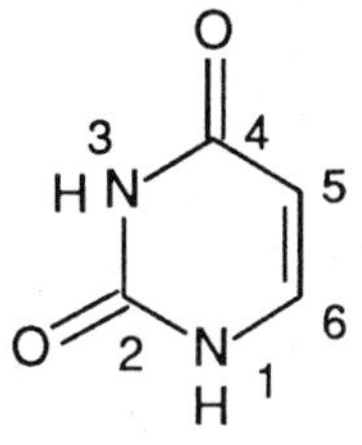   Uracil 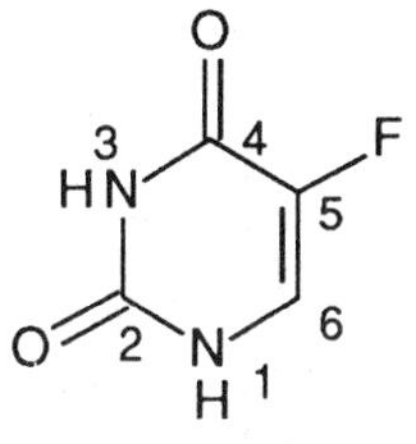   5-Fluorouracil

18.2   (b) Cytidine is present in ribonucleic acid (RNA) and so is a nucleoside of D-ribose. The base is cytosine.

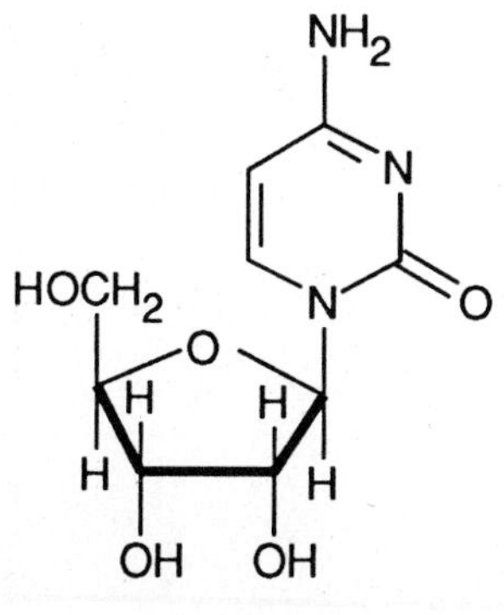

(c) Guanosine is present in RNA and so is a guanine nucleoside of D-ribose.

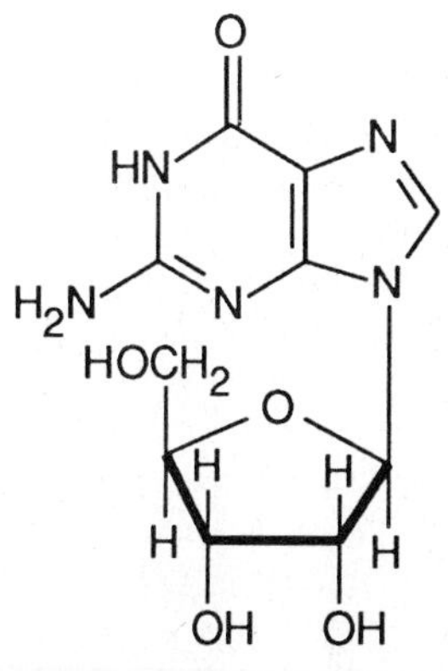

323

18.3 The complementary base pairs in DNA are A···T and G···C. As in the solution to part (a), write the symbol for the complementary base opposite each symbol in the original strand.

(b)

Original strand: —A—T—G—G—A—C—T—

Complementary strand: —T—A—C—C—T—G—A—

(c)

Original strand: —G—G—G—C—C—A—G—T—T—

Complementary strand: —C—C—C—G—G—T—C—A—A—

18.4 The anticodon is the set of bases complementary to the codon base sequence. The complementary base pairs are G···C and A···U. The amino acid corresponding to each codon is found in Table 18.1 of the text.

	(b)	(c)	(d)
Codon:	GAC	UUU	AGG
Anticodon:	CUG	AAA	UCC
Amino acid:	aspartic acid	phenylalanine	arginine

18.5 A nucleoside is a combination of a purine or pyrimidine base with ribose or 2-deoxyribose. Nucleotides are phosphoric acid esters of nucleosides.

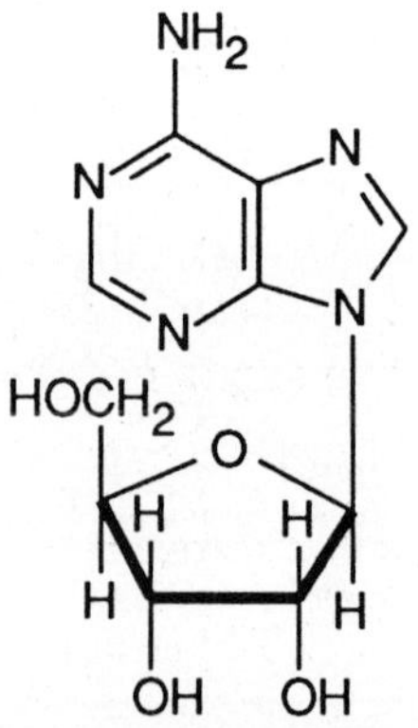

Adenosine

(a nucleoside)

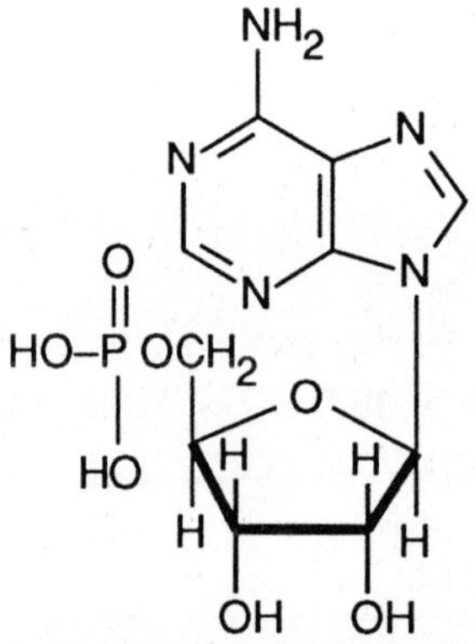

Adenosine 5'-monophosphate

(a nucleotide)

18.6 Purine and is numbering system are as shown:

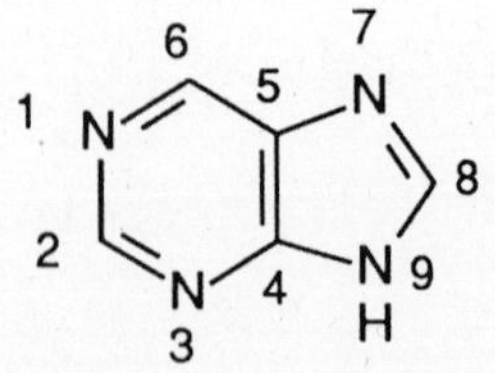

324

In nebularine, D-ribose in its furanose form is attached to position 9 of purine. The stereochemistry at the anomeric position is β.

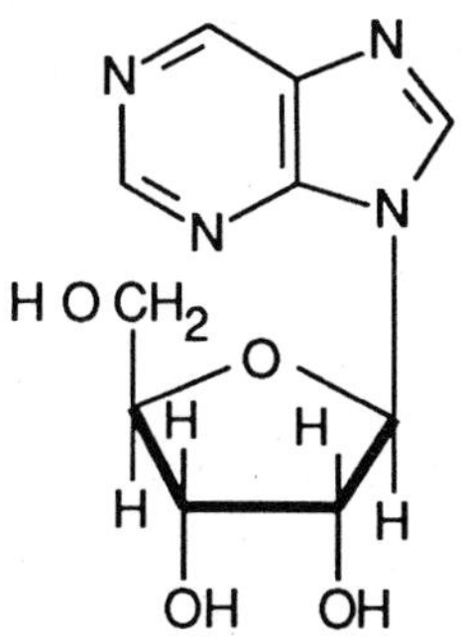

9-β-D-Ribofuranosylpurine (nebularine)

18.7 The problem states that vidarabine is the arabinose analog of adenosine. Arabinose and ribose differ only in their configuration at C-2 in the carbohydrate portion of the molecule. The hydroxyl group at C-2 is "down " in adenosine and "up" in vidabarine.

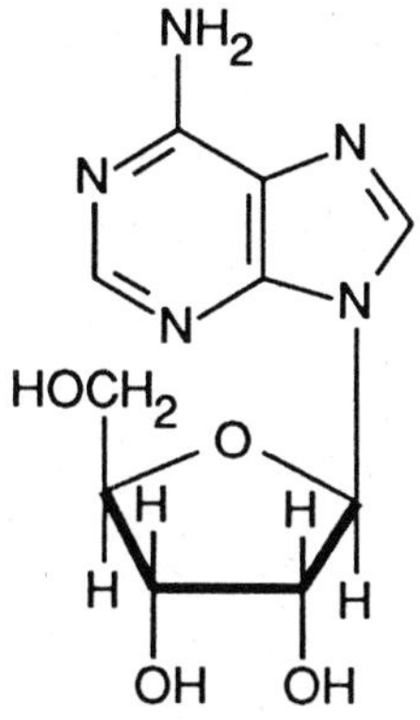

Adenosine

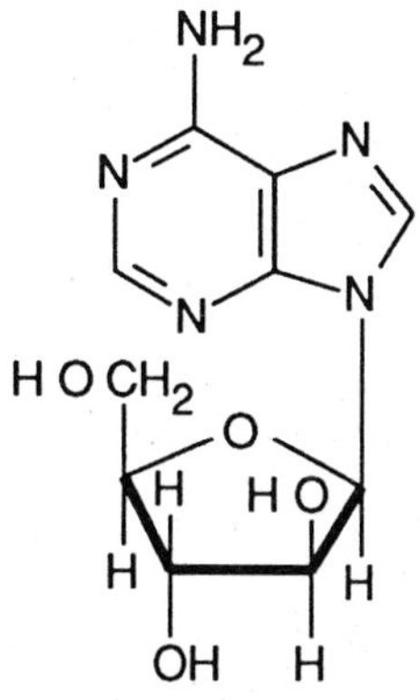

Vidabarine

18.8 Write the symbol for the complementary bases opposite each nucleotide segment in the original strand. Recall that the complementary base pairs in DNA are A⋯T and G⋯C.

(a)

Original strand:

—A—A—A—G—G—T—C—C—C—G—T—A—

Complementary strand:

—T—T—T—C—C—A—G—G—G—C—A—T—

(b)

Original strand:

—T—A—C—T—C—G—C—G—G—A—T—G—

Complementary strand:

—A—T—G—A—G—C—G—C—C—T—A—C—

18.9 The pairing of RNA bases with the DNA strand is the same as DNA base-pairing, except that uracil (U) replaces thymine (T) in RNA.

(a)

Original strand:

—A—A—A—G—G—T—C—C—C—G—T—A—

mRNA strand:

—U—U—U—C—C—A—G—G—G—C—A—U—

(b)

Original strand: —T—A—C—T—C—G—C—G—G—A—T—G—

mRNA strand: —A—U—G—A—G—C—G—C—C—U—A—C—

18.10 The codons are triplets of nucleotides in the mRNA strand.

(a)   UUU   CCA   GGG   CAU

(b)   AUG   AGC   GCC   UAC

18.11 Each mRNA codon codes for a specific tRNA and each tRNA delivers a particular amino acid to the growing protein. The codons and the amino acids for which they code are listed in Table 18.1.

(a)	Codon	Amino Acid		(b)	Codon	Amino Acid
	UUU	Phenylalanine			AUG	Methionine
	CCA	Proline			AGC	Serine
	GGG	Glycine			GCC	Alanine
	CAU	Histidine			UAC	Tyrosine

The amino acid sequences are:

(a)   Phe-Pro-Gly-His   (b)   Met-Ser-Ala-Tyr

18.12 An anticodon is a group of three bases in a particular section of tRNA complementary to the bases of each mRNA codon. There is a different tRNA for each amino acid.

(a)	Codons:	UUU	CCA	GGG	CAU
	Anticodons:	AAA	GGU	CCC	GUA

(b)	Codons:	AUG	AGC	GCC	UAC
	Anticodons:	UAC	UCG	CGG	AUG

These are the base sequences deduced using the customary relationships between purine and pyrimidine bases. It should be pointed out that tRNA often has a modified base as the first base in the sequence. These bases are derived from the usual A, U, G, and C but have been enzymatically modified such as, for example, by introduction of an methyl group.

18.13 (a)   Each of the amino acids in Problem 18.11 except methionine are specified by more than one mRNA codon. These are tabulated below for the amino acid sequence Phe-Pro-Gly-His.

Amino acid:	Phe	Pro	Gly	His
Codons:	UUU	CCU	GGU	CAU
	UUC	CCC	GGC	CAC
		CCA	GGA	
		CCG	GGG	

Thus, the sequence of 12 bases is:

UU(U or C)   CC(U, C, A, or G)   GG(U, C, A, or G)   CA(U or C)

(b)   For the sequence Met-Ser-Ala-Tyr:

Amino acid:	Met	Ser	Ala	Tyr
Codons:	AUG	UCU	GCU	UAU
		UCC	GCC	UAC
		UCA	GCA	
		UCG	GCG	
		AGU		
		AGC		

The sequence of 12 bases may be any of the combinations:

AUG UC(U, C, A, or G) GC(U, C, A, or G) UA(U or C)

or   AUG AG(U or C)   GC(U, C, A, or G) UA(U or C)

18.14   As in previous problems, align complementary base pairs.   Recall that thymine (T) is present in DNA instead of uracil (U).

(a)   The various mRNA codons from problem 18.13 (a) are:

UU(U or C) CC(U, C, A, or G) GG(U, C, A, or G) CA(U or C)

These translate to the DNA sequence of bases:

AA(A or G) GG(A, G, T, or C) CC(A, G, T, or C) GT(A or G)

(b)   The various mRNA codons from problem 18.13 (b) may be:

AUG UC(U, C, A, or G) GC(U, C, A, or G) UA(U or C)

which translate to:

TAC   AG(A, G, T, or C)   CG(A, G, T, or C) AT(A or G)

or: AUG AG(U or C)   GC(U, C, A, or G) UA(U or C)

which translate to:

TAC   TC(A or G)   CG(A, G, T, or C) AT(A or G)

18.15   Table 18.1 in the text lists the messenger RNA codons for the various amino acids. The codons for valine and for glutamic acid are:

Valine:	GUU	GUA	GUC	GUG
Glutamic acid:		GAA		GAG

As can be seen, the codons for glutamic acid (GAA and GAG) are very similar to two of the codons (GUA) and (GUG) for valine. Replacement of adenine in the glutamic acid codons by uracil causes valine to be incorporated into hemoglobin instead of glutamic acid and is responsible for the sickle-cell trait.

18.16 Leucine enkephalin has the amino acid sequence Tyr-Gly-Gly-Phe-Leu. Each of the amino acids in leucine enkephalin is specified by more than one mRNA codon (Table 18.1).

Amino acid:	Tyr	Gly	Gly	Phe	Leu
Codons:	UAU	GGU	GGU	UUU	CUU
	UAC	GGC	GGC	UUC	CUC
		GGA	GGA		CUA
		GGG	GGG		CUG
					UUA
					UUG

The sequence of 15 bases may be any of the combinations:

UA(U or C)   GG(U, C, A, or G)   GG(U, C, A, or G)   UU(U or C)   CU(U, C, A, or G)

or   UA(U or C)   GG(U, C, A, or G)   GG(U, C, A, or G)   UU(U or C)   UU(A or G)

18.17 As in Problem 18.14, align complementary base pairs to describe the DNA sequence which would serve as a template for the leucine enkephalin mRNA sequence.

mRNA: UA(U or C)   GG(U, C, A, or G)   GG(U, C, A, or G)   UU(U or C)   CU(U, C, A, or G)

DNA: AT(A or G)   CC(A, G, T, or C)   CC(A, G, T, or C)   AA(A or G)   GA(A, G, T, or C)

or: *mRNA:*   UA(U or C)   GG(U, C, A, or G)   GG(U, C, A, or G)   UU(U or C)   UU(A or G)

DNA: AT(A or G)   CC(A, G, T, or C)   CC(A, G, T, or C)   AA(A or G)   AA(T or C)

18.18 The accuracy of DNA replication depends on the paired bases adenine and thymine, and guanine and cytosine (Figure 18.2). In the guanine-cytosine base pair, one of the ring nitrogens of guanine acts as the donor of a proton in a hydrogen bond to cytosine.

proton donor in a hydrogen bond

Deoxyribose

In *O*-methylguanine, the corresponding ring nitrogen has no proton attached to it. This nitrogen can only act as a proton acceptor in hydrogen bond formation. It cannot form a stable base pair with cytosine.

proton acceptor in a hydrogen bond

Deoxyribose

Some research studies have suggested that *O*-methylguanine base-pairs with thymine, instead of the correct base-pairing with cytosine.